Technical and Risk Management Reference Series

Security, Audit and Control Features

Oracle® E-Business Suite

3rd Edition

ISACA®

With more than 86,000 constituents in more than 160 countries, ISACA (*www.isaca.org*) is a leading global provider of knowledge, certifications, community, advocacy and education on information systems (IS) assurance and security, enterprise governance of IT, and IT-related risk and compliance. Founded in 1969, ISACA sponsors international conferences, publishes the *ISACA® Journal*, and develops international IS auditing and control standards. It also administers the globally respected Certified Information Systems Auditor™ (CISA®), Certified Information Security Manager® (CISM®), Certified in the Governance of Enterprise IT® (CGEIT®) and Certified in Risk and Information Systems Control™ (CRISC™) designations.

ISACA offers the Business Model for Information Security (BMIS) and the IT Assurance Framework™ (ITAF™). It also developed and maintains the COBIT®, Val IT™ and Risk IT frameworks, which help IT professionals and enterprise leaders fulfill their IT governance responsibilities and deliver value to the business.

Disclaimer

ISACA has designed and created *Security, Audit and Control Features Oracle® E-Business Suite, 3rd Edition* (the "Work") primarily as an educational resource for control professionals. ISACA makes no claim that use of any of the Work will assure a successful outcome. The Work should not be considered inclusive of all proper information, procedures and tests or exclusive of other information, procedures and tests that are reasonably directed to obtaining the same results. In determining the propriety of any specific information, procedure or test, readers should apply their own professional judgment to the specific control circumstances presented by the particular systems or information technology environment.

Oracle is a registered trademark of Oracle Corporation. Oracle Corporation is not the publisher of this book and is not responsible for it under any aspect of press law.

ISACA
3701 Algonquin Road, Suite 1010
Rolling Meadows, IL 60008 USA
Phone: +1.847.253.1545
Fax: +1.847.253.1443
E-mail: *info@isaca.org*
Web site: *www.isaca.org*

ISBN 978-1-60420-106-2
Security, Audit and Control Features Oracle® E-Business Suite, 3rd Edition (Technical and Risk Management Reference Series)
Printed in the United States of America

Acknowledgments

ISACA wishes to recognize:

Researchers
Primary Research Team
Mark Sercombe, CISA, CA, CIA, Sponsoring Partner, Deloitte, Australia
Duncan Auty, CISA, Deloitte, Australia
Najeeba Hossain, Deloitte, Australia
Ryan Lee, Deloitte, Australia
Research Support Team
Vicky Vargas, CISA, Deloitte, Australia,
Gerardo Lopez, CISA, CISSP, Deloitte, Australia
James Mann, CISA, Deloitte, UK
Iain Muir, CISA, CA, CISSP, Deloitte, Australia

Expert Reviewers
Munsha Ahmed, KPMG LLP, Canada
Akin Akinbosoye, CISA, CISM, CGEIT, PMI-RMP, Gizmosearch Inc, USA
Mustapha Benmahbous, Ph.D., CISA, CISM, XPertics Solutions Inc., Canada
Madhav Chablani, CISA, CISM, TippingPoint Consulting, India
Stephen Coates, CISA, CGAP, CIA, Coates Associates Pty Ltd, Australia
Pinaki Das, SOAProjects, Inc., Canada
Mayank Garg, CISA, USA
Abdus Sami Khan, BE, MIE (PAK), MS, PE, Sami Associates, Pakistan
Prashant A. Khopkar, CISA, CA, USA
Stephen Kost, Integrigy Corp., USA
Larry Marks, CISA, CGEIT, CFE, CISSP, PMP, USA
Lucio Augusto Molina Focazzio, CISA, CISM, ITIL V3, Independent Consultant, Colombia
Jean-François Oligny, CA.IT, XPertics Solutions Inc., Canada
Megah Santio, Australian Taxation Office, Australia
Vinoth Sivasubramanian, ABRCCIP, CEH, ISO 27001 LA, UAE Exchange Center LLC, UAE
Vikrant V. Tanksale, ACWA, CMA, AlBahja Industrial Holdings LLC, Sultanate of Oman
John Tannahill, CISM, CGEIT, CA, J. Tannahill & Associates, Canada
William G. Teeter, CISA, CGEIT, PMP, USA
Andre van Winssen, CISA, CISSP, Oracle 10*g* Certified Master, Acision, The Netherlands
Chakri Wicharn, Fuji Xerox Co., Ltd., Thailand

ISACA Board of Directors
Emil D'Angelo, CISA, CISM, Bank of Tokyo-Mitsubishi UFJ Ltd., USA, International President
Christos K. Dimitriadis, Ph.D., CISA, CISM, INTRALOT S.A., Greece, Vice President
Ria Lucas, CISA, CGEIT, Telstra Corp. Ltd., Australia, Vice President
Hitoshi Ota, CISA, CISM, CGEIT, CIA, Mizuho Corporate Bank Ltd., Japan, Vice President
Jose Angel Pena Ibarra, CGEIT, Alintec S.A., Mexico, Vice President
Robert E. Stroud, CGEIT, CA Technologies, USA, Vice President
Kenneth L. Vander Wal, CISA, CPA, Ernst & Young LLP (retired), USA, Vice President
Rolf von Roessing, CISA, CISM, CGEIT, KPMG Germany, Germany, Vice President
Lynn C. Lawton, CISA, FBCS CITP, FCA, FIIA, KPMG Ltd., Russian Federation, Past International President
Everett C. Johnson Jr., CPA, Deloitte & Touche LLP (retired), USA, Past International President
Gregory T. Grocholski, CISA, The Dow Chemical Co., USA, Director
Tony Hayes, CGEIT, AFCHSE, CHE, FACS, FCPA, FIIA, Queensland Government, Australia, Director
Howard Nicholson, CISA, CGEIT, CRISC, City of Salisbury, Australia, Director
Jeff Spivey, CPP, PSP, Security Risk Management, USA, ITGI Trustee

Acknowledgments *(cont.)*

Knowledge Board
Gregory T. Grocholski, CISA, The Dow Chemical Co., USA, Chair
Michael Berardi Jr., CISA, CGEIT, Nestle USA, USA
John Ho Chi, CISA, CISM, CBCP, CFE, Ernst & Young LLP, Singapore
Jose Angel Pena Ibarra, CGEIT, Alintec S.A., Mexico
Jo Stewart-Rattray, CISA, CISM, CGEIT, CSEPS, RSM Bird Cameron, Australia
Jon W. Singleton, CISA, FCA, Auditor General of Manitoba (retired), Canada
Patrick Stachtchenko, CISA, CGEIT, CA, Stachtchenko & Associates SAS, France
Kenneth L. Vander Wal, CISA, CPA, Ernst & Young LLP (retired), USA

Guidance and Practices Committee
Kenneth L. Vander Wal, CISA, CPA, Ernst & Young LLP (retired), USA, Chair
Christos K. Dimitriadis, Ph.D., CISA, CISM, INTRALOT S.A., Greece
Urs Fischer, CISA, CRISC, CIA, CPA (Swiss), Switzerland
Ramses Gallego, CISM, CGEIT, CISSP, Entel IT Consulting, Spain
Phillip J. Lageschulte, CGEIT, CPA, KPMG LLP, USA
Ravi Muthukrishnan, CISA, CISM, FCA, ISCA, Capco IT Service India Pvt. Ltd., India
Anthony P. Noble, CISA, CCP, Viacom Inc., USA
Salomon Rico, CISA, CISM, CGEIT, Deloitte, Mexico
Frank Van Der Zwaag, CISA, CISSP, Westpac, New Zealand

To ISACA member Lily M. Shue, CISA, CISM, CGEIT, CCP, LMS Associates LLP, USA, for her financial support

Table of Contents

Page intentionally left blank

Preface

This book is the third edition of *Security, Audit and Control Features Oracle® E-Business Suite* (Technical and Risk Management Reference Guide). Oracle Corp. is one of the leading developers of enterprise resource planning (ERP) applications, which are applications that integrate an enterprise's operations. Although Oracle markets JD Edwards EnterpriseOne and Oracle E-Business Suite as ERP solutions, its primary ERP product is the Oracle E-Business Suite (EBS). This third edition of the technical reference guide is one in a series of three technical reference guides providing information relating to the world's three major ERP systems. The other guides in the series focus on SAP and PeopleSoft. A related publication in the technical reference guide series is *Security, Audit and Control Features Oracle® Database, 3rd Edition*.

The purpose of this guide is to provide an update on current industry standards and identify future trends in Oracle EBS risk and control. The objective is to enable audit, assurance, risk and security professionals (IT and non-IT) to evaluate risks and controls in existing ERP implementations, and facilitate the design and implementation of better practice controls into system upgrades and enhancements. This book also aims to assist system architects, business analysts and business process owners who are implementing Oracle EBS, as well as people responsible for managing it in live production to maintain the appropriate level of control and security according to business needs and industry standards. This publication is designed to be a practical how-to guide based on Oracle EBS 12.1, with a primary focus on the Oracle EBS Financials applications.

The popularity of the earlier editions of this guide confirmed the need for a series of definitive audit guides for these products. Using a definitive approach, the authors sought to provide detail on testing techniques within the ERP products and their execution, rather than generic descriptions of the audit tests to be performed.

Page intentionally left blank

1. Executive Introduction

Prior to ERP systems, an enterprise's applications were typically organized around functions or departments (e.g., sales, purchasing, inventory and finance), rather than business processes (e.g., purchase to pay, order to cash). These applications evolved independently of each other, which resulted in data redundancy and data errors across the disparate systems. More often than not, these systems had been developed on different platforms, resulting in technical diversity and the need for complex interfaces between systems.

ERP systems, on the other hand, have a business process focus. An ERP system is a packaged business software system that allows an enterprise to:

- Automate and integrate its core business processes.
- Share common data and practices across the entire enterprise.
- Produce and access information in a real-time environment.
- Provide single access to enterprisewide data.

Their relational database tables are designed around a complete set of core functions, rather than disparate modules that merely pass transaction data from one module to another. While traditional paper-based audit trails are replaced by electronic need-based logs, a unique feature of ERP systems is the standard exception reports that can be relied upon to monitor the health of a system in a real-time mode. Controls shift from detective to preventive, and traditional matching reconciliation controls are automated in the ERP software. Consequently, it makes business sense to ensure that adequate controls are properly integrated into the reengineered ERP-enabled processes.

Key factors to consider when implementing an ERP are:

- Senior management buy-in
- Data ownership
- Staff training and communication
- Configuration options
- Data conversion

The implementation of an ERP system can introduce new risks and alter an enterprise's risk profile. The first step in embarking upon an ERP initiative is to carry out a business process risk and control assessment followed by a detailed evaluation of available ERP options. A risk and control assessment requires a framework covering the areas of business process controls, application security, program change controls, data conversion controls, technology infrastructure and project management. When control issues are identified, the auditor should attempt to uncover and report to management the causes of the problem together with recommendations. In this respect, ISACA's COBIT helps meet the multiple needs of management by bridging the gaps among business risks, control needs and technical issues. COBIT 4.1 provides guidance across a domain and process framework and presents activities in a manageable and logical structure. It provides a measure against which to judge when things do go wrong and can assist in identifying the root causes of problems. Once implemented, the ERP environment should be subjected to regular risk and control assessments scheduled at planned intervals. This process should be aligned with the enterprise's risk management framework, e.g., *The Risk IT Framework* issued by ISACA.

The first-year audit of enterprises that have implemented ERP systems needs to be carefully scoped since the enterprise may be using a combination of centralized accounting controls and decentralized operational controls. A detailed knowledge of the ERP system is necessary to effectively understand security and control issues over application areas and the technical environment prior to the use of automated diagnostic tools to review security configurations and data integrity.

In the web-enabled ERP environment, control solutions for risks associated with e-business must be developed. The traditional control framework that focused on end users accessing the system through traditional workstations must be extended to include identity management, content quality, privacy, collaborative commerce and integrity.

What Is New in This Edition

The second edition of this guide was based on release (R) 11*i*.10. Since then, two releases of Oracle EBS have been made available. The new releases contain additional functionality, which are explained in more detail in chapter 2. This guide covers the core financial modules (Financial Accounting and Expenditure) of Oracle EBS Release 12.1. It also contains audit programs updated for release 12.1, which contain references to COBIT 4.1. Future directions for Oracle EBS are explored, and a new chapter on continuous monitoring in the Oracle EBS environment is also included.

How This Book Is Organized

Introduction to Oracle EBS and ERP Systems

The evolution of ERP software is described—from its early beginnings in material requirements planning packages centered around manufacturing, to the present-day systems that provide enterprisewide integrated solutions. The benefits and characteristics of ERP systems are discussed. Oracle Corp. and the innovation surrounding the development of Oracle EBS are introduced. Major Oracle EBS modules and functionality are overviewed. The publication outlines the manner in which an ERP implementation and its associated business process changes transform critical elements of the business, including the control environment.

Risk Management in an ERP Environment

Business risks (e.g., business process, application and technical infrastructure security, data conversion, program interface, and project management risks) and key management controls for ERP implementations are outlined, leading to a discussion of the importance of establishing a control framework for ERP environments. The impacts on the audit following the implementation of an ERP system are also described. The purpose of this chapter is to help enterprises minimize the risk of not obtaining the significant benefits that can flow from a well-executed ERP implementation.

ERP Audit Approach

This section details how the implementation of an ERP system affects the audit process, and provides frameworks and methodologies for auditing and testing the Financials module in an Oracle EBS environment. These include a recommended Oracle EBS audit framework, how to adopt a risk-based audit approach to ERP, an overview of the Oracle EBS security concept, configurable controls and segregation of duties/excessive access. The need to identify the causes of issues arising from audit or control testing and a technique to assist in identifying the cause of issues using the COBIT framework are also described.

Auditing Oracle EBS—Core Financial Business Cycles (Financial Accounting and Expenditure) and Security

The relationship between Oracle EBS modules and the major business cycles operating within enterprises is explained. An overview of the core financial business cycles for an enterprise (i.e., financial accounting and expenditure) and their integration is provided. For each of these business cycles, the functionality of the Oracle EBS business process and their subprocesses are described from a controls perspective. Specific risks are identified, potential automated controls are outlined and sample testing techniques are suggested. Techniques for testing user access to business cycle functions and segregation of duties (SoD) are also covered.

The Oracle EBS Security sections provide an overview of the Oracle EBS Internet computing infrastructure and associated Security Administration functionality. They provide details of the risks, key controls and techniques to consider when testing security administration.

Continuous Monitoring in an Oracle EBS Environment

The tools available for continuous control monitoring within an Oracle EBS environment are explored. The main features of the Oracle Governance, Risk and Compliance (GRC) solutions are provided as an example of continuous monitoring tools. The key risks associated with these tools that should be considered as part of an audit are also discussed.

New Directions in Oracle EBS and ERP Audit

The guide concludes by looking at the various stages through which ERP auditing has progressed and looks ahead to major emerging directions for ERP auditing (e.g., improvements in application security assurance and data security assurance, and the changing compliance landscape). Oracle's Fusion Middleware products are also described with regard to the current and anticipated developments of this product, particularly in light of the current fluid business climate, where the direction for software vendors is to engage in partnerships, collaborations, acquisitions and mergers.

Who Should Read This Book

This publication has been written with the business manager in mind. IT, audit and assurance professionals and security, compliance and risk management professionals will also find this publication to be highly informative and helpful. Parts of the publication are written for those looking to learn more about how Oracle EBS Financials works, as well as the strategic and risk management issues. However, for the most part, the book assumes that the reader has a fundamental working knowledge of Oracle EBS.

What Makes This Book Different

Although there are many books that have been written about Oracle EBS, they are focused more narrowly on implementation, business aspects or how one of the Oracle EBS modules works. This publication is unique in that it deals with aspects of risk management, audit, security and control over Oracle EBS. It contains audit/assurance programs, audit suggestions and internal control questionnaires (ICQs) for the business cycles addressed within the publication.

2. Introduction to Oracle E-Business Suite and ERP Systems

This chapter provides an introduction to ERP systems in general and Oracle EBS in particular. Key changes from previous versions of Oracle EBS are described, as well as navigation techniques and key functionality.

Before ERP systems were developed, an enterprise's applications were typically designed around functions or departments (e.g., sales, purchasing, inventory and finance), as shown in **figure 2.1**, and not by business processes (e.g., purchase to pay, order to cash). A function evolved independently and might have had support from an individual application system or a number of systems by manual or system interface. This approach resulted in delays, additional costs, data redundancy and the need for reconciliation. Frequently, business controls had a significant number of manual processes. For example, when the invoice arrived, the purchase order (PO) was either printed out again or retrieved from the files, and then stapled to the invoice. The invoice was then approved for payment. The documents may have been scrutinized once again and approved during the check payment process.

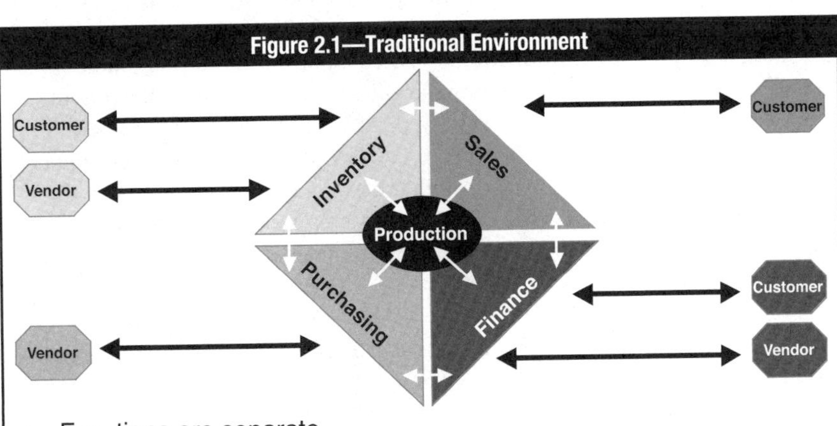

Figure 2.1—Traditional Environment

- Functions are separate.
- Interface and communication links with other functions.
- Variety of systems
- Each function maintains its own master data.
- Duplication of data entry

Non-ERP systems are typically designed around disparate and independent modules that transmit transaction data among themselves by means of interfaces, where the information is normally summarized (e.g., totals or balances only). In such cases, details of transactions are often difficult to ascertain, unlike the ability to drill down as provided by ERP systems.

ERP systems have a business process focus. They grew out of the need to integrate separate material resource planning (MRP) systems (used to integrate material requirements to production, demand and capacity) and financial accounting systems in manufacturing organizations. The integration of these functional capabilities into an online and real-time application system designed to support end-to-end business processes enables organizations to plan and optimize their resources across the enterprise. Oracle EBS is designed using a Relational Database Management System (RDBMS), allowing information to be shared among modules rather than passing transactional data from one module to another. The financial module in Oracle EBS is integrated with other modules, such as supply chain, sales and manufacturing. With correct configuration, the data input in the other modules is updated automatically in the financial module through seamless system integration configuration. For example, when a clerk at a shipping dock records that goods have been shipped to a customer, changes are made in order management, billing and inventory. The new transaction is used to update the revenue recognition and the cost of goods sold systems.

An ERP environment operates in line with the business—online and in real time. Oracle EBS provides online transaction processing (OLTP) with the flexibility to perform high-volume tasks in a batch process mode. Management has access to online and up-to-date information on how the business is performing. Common and consistent information is shared among application modules and users from different departments simultaneously. For example, following the implementation of an ERP, enterprises typically report completion of period or year-end closes in one or two days, as opposed to two to three weeks under their legacy system environments. Another key change brought about by the implementation of ERP systems is that the systems are owned and driven by business process owners/end users, with the technical support of IT, rather than being owned and driven by IT alone.

Enterprises implementing ERP systems can achieve significant benefits, such as:

- Improvements in:
  - Personnel deployment to more value-producing activities
  - Productivity
  - Order management cycle
  - Cash management
  - On-time delivery
- Reductions in:
  - Inventory
  - Financial close/cycle
  - IT cost
  - Procurement cost
  - Transportation/logistics cost
  - Hardware and software maintenance

The intangible benefits of an ERP implementation—while difficult to quantify—can deliver significant business value through capability improvements, including:

- Information visibility (e.g., drill-down capability, consistent and reliable information across business areas)
- New/improved processes
- Improved customer responsiveness
- Integration and standardization
- Flexibility
- Globalization

Oracle Software

Oracle Corp. products can be categorized in two broad areas: systems software and applications.

Systems software includes a platform for developing and deploying applications on the Internet and corporate intranets. Systems software products include database management software; application server software; and development tools that allow users to create, retrieve and modify the various types of data stored in a computer system. Oracle Corp. applications, which are now designed to be accessed with a standard web browser on a client computer, automate the performance of business processes and functions for Asset Lifecycle Management, Customer Relationship Management, ERP, Supply Chain Management, Product Lifecycle Management, Procurement and Manufacturing.

The software runs on a broad range of computers, including central and distributed processing servers, mainframes, workstations, personal computers, laptop computers and information appliances (such as handheld devices and mobile phones), and is supported on numerous operating systems, including Windows and variants of UNIX, such as Linux, Solaris, AIX and HP-UX.

Systems Software

The Oracle RDBMS, the key component of Oracle's Database platform, enables storing, manipulating and retrieving relational, object-relational, multidimensional and other types of data. In 2004, the Oracle Corp. introduced Oracle RDBMS 10*g* ("*g*" standing for "grid") as the then latest version of Oracle Database. Oracle Application Server 10*g* using Java 2 Platform Enterprise Edition (J2EE), integrates with the application server part of that database version, making it possible to deploy web-technology applications. The application server comprised the first middle-tier software designed for grid computing. The interrelationship between Oracle RDBMS 10*g* and Java has enabled Oracle Corp. to allow developers to set up stored procedures written in the Java language, as well as those written in the traditional Oracle Database programming language (PL)/Structured Query Language (SQL). The latest release replacing Oracle RDBMS 10*g* is 11*g*, which is now certified with Oracle EBS. Introduced in 2007, Oracle RDBMS 11*g* includes additional features such as built-in testing for changes, the capability for viewing table logs, extended compression of all types of data types and enhanced disaster recovery functions.

The Oracle Developer Suite is an integrated suite of development tools for rapidly developing Internet database applications and web services. Built on Internet standards such as Java, Extensible Markup Language (XML), Common Object Request Broker Architecture (CORBA) and Hypertext Markup Language (HTML), Oracle Developer Suite contains application development tools, business intelligence tools, and database and data warehouse design tools.

Application development tools include Oracle Designer, Oracle Forms Developer, Oracle JDeveloper and Oracle Software Configuration Manager. Oracle Designer allows developers to model business processes and automatically generate enterprise database applications. Oracle Forms Developer allows for the building of database applications that can be deployed unchanged in Internet and client/server-based environments. For Java programmers, Oracle JDeveloper provides a Java development tool suite for building applications for use on the Internet. Oracle Software Configuration Manager helps manage structured and unstructured data and different file types throughout the software development life cycle.

Oracle Corp. business intelligence tools and database and data warehouse design tools are designed for the Internet and provide a comprehensive and integrated suite of products that enable enterprises to address the full range of user requirements for information publishing, data exploration, advanced analysis and data mining. Business intelligence tools include Oracle Business Intelligence Beans and Oracle Reports Developer. Oracle Database and data warehouse design tools include Oracle Warehouse Builder and Oracle Discover.

Oracle EBS

Oracle EBS R12.1 is a fully integrated and Internet-enabled set of applications. Oracle EBS offers business flow applications, enabling enterprises to automate discrete business flows, such as procurement to payment or order to cash. The applications combine business functionality with technologies such as workflow and self-service applications, and enable customers to lower the cost of their business operations by providing their customers, suppliers and employees with self-service access to transaction processing and selected business information using the Internet platform. Self-service applications automate a variety of business functions, such as customer service and support, procurement, expense reporting, and reimbursement.

Main Updates in Releases 12 and 12.1

Oracle Corp. released EBS R12 in 2007, which introduced changes from R11*i*.10 to the application technology platform, provided changes to the security authorization concept and a new architecture for Oracle EBS Financials. The latest version, Oracle EBS R12.1, was released in 2009 and introduced changes to the other enterprise application areas (e.g., Supply Chain Management, Procurement, Customer Relationship Management and Human Capital Management). Although this book is based on R12.1, the section below briefly summarizes the major changes introduced in both R12 and R12.1.

New R12 Technology Stack

Having acquired a variety of software application vendors and their application suites, Oracle Corp. has embarked on the ambitious software engineering exercise of combining each of the disparate application suites into a consolidated application suite. The name given by Oracle Corp. to this next-generation enterprise application suite is Fusion Applications. Fusion Applications are planned to combine the best features of their current successful applications, including Oracle EBS, Siebel's Customer Relationships Management (CRM) Suite, PeopleSoft Enterprise Applications, and JD Edwards' World and EnterpriseOne Application Suites among other smaller software acquisitions and developments. With the introduction of Oracle Fusion Middleware, the previous Application Server Architecture was replaced with a Service Oriented Architecture (SOA), allowing for a more web-based, interoperable and integrated technology platform for Oracle EBS. This has also resulted in conformance of the user interface for Oracle EBS to Web 2.0 standards from R12 onward.

New Security Authorization Concept With Role-Based Access Control

With Oracle EBS R12, the core security area of Oracle User Management (UMX) comes standard with a role-based access control (RBAC) model that builds on existing function security and data security models. RBAC was first implemented in EBS R11.5.10, but fully integrated and supported with a full set of features in R12. RBAC was briefly mentioned in the previous edition of this guide, but detail has been included in this version. More information regarding the security authorization concept is provided in chapters 4, 9 and 10.

New Oracle EBS Financials R12 Architecture

There are more than 300 additional features introduced in Oracle EBS R12, including several changes to the Financial Application Architecture:[1]

- **Ledger sets**—Allow processing and reporting of multiple ledgers simultaneously, including the ability to view and report, open and close

[1] Melatti, Annette; "The Business Value of Upgrading to Oracle E-Business Suite Financials Release 12," Oracle White Paper, USA, 2008

periods, create journal entries and perform allocations across ledgers

- **Multi-organizational access control (MOAC)**—Provides role-based access to operating units, enabling the performance of multiple tasks across operating units without having to change responsibilities
- **Subledger accounting**—Provides a rules-based accounting engine aimed at improving the monitoring, controlling, auditing and reconciling of subledger accounting entries
- **Tax engine**—Provides a centralized repository, Oracle E-Business Tax, for managing transaction tax rules and transactions
- **Advanced global intercompany system**—Aims to streamline the intercompany and trading reconciliation process across ledgers
- **Bank model**—Associates bank accounts with legal entity rather than operating unit, resulting in more centralized banking options

Rapid Value Stand-alone Solutions

Another of the changes introduced by Oracle EBS R12.1 is the ability for enterprises to implement rapid value stand-alone solutions that are compatible with existing Oracle EBS R11*i* or R12 environments, without having to perform a major upgrade to R12.1.

Global Business Platform

A global business platform offering centralized administration of global shared services means that responsibility for functional tasks (e.g., financial reporting) across all organizational entities can now be managed centrally. This is aimed to help improve standardization and simplification throughout the enterprise.

Major Oracle EBS R12.1 Modules and Functionality

This section introduces Oracle EBS and the major applications within it. Oracle EBS applications are an integrated suite of modules identified by major business functions.

Oracle EBS applications are as follows:

- Asset Lifecycle Management
- Customer Relationship Management
- Enterprise Resource Planning
  - Channel Revenue Management
  - Financial Management (referred to also as Oracle EBS Financials)
  - Human Capital Management
- Procurement

- Product Lifecycle Management
- Supply Chain Management
  - Supply Chain Planning
  - Logistics and Transportation Management
  - Order Management
  - Price Management
- Manufacturing

Oracle EBS Financials Management

The modules contained within the Financial Management application are as follows:

- Asset Lifecycle Management
- Cash and Treasury Management
- Credit-to-Cash
- Financial Control and Reporting
- Financial Analytics
- Governance, Risk and Compliance
- Internal Controls Manager
- Lease and Finance Management
- Procure-to-Pay
- Travel and Expense Management

As the focus of this guide is on the Oracle EBS Financials application, the modules contained in the other application areas have not been specified. More information regarding the other application areas can be found on the Oracle Corp. web site (*www.oracle.com*).

This technical reference guide is part of a series of three guides. The series is intended to be considered collectively; therefore, common business processes and related risks and control features are not covered in every guide. For example, risks and typical controls associated with human resources and payroll are considered in detail in *Security, Audit and Control Features PeopleSoft,® 2nd Edition*, and inventory risks and controls are considered in *Security, Audit and Control Features SAP® ERP, 3rd Edition*. This guide covers in detail the key risks and controls associated with the core financial modules (Financial Accounting and Expenditure) of Oracle EBS R12.1. However, many risks dealt with in the other guides may be applicable to the core functional modules not covered in detail in this guide. For example, this publication is not focused on some of the risks associated with the more common applications, such as manufacturing and human resources, and these could be considered when planning an audit. These lists should not be considered exhaustive and will be dependent on the implementation and processes within the enterprise.

Navigating the Oracle EBS R12.1 System

This section provides an overview of Oracle EBS basic navigation techniques and methods. Oracle EBS is an integrated group of applications that look and feel the same. The login process is described in the following section.

Logging In—Oracle EBS

The login screen for the EBS is displayed in **figure 2.2**.

Figure 2.2—Oracle EBS Login Screen

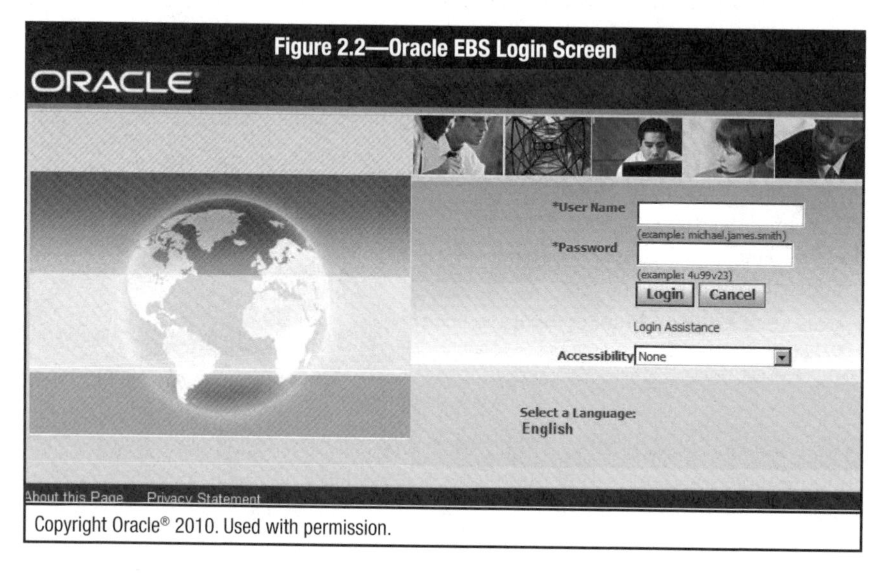

Once users successfully log in by entering their usernames and passwords, they are presented with the EBS home page. This page allows users to access the EBS functions, grouped by responsibility, from the applications menu. Appearance of the login screen may vary between versions of the EBS, such as the CRM and the Oracle Access or Applications Manager (OAM) login pages. Users may also set preferences and navigate to frequently used functions or self-service pages from their favorites. The EBS home page can be seen in **figure 2.3**.

Responsibilities listed on the EBS home page are a level of authority assigned to a user in Oracle EBS that enables user access to functions that may be appropriate to the user's organizational role. A user may have one or more responsibilities. The example functions list illustrated in **figure 2.3** was populated as a result of selecting the System Administrator profile.

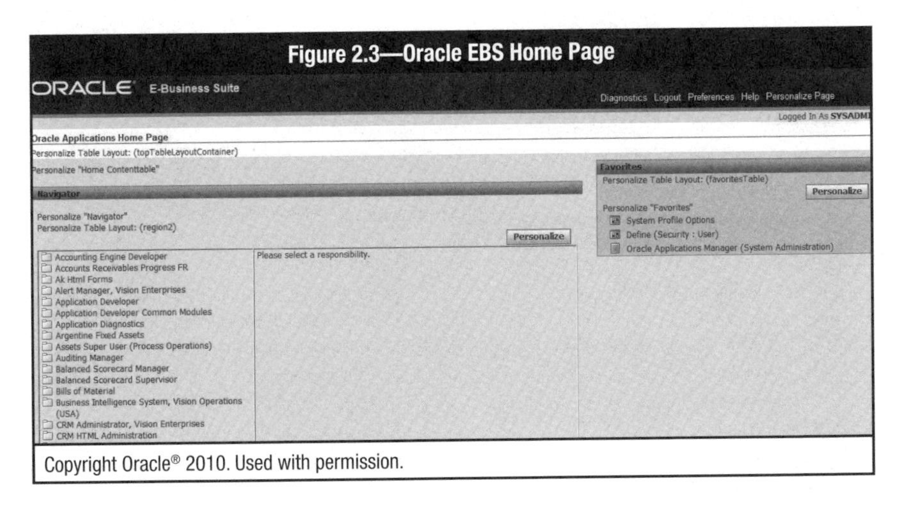

Figure 2.3—Oracle EBS Home Page

Copyright Oracle® 2010. Used with permission.

To access a specific function, select a responsibility and then select the function to launch it. Depending on the function selected, two types of interface may be launched. These include forms-based applications that are designed to process a large volume of transactions and HTML-based applications, which may also be referred to as self-service applications. The HTML-based interface is used for low-volume and broad-use audiences because no software is required to be installed on the end user's computer. Forms-based views and HTML-based views may be seen in **figures 2.4** and **2.5**, respectively. Forms-based views and HTML-based views may also be differentiated by the icons listed on the EBS home page. These icons are shown in **figure 2.6**.

Forms-based View—Oracle EBS

When selecting a forms-based function, the Oracle Applications Navigator is launched. From the Navigator, a user can do one of the following:

- **Open forms from the Functions tab**—The Functions tab provides links to the forms that are accessible to a user's responsibility. To open a form, the function heading can be expanded and the form name can be selected.
- **Link to documents from the Documents tab**—The Documents tab can be customized to allow the user to create links to frequently used documents, e.g., POs, invoices, sales orders, employee information or plans. This feature allows the user to access these documents later. The user may create as many links as desired. Any documents opened using this feature are opened in the appropriate form window.
- **Launch business processes from the Processes tab**—The Processes tab provides graphical maps of the business processes, allowing users to view business flows across form screens and, if permitted by their responsibilities, to initiate action. The Processes tab guides users through each required function in the business process. The Process Navigator tab also launches the appropriate forms and standard reports in each step of the business process.

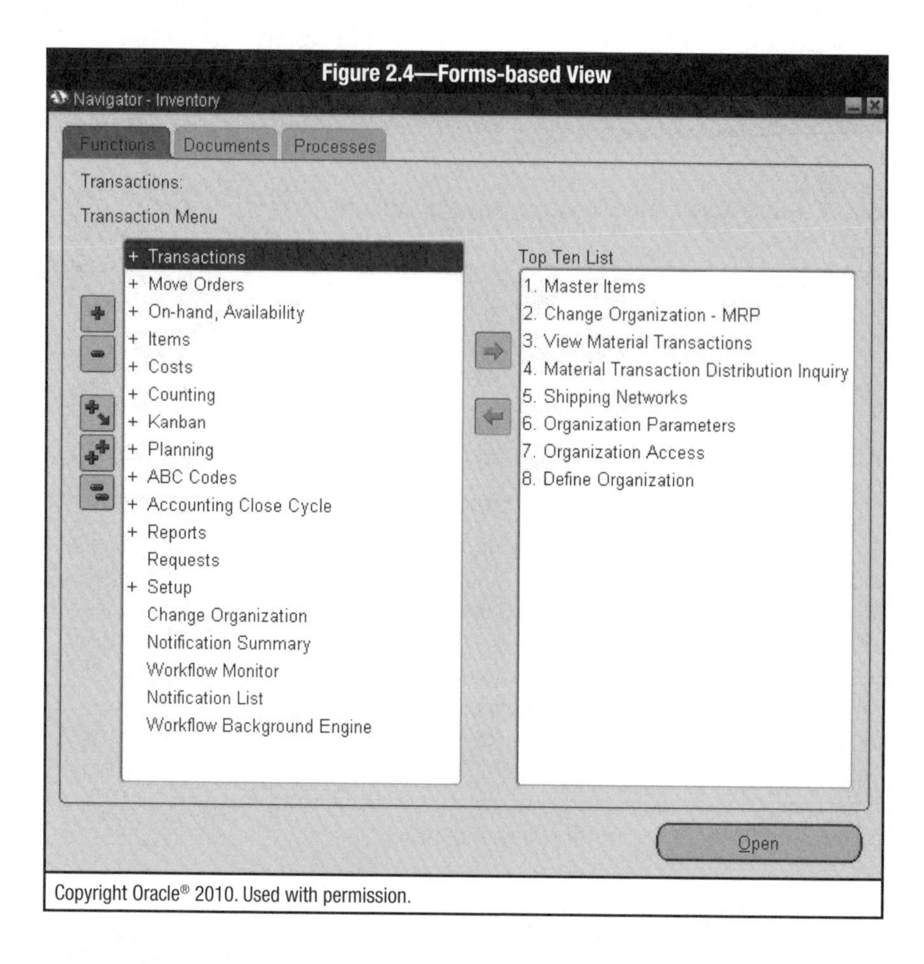

Figure 2.4—Forms-based View

Figure 2.5—HTML-based View

ORACLE

Diagnostics Preferences Help Personalize Page Close Window

Notifications

Enter filter criteria such as Notification ID, Owner, To, From, Workflow Type, Type Internal Name, Subject

* Indicates required field

Search

Personalize "Search"

Notification ID

Find the Notification matching this Notification ID only, other search parameters will be ignored

Owner: All Employees and Users

To: All Employees and Users

From: All Employees and Users

* Status: Open

Workflow Type

Type Internal Name

Subject

* Sent Date: Any Time

* Due Date: Any Time

* Priority: All

Personalize "Button Region"

Go Clear

Figure 2.6—Icons

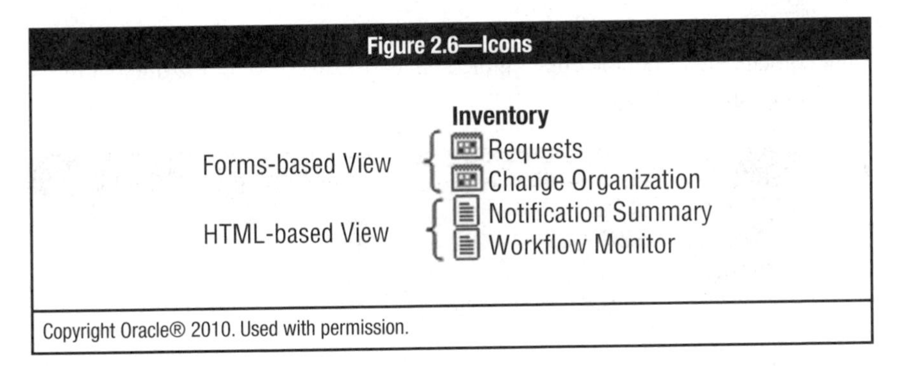

Expand or Collapse the Navigator Window List

Select one of the following to expand an expandable item to its next sublevel:

- Double click on the menu option.
- Select the menu option, and click on the Open button.
- Select the menu option, and click on the + button.

Select one of the following to collapse an expanded item:

- Double click on the Menu option.
- Select the Menu option, and click on the - button.

To expand or collapse several items simultaneously, use one of the following:

- **Expand all children** +->—Expands all the sublevels of the selected item
- **Expand all** ++—Expands all the sublevels of all expandable items in the navigation list
- **Collapse all** --—Collapses all currently expanded items in the navigation list

Open a Form

Either of the following can be used to open a form:

- Select the desired menu option and click on the Open button.
- Double click on the desired menu option.

Open a Form Using a List of Values (LOV) Window

To open a form using an LOV window:

- Press the Ctrl and L keys to open an LOV window, which will bring up a list of form functions, as shown in **figure 2.7**.
- Select the desired form and click on the OK button, or create a short list by keying in a partial form name.

Top Ten List

Frequently used forms can be copied into the Top Ten List. Forms are displayed numerically. A maximum of 10 forms can be selected for each responsibility assigned to a user. To add an item to the Top Ten List, select the form and then select the Add to List button. To remove a form, select the form and then select the Remove from List button. To open a form from the Top Ten List on the Function tab, either double click on the item on the list or enter the number of the item on the list.

Figure 2.7—List of Values

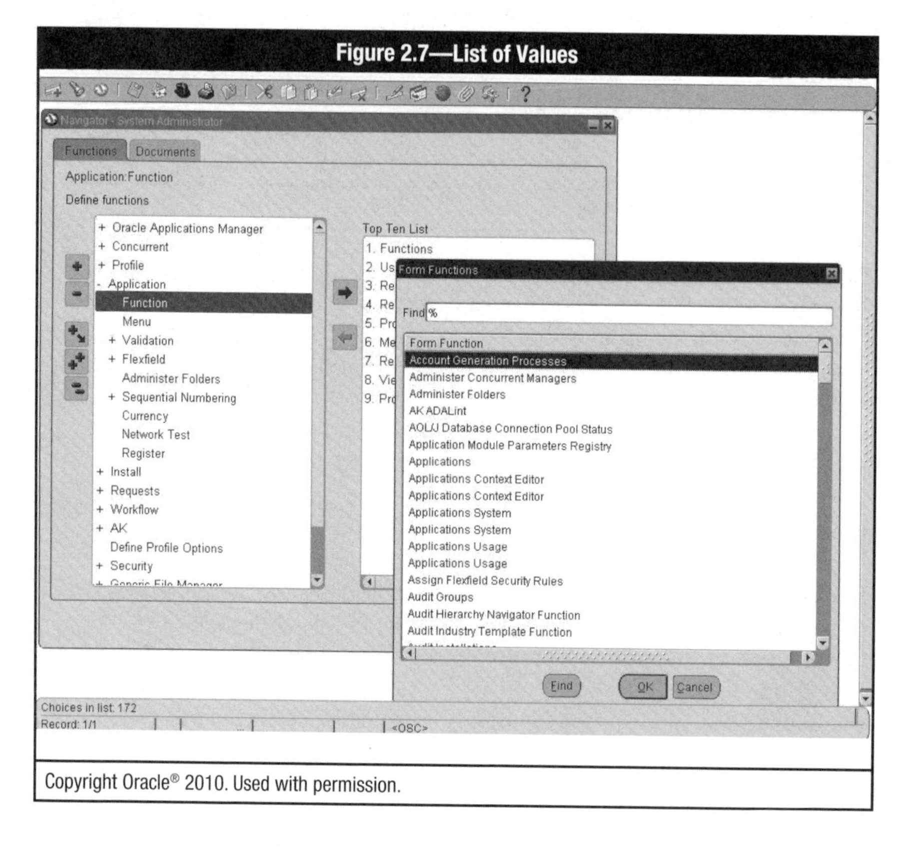

HTML-based View—Oracle EBS

When selecting an HTML-based interface within Oracle EBS, an HTML-based window is launched, as shown in **figure 2.8**. Common attributes of an HTML-based form include:

1. Navigation links
2. Date picker
3. Global navigation

Generally, the screens shown in **figures 2.8** and **2.9** are secured to prevent nonexpert users from changing their regional settings. Entry or validation errors may occur if the regional settings are set up inadequately for some users.

Figure 2.8—HTML-based Window

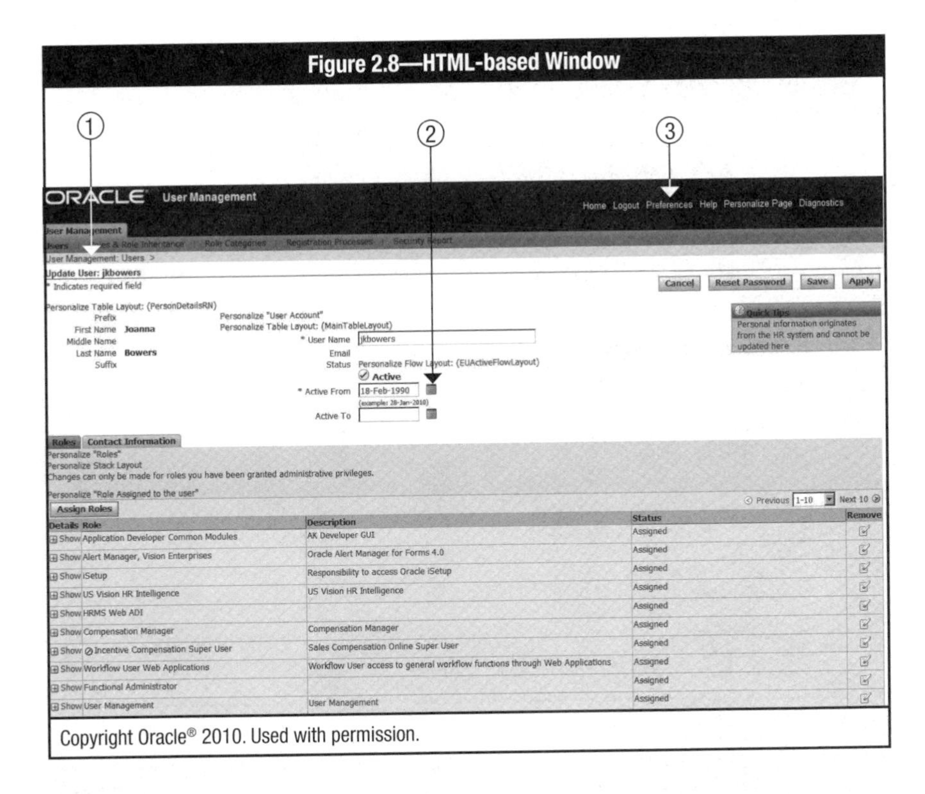

Switching Responsibilities—Oracle EBS

Users may change the responsibility that they are logged into in two ways:

1. By selecting a different responsibility from the Oracle EBS home page, as shown in **figure 2.9**
2. By using the Switch Responsibility item from the File menu, as shown in **figure 2.9**. Once it is selected, users will then be presented with a list of responsibilities from which they can select within Oracle EBS. This list will be dependent on responsibilities assigned to those particular users, as shown in **figure 2.10**.

Figure 2.9—Switching Responsibilities

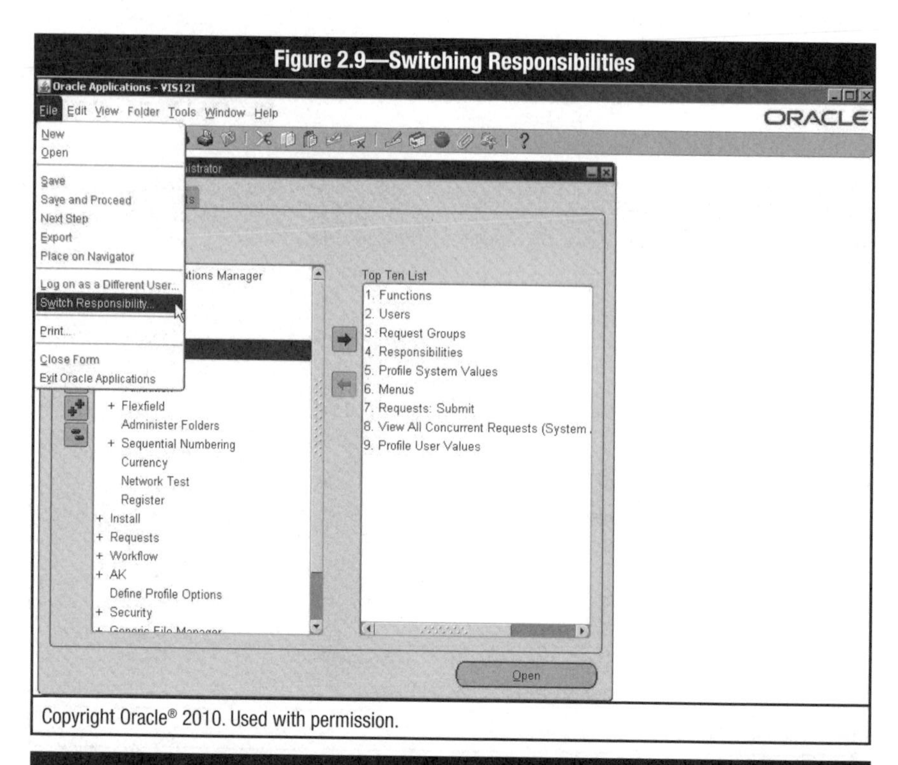

Copyright Oracle® 2010. Used with permission.

Figure 2.10—Responsibilities

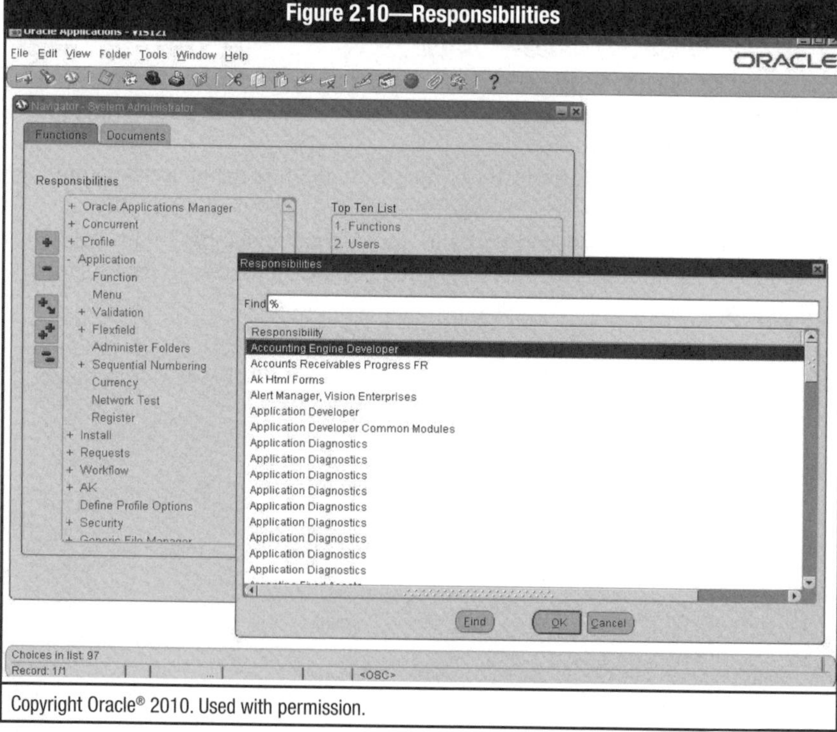

Copyright Oracle® 2010. Used with permission.

Home Page Preferences

The ability to change certain preferences exists within a user's personal home page. Preferences include languages, date formats and number formats, notification style, and password changes. The preferences can be changed by selecting the Preferences link in the global navigation, as shown in **figure 2.3**, on the EBS home page. Once the changes have been made, the Apply button should be clicked for the changes to take effect. The Preferences screen used to choose personal preferences is shown in **figure 2.11**.

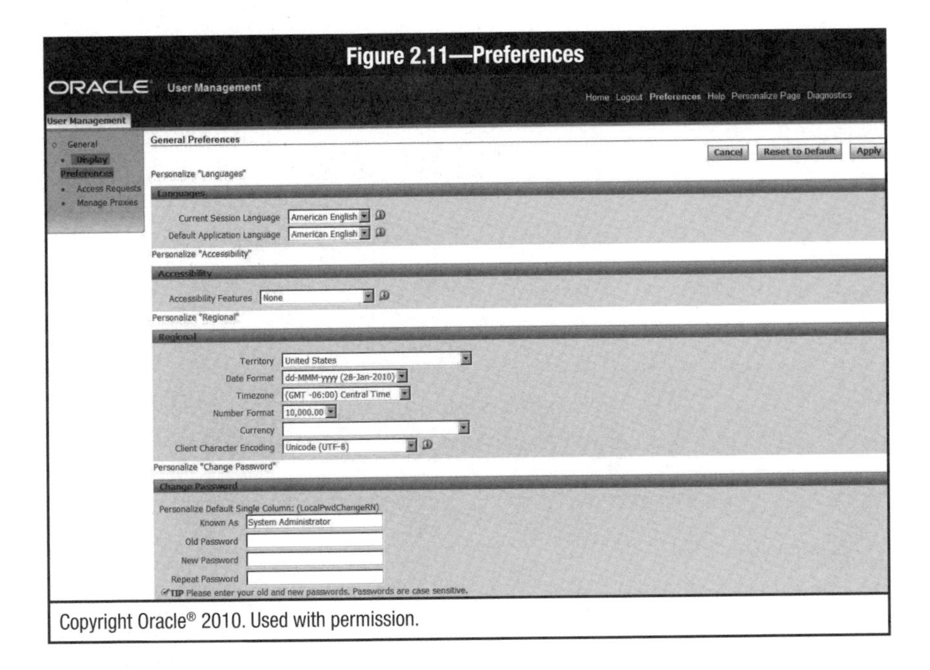

Figure 2.11—Preferences

Keyboard Shortcuts

Users can bypass the menu by using keyboard shortcuts. Keyboard shortcuts are keystrokes that perform the same functions as corresponding menu items. A list of keyboard shortcuts can be displayed by selecting Keyboard Help from the Help menu, as shown in **figure 2.12**. Many common keyboard shortcuts can be selected from the toolbar.

Figure 2.12—Keyboard Shortcuts

Keys

Function	Key
Actions LOV	Shift+Ctrl+F8
Block Menu	Ctrl+B
Clear Block	F7
Clear Field	F5
Clear Form	F8
Clear Record	F6
Commit	Ctrl+S
Count Query	F12
Delete Record	Ctrl+Up
Display Error	Shift+Ctrl+E
Down	Down
Duplicate Field	Shift+F5
Duplicate Record	Shift+F6
Edit	Ctrl+E
Enter Query	F11
Execute Query	Ctrl+F11
Exit	F4
Function 0	Shift+Ctrl+F10

OK

Oracle Help

An online help feature is available in Oracle EBS. The help feature is accessed by clicking on the Ctrl and H keys or by clicking on Window Help from the Help menu. Help topics can be selected using the Contents tab or the Search tab.

Oracle Report Manager

Oracle Report Manager is an online report distribution system that provides a secure and centralized location to produce and manage real-time reports. The Oracle Report Manager can be used to submit, publish, set security for reports, and view and approve reports. There are four types of reports that can be accessed in Oracle EBS:

- **Single Report**—A complete report for a specific time period
- **Single Report Over Time**—A single report produced for different periods
- **Expanded Report**—One report displayed in a variety of ways depending upon a specific variable
- **Expanded Report Over Time**—An expanded report available for different time periods

The system administrator assigns access to the Oracle Report Manager menu items to various responsibilities.

Submitting Requests

Standard request submission is a feature that provides a common interface to run Oracle EBS reports and programs. This feature provides a set of windows for running reports and programs, as well as control over the submission and output of reports and programs. Reports can be viewed online or printed. Reports run as concurrent programs in Oracle EBS.

Concurrent processing options allow the user to:
- Specify the number of copies to print
- Select print style
- Select the printer
- Hold a request
- Specify dates and times to run a request
- Save the results in a standard file format

Requests can be run as either a:
- **Single Request**—A single report and/or program to be run. To submit a Single Request, select the Requests menu option in the Navigator window. Select the Run option. Click on Single Request in the pop-up window, and click the OK button.
- **Request Set**—A collection of reports and/or programs that the user can group together. All the reports and/or programs of a Request Set can be submitted in a single step using a single transaction. To submit a Request Set, select the Requests menu option in the Navigator window. Select the Run option. Click on Request Set in the pop-up window, and click the OK button.

Figure 2.13 shows the step of selecting between a Single Request and Request Set.

To view submitted requests, select the Requests menu option in the Navigator window. Users are then directed to the Find Requests window in which they are able to define the search criteria as shown in **figure 2.14** on page 23. If All My Requests is selected, a list of all reports will be generated, as shown in **figure 2.15** on page 24.

Oracle EBS allows the security of output requests to be set for sharing by users who have the same responsibility or to not be shared at all (Profile option). If not shared (default behavior), even the System Administrator does not have access to the output of the request by the application. The screen in **figure 2.15** is available in the administrative mode or end user's mode.

Searching in the Application

To search within the application, click on the Find icon on the toolbar. This will invoke a search within a given field. Only certain fields allow searches to be performed. The number of options available will depend on the field selected and the application setup. The Search forms allow the user to enter % (for all)

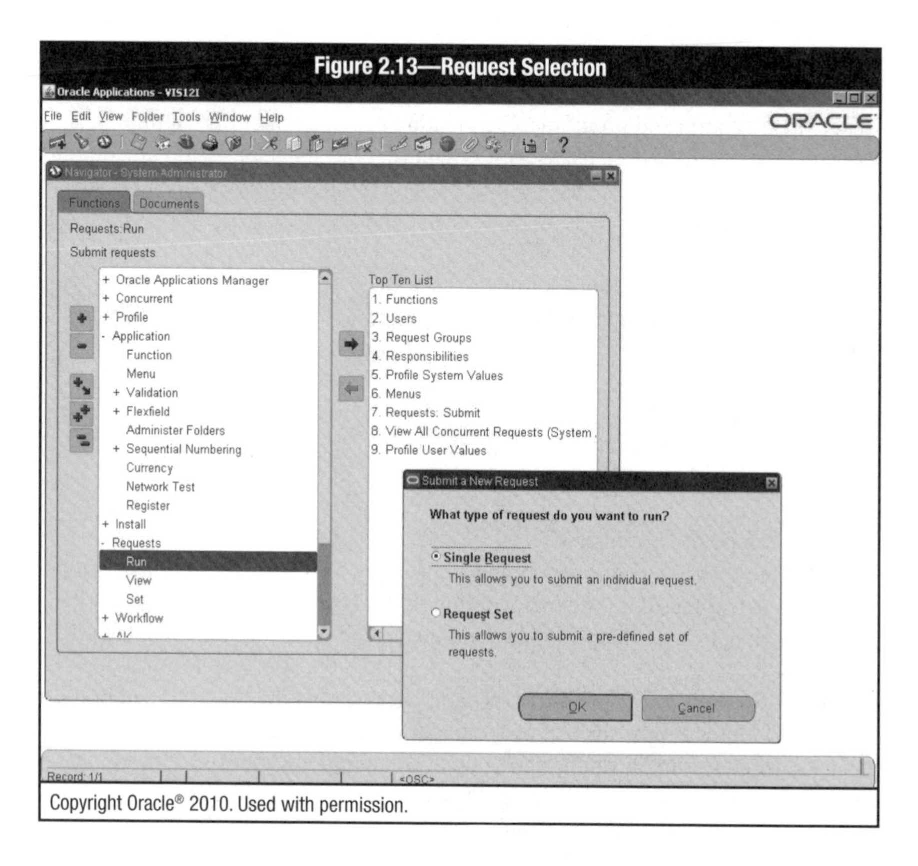

Figure 2.13—Request Selection

or a combination of letters (e.g., ora%) for which to search. **Figure 2.16** on page 25 shows an example of using the Search function to find a responsibility. To retrieve a group of records based on more sophisticated search criteria, the Query by Example function can be used. Query by Example allows the user to specify search criteria in any field in the current block that can be queried. The search criteria can contain specific values, wildcard characters or query operators.

Printing a Window

To print the current window:

- Select Print from the toolbar.
- Change any of the options required within the standard Microsoft Windows Print window, and execute the print job by clicking the OK button.

Saving Work

When work is saved in Oracle EBS, the underlying database is updated with the new information. Oracle EBS also performs a validation of the work when saving it. Any incomplete or invalid data are flagged, and a message is displayed. The incomplete or invalid information will not be saved.

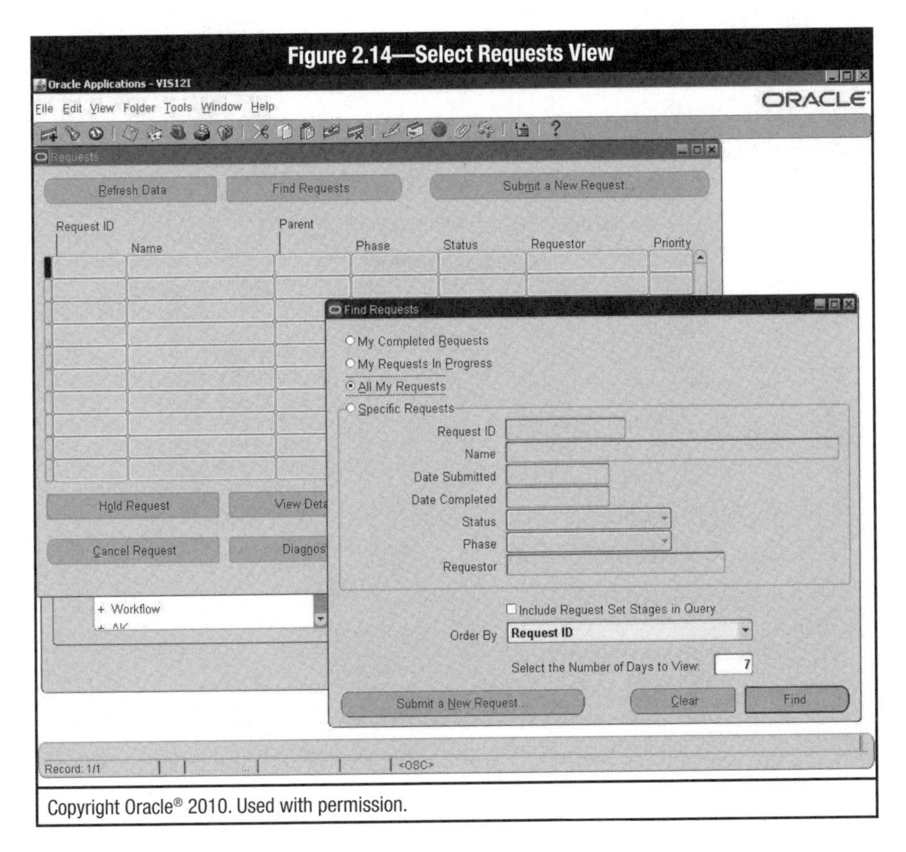

Figure 2.14—Select Requests View

To save work, select Save from the toolbar. To save work and update the process in the Navigator window, select Save and Proceed from the File menu.

Query and Export

This feature allows records to be exported in a multiple row block to a tab-delimited file (shown in **figure 2.17** on page 26) that can be opened using many standard desktop applications. The data to be exported can be controlled using the Query functions of the form. Columns can be reordered or removed in folder forms before export. Depending on how the application has been configured and implemented, this functionality may not be available in some enterprises.

To export records, the following is necessary:

- Run a query on the records. Queries can be run in a form/function by using the Find or Query by Example options as discussed previously.
- Ensure that the cursor is in the multirow block that contains the records to be exported.
- Select Export from the File menu in the taskbar.
- After the extraction process has finished, select either Open This File From Its Current Location or Save This File to Disk. The file is saved in a .tsv format that can be opened by, for example, Microsoft Excel.

Figure 2.15—View Submitted Requests

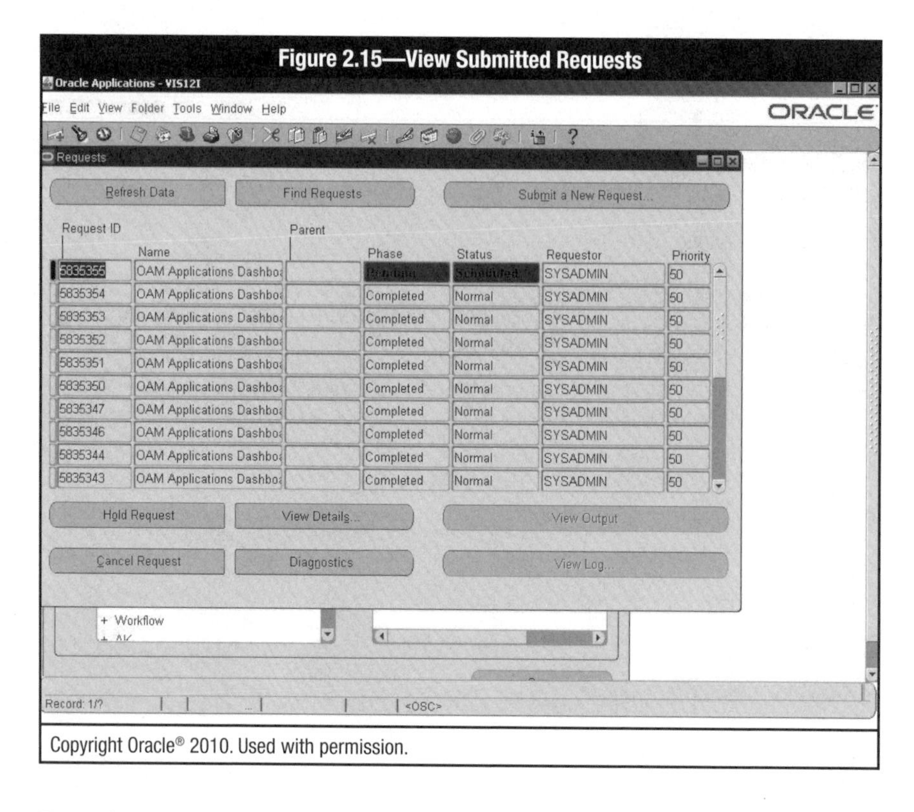

Exporting more than 100 records results in a prompt asking users whether they would like to:

- Stop (stops after first 100)
- Continue (selects the next 100)
- Continue to End (selects all the records matching the query criteria)

The Export feature should not be used to export large numbers of records (greater than 1,000) since system performance can be affected. The number of records that the query brings up can be seen by the record count in the bottom left corner of the screen.

The Oracle EBS Financials RXi Reports Administration Tool is a tool that can be used to design the content and layout of the RXi reports without changing the underlying report code. There are approximately 45 reports in the General Ledger (GL), Payables, Receivables, Fixed Assets, Cross-Product and Globalizations modules where the RXi tool can be used. The RXi Reports Administration Tool allows the same report to be printed using a number of different layouts and contents. This means the data items that are included in the report can be selected.

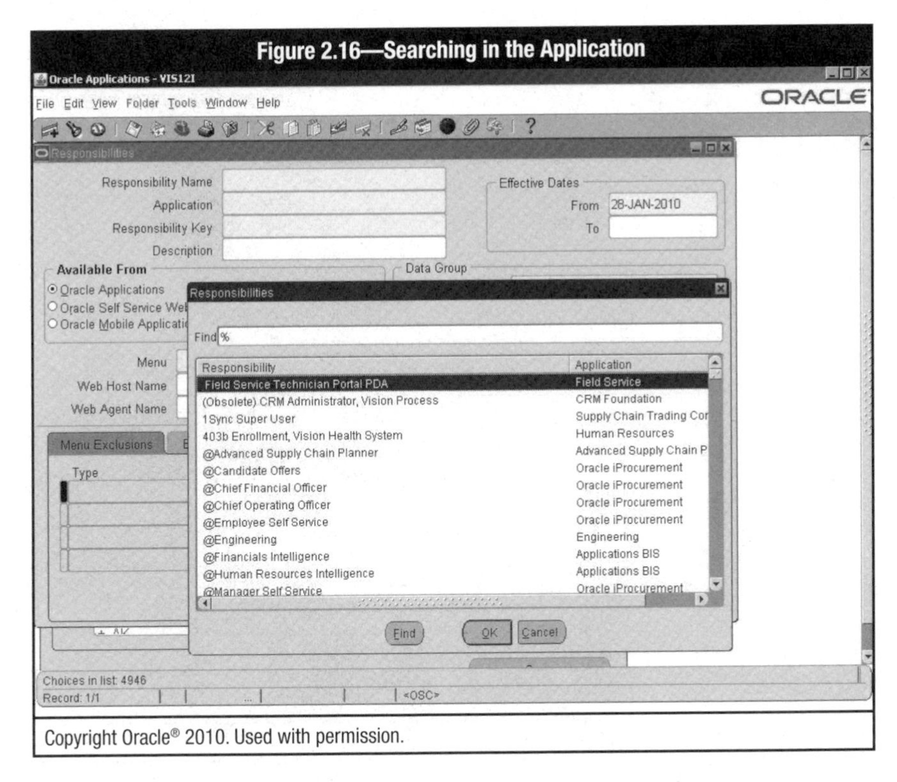

Figure 2.16—Searching in the Application

The following requirements may be addressed by the RXi tool:

- Provide multiple layouts with different contents for the same report.
- Remove column(s) from a given layout.
- Modify column formatting features, such as length, title, display, sequence and amount format.
- Modify page formatting features, e.g., report titles, page number display and parameters display.
- Modify grouping and summarizing features.
- Generate reports in text, HTML, comma-separated values (CSV) and tab-delimited formats.

Exiting From Oracle EBS

There are two ways to exit from Oracle EBS:

- If using a Forms-based application, select Exit Oracle Applications from the File menu, or click on the Close button (closes the active program or file) in the top right corner of the Oracle EBS window. Select Save or Discard Changes when exiting from Oracle Applications.
- If using an HTML-based application, click on the Logout link in the global navigation area.

Figure 2.17—Exporting Records

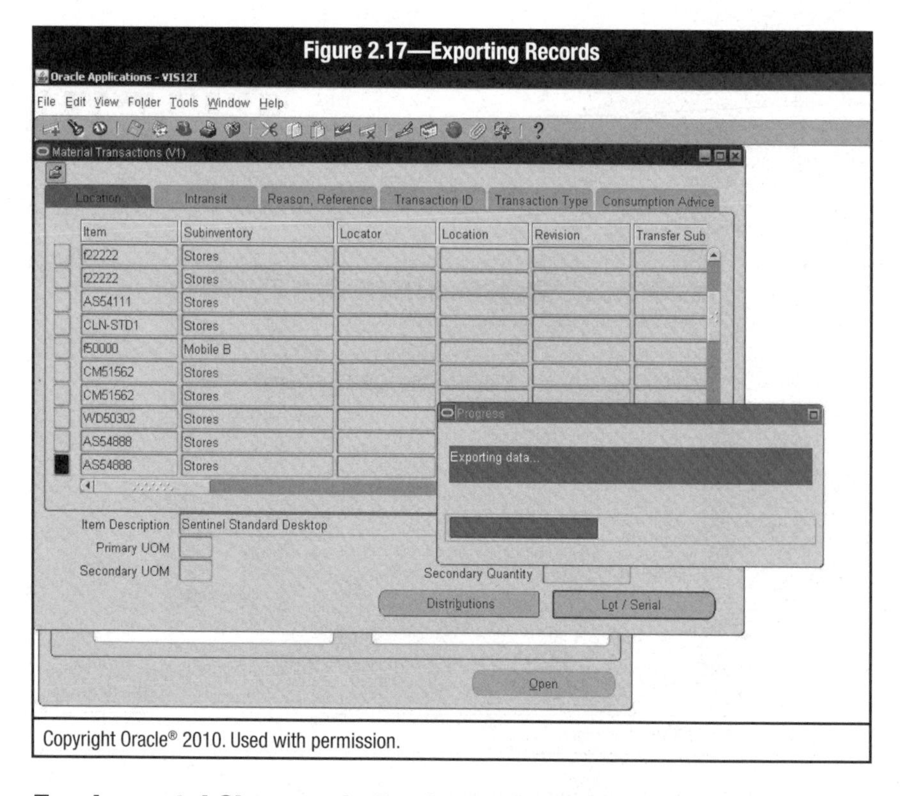

Fundamental Changes in Business Controls

An ERP implementation and its associated business process changes affect critical elements of the enterprise, resulting in increased business, process, system and project risk. Some reasons for the increased risk include:

- Decisions taken on erroneous real-time information are often irreversible or costly to set right.
- Batch-oriented controls are not the focus in an online and real-time environment.
- Traditional (paper-based) audit trails are lost.
- Access requirements have expanded vastly to include field personnel and, increasingly, suppliers and customers.
- Master data changes can have a significant impact on transactional data across multiple business processes and business units, due to the integrated nature of Oracle EBS.
- A single significant point of failure exists.
- There is potential erosion of SoD since end-to-end business processing now occurs in one system.

As a result, the integrity and control structure supporting ERP-enabled business processes must also be transformed. ERP systems can change internal controls in three fundamental ways:

- **The method of control**—From rechecking and revalidating paper-based records to online monitoring and measurement
- **The point of control**—From multiple validations of transactions, often based on printed outputs and source documents, to a single validation at the point of creation, often an online approval
- **The amount of control**—From many redundant, process-impeding controls to fewer automated and strategic controls

Consequently, it makes business sense to ensure that these enhanced controls are integrated into the reengineered and ERP-enabled processes. **Figure 2.18** shows the four main steps in the expenditure (noninventory) business cycle for an ERP-enabled enterprise. Some examples of fundamental changes in business controls are described in the following sections.

Figure 2.18—Expenditure Cycle (Sample)

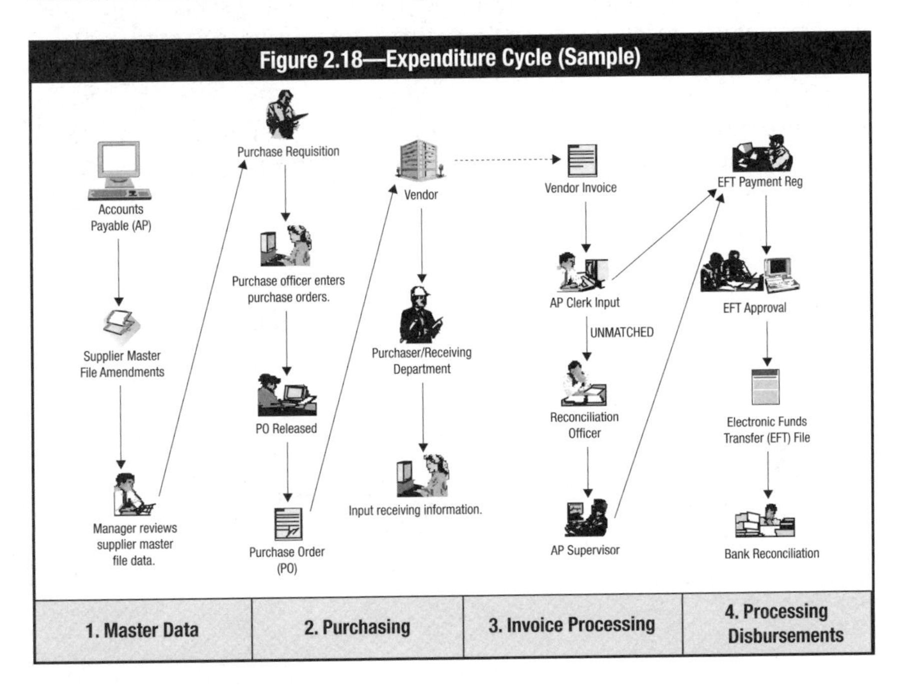

Case 1—Three-way Match

An educational institution employs three-way matching on POs, goods receipts and invoice processing. POs are entered and approved online under delegated authority. Goods receipt information entered online is matched to the PO quantity and tolerances set for over/under receipt. When the invoice arrives, a three-way matching process occurs, whereby the quantity on the PO is matched to the quantity on the goods receipt, and the amount on the invoice is compared to the amount on the PO. If these fields match within preset tolerance levels, the transaction passes for payment. This matching process effectively reduces the extensive manual and paper-based control activities, such as printing already approved POs and stapling them to the back of invoices for reapproval at invoice receipt and again at the payment stage. Controls have moved to the beginning of the process in an electronic form.

Case 2—Evaluated Receipt Settlement

When a food company loads POs into its system, orders are validated against vendor master data details already loaded into the ERP application system. Details of contractual arrangements and prices with vendors have also been loaded into the system. When the goods are received, there is a two-way match on quantity and, providing it matches within preset tolerance limits, the vendor is paid based on the set payment terms. There is no need to process a physical invoice. Oracle EBS, for purposes of accounts payable (AP) processing, can generate an invoice within the system. The controls in this example reside at the beginning of the process cycle—at the approval of requisitions or orders and receiving of the goods/services.

These two cases demonstrate that controls are being shifted from detective to preventive and traditional matching reconciliation controls are automated in the Oracle EBS software.

3. Risk Management in an ERP Environment

While implementation of an ERP package may provide significant benefits, it may also introduce new risks and changes in the enterprise's risk profile. This chapter outlines:
- Strategic risks and controls
- The importance of implementing a control framework

Risks and Key Management Controls

The key project and operational risk areas associated with the implementation of an ERP system are:
- Program and project governance
- Business processes and functions
- Security (applications and infrastructure) and business continuity

Immediately following each risk area is a summary of controls that, when implemented, may manage the risk. The risks may apply to any ERP implementation and are not specific to an Oracle EBS implementation.

Program and Project Governance

The major concerns for ERP implementations involve organizational rather than technological issues.[2] This section discusses the risks and key controls for an ERP project, including:
- Organizational change management and training
- Planning and project management
- Executive sponsorship
- Reliance on third parties
- Project cost

Organizational Change Management and Training: Risks

Organizational change management and training are often primary areas of risk for enterprises implementing an ERP system. During the initial budget and business case phase of a project, change management and training are often descoped or downscaled to reduce project costs. ERP implementations typically bring with them significant amounts of change, and insufficient effort on change management and training are often major causes of project failure. Employees often defer their involvement in the ERP development, even though it may significantly affect the way they perform their roles in the new ERP-enabled enterprise, and employees require considerable training on changed business

[2] Manual, J.E.; *Definition and Analysis of Critical Success Factors for ERP Implementation Projects*, 2004, *http://profesores.ie.edu/jmesteves/thesis_research.htm*

processes and hands-on exposure to the system to adapt to the new processes and systems. An important aspect of change management that is sometimes underemphasized is the role of the users and the impact their actions have on a single, integrated, enterprisewide application. Another key risk is the retention of employees once they are trained in new processes and systems.

Organizational Change Management and Training: Key Controls

The project sponsor must ensure that the enterprise has the same vision as the original motivations for implementing ERP-enabled processes: the targeted capabilities as well as the targeted benefits. Aligning with the true destination (as opposed to the initial go-live phase) is a hearts-and-minds issue that requires special focus on people: communicating, managing expectations, offering education and providing senior management support. The change management and training program must:

- Provide relevant users with the skills and knowledge required to participate appropriately in the development of the system
- Understand the changes to their job roles in the post-go-live environment
- Contribute to the next milestone

Typically, this is not an area where the budget can be trimmed successfully. Enterprises need to establish business process owners and champions who own the business processes and understand the impact of the actions of one group on another. A key success factor for the business process owners often revolves around how early and to what extent they obtain hands-on experience on the new or redesigned processes and the new ERP system. These owners and champions should ensure that an understanding of the dependencies among processes and modules is communicated effectively. Retention plans are required to ensure that employees, once trained and marketable, remain with the enterprise. Each key member of the team should have a backup staff member with similar training and experience.

Planning and Project Management: Risks

Key challenge areas for an enterprise implementing an ERP system include detailed planning and project management of the people, processes and technology factors. Often the issues and obstacles facing an ERP implementation concern project factors. These factors include prioritization, resource allocation, team and project structure, discipline, ownership, and communication. Failure to place sufficient emphasis on these factors, as opposed to process and technology factors, often leads to disappointment with the implementation outcome.

While this publication focuses on Oracle-EBS-specific operational risk, those who are interested in implementation and project management risk should refer to the audit/assurance program for systems development and project management.[3]

[3] ISACA, *Systems Development and Project Management Audit/Assurance Program*, USA, 2009, *www.isaca.org/Template.cfm?Section=Research2&CONTENTID=53961&TEMPLATE=/ContentManagement/ContentDisplay.cfm*

Planning and Project Management: Key Controls

Enterprises need a strong business imperative to implement ERP so their projects do not stop midstream and end in disillusionment. This imperative needs to be embodied in the business case and carried through to an effective implementation plan and design with appropriate user involvement. Successful projects are guided by detailed work plans, milestone plans and rollout plans. Key dates and deliverables are spelled out and dependencies are synchronized, while benefit scoreboards are created and results are tracked and communicated. There is a need for a professional project manager with the ability to:

- Integrate IT and business users into joint decision making
- Facilitate significant and difficult decisions, such as whether to reengineer existing business processes to accommodate a standard ERP package or to redesign/customize the ERP itself to fit existing processes

Executive Sponsorship: Risks

Project management and users may become frustrated and effective change may not be achieved if there is no sponsorship or the active involvement of executive management is lacking. Project resources may be redirected to other priorities and the project may stall. Conflicts may arise between the business areas and IT (or between business areas), and effective resolution may not be achieved. The right level of investment may not be maintained, and the project may lose its purpose.

Executive Sponsorship: Key Controls

Senior executive buy-in and sponsorship are needed to achieve the right mix of business and IT involvement in the project and to resolve conflicts. Business process reengineering needs championing and the systems architecture needs investment. To succeed, these aspects need executive support. During implementation, the responsibility for going live on time and on budget usually rests with the project leader. However, going live is only an interim destination. In successful enterprises, there is no mystery about the accountability for results. An unambiguous responsibility and accountability structure should be set up for the benefits of the project and should be communicated to the entire enterprise.

Reliance on Third Parties: Risks

While consultants bring valuable experience and methodologies to a project, their presence alone does not guarantee success. The enterprise may over-delegate to consultants or third-party vendors of ERP solutions, expecting them to intuitively know their business requirements and effectively test and implement the solution. A major pitfall may also involve the payment of large sums of money on the delivery of documents, such as designs and flowcharts, without any tangible delivery of the computer system solution. Often enterprises focus so much on going live as the end product that the need for a postimplementation stabilization phase and benefits realization phase are ignored or not well understood. This results in the consultants or the project team being let go too early, skills and knowledge not being transferred effectively to the enterprise, and inadequate support in the postimplementation environment. Another consequence of remunerating consultants upon going live is that the enterprise may be pushed to go live even if it is not ready.

Reliance on Third Parties: Key Controls
Business process owners who understand the enterprise and its business requirements need to be appointed. These business process owners must gain hands-on experience with the solution and must champion the cause, ensuring effective testing and implementation of the solution. The enterprise needs to contract effectively with its suppliers to manage the quality of deliverables and effective postimplementation support. This may take the form of warranties or retainers until the delivered product is proven in production. Effective transition planning and training are required to transfer skills from vendors or consultants to appropriate staff.

Project Cost: Risks
Some of the major causes of project cost overruns have already been discussed. These include change management, training and a lack of software functionality. In addition, the customization and integration of software packages can make up a considerable component of total implementation costs. Changes to the vendor-supplied software or customizations usually build in upgrade costs since additional testing of the changes is required during the upgrade. Generally, it is better not to customize; however, the enterprise needs to be sure that the standard solution can handle the major parts of its business. Other areas that are often underestimated include program interfaces, data conversions, report changes, integration testing, process rework and consequent increases in consulting fees. Unexpected project costs may also be hidden in business-area desktop computing budgets or in other IT infrastructure budgets.

Project Cost: Key Controls
Project cost overruns, particularly in relation to change management and training, need to be identified early through effective reporting to the appropriate governance mechanism. Change management and training challenges involve an understanding of the integration between business areas, the data flow through the enterprise and the impact of one area's actions in the system on another. The change management process and training program that accompany an ERP implementation need to be presented to (and understood by) executive management. The program needs to be targeted and funded appropriately.

The business case should not be a static, one-time exercise intended to secure funding. On the contrary, the business case needs to be a dynamic and evolving management tool—one that should last beyond the go-live phase to the benefits realization phase. Successful enterprises use the business case tool in a variety of ways including:

- Justifying the program
- Validating the design

- Setting and managing postimplementation targets
- Prioritizing postimplementation change initiatives

Too often, the business case for an ERP implementation consists of a high-level mission statement or description of intangible, unquantified business benefits. A proven business case template should be employed and tailored to the enterprise's environment. Factors that should be considered include:

- The total cost of ownership, factoring in, for example, the additional cost of upgrading as a result of making software customizations
- The appropriate due diligence in determining benefit and cost items involving the input of variables and formulas for determining inventory, people savings, and conversion and integration costs
- A discounted cash flow analysis, including appropriate risk factors and cost of capital

Measurements need to be initiated in the legacy systems environment to baseline costs and benefit streams so the improvements in the postimplementation ERP environment can be measured effectively.

Business Processes and Functions

Users who are familiar with the functional orientation of a legacy system environment can find it challenging to embrace the notion of an integrated ERP environment based on business processes. As with any integrated environment, errors in one part of the process may affect steps throughout the processes. This section discusses the risks and key controls for business process reengineering and software functionality.

Business Process Reengineering: Risks

Reengineering of the business processes will most likely also result in structural and job role changes within the enterprise. Staff members who had worked within the legacy environment for an extended period may find it difficult to adapt to new roles, and as a result, certain business functions may not be performed properly in the postimplementation environment. There is a risk that the reengineered business processes may not have been configured properly, resulting in incorrect processing (e.g., incorrect tax indicators) or inadequate business controls (e.g., bypassing the three-way match on purchases).

Business Process Reengineering: Key Controls

Before any ERP implementation takes place, a requirements analysis is necessary for determining user needs and system constraints. This analysis is useful for determining functional specifications for the implementation.

The change management process and training program need to provide users with an appropriate overview and understanding of the impact of their actions on the process, system and other users. Users need to be trained sufficiently, and the appropriate procedural controls need to be defined so users are able to execute their new roles in the new, integrated processes and system on the first day of going live.

Enterprises, even those successful at implementing ERP, usually experience a temporary dip in performance after going live. Going live with an ERP is a significant change for any enterprise. The dip varies among enterprises, depending on how well they were prepared for the introduction of this new system. Most users need to walk before they run, and after mastering the basics on the live environment, they may require refresher training on the more advanced topics. The enterprise needs to be prepared for contingencies and the considerable effort often involved in correcting errors made in an online, real-time environment. This may require additional trained data entry or programming resources to correct data errors.

Configurable options need to be explained thoroughly to users and documented appropriately in the business requirements, design or blueprint documentation. Changes in the system of business controls need to be considered early in the implementation process and included in the design to minimize the cost of retrofitting controls at a later date.

Software Functionality: Risks

When enterprises get down to the detail or, worse still, when they are in production, they often find that the ERP solution cannot handle the major parts of the business. While the enterprise perceived that the vendor or reseller presented the solution as being capable of meeting the business requirements, the requirements may not have been specified clearly or detailed effectively. Whatever the reason for the misunderstanding, if there is a fundamental mismatch between the system and the business need, the consequences may be costly.

Software Functionality: Key Controls

Management needs to take the necessary time to complete effective due diligence on the new system. This includes translating business requirements into a high-level design specification for software acquisition, taking into account the enterprise's technological direction and information architecture. Management should review and approve the design specifications to ensure that the high-level design addresses the requirements. When significant technical or logical discrepancies occur during development or maintenance, the document will need to be reassessed. Appropriate software selection guidelines should be used, and factors to consider include:

- Others in the industry using the solution
- Particular local requirements
- Legislative or compliance requirements (e.g., tax, statutory reporting, industrial awards/agreements)

- Foreign currency handling (e.g., financial vs. management accounting treatment, and reporting through time and across country borders on a transaction)
- Particular reporting requirements (e.g., external reporting and reconciliation needs, compliance with International Financial Reporting Standards [IFRS])
- The impact that the loss of specific legacy system functionality may have on customer service (However, caution needs to be exercised prior to replicating legacy functionality; only if it is the best, or at least the better, process should the enterprise consider replicating it. It should not be replicated only because the enterprise is used to it.)
- Stability of the current software release
- Specific operational needs (e.g., handling fresh produce or livestock)
- Marketing needs (e.g., bulk discounting across product lines)

Further along in the implementation, adequate user acceptance and system and integrated testing need to be performed to ensure that the system performs as anticipated. System performance is another critical area that must be tested to ensure that the application and the related infrastructure can handle the typical transaction loads processed by the enterprise.

Security (Applications and Infrastructure) and Business Continuity

For many enterprises implementing Oracle EBS, the focus lies in the implementation of business functions and the automation of business processes. In such cases, there may be a potential risk that a lower level of priority is placed on ensuring that appropriate functional security controls are in place, such as the activation of audit trails and monitoring security functions, restriction of access by default, and restriction of access to high-privilege accounts. Therefore, it is important for information security and administration groups and internal audit groups within the enterprise to be involved to ensure that appropriate security control features are designed into the Oracle EBS implementation.

This section discusses the specific risks and key controls related to the security of the applications and infrastructure implemented for an ERP project, including:

- A single point of failure
- Distributed computing experience
- System access
- Data quality
- Program interfaces

A Single Point of Failure: Risks

Within the legacy environment, the impact of a component failure within a system has a limited impact, if any, on other systems. This is true even in the case of a total loss of a particular application system. For example, in most cases, a purchasing system could be managed through manual workarounds. In an ERP environment, where the entire enterprise may rely on the system, the loss of the system for any extended period is likely to have significant effects on the enterprise's operations, as well as significant financial implications. In the legacy environment, systems typically can be unavailable for a few days before offsite and contingency facilities have to be invoked. In an ERP environment, the period between when the system is unavailable and when the contingency plan needs to be invoked is typically measured in hours, not days. Because the enterprise has moved to operating in an online, real-time mode, its business operations may be disrupted when the system is unavailable. For example, in one case, a distributor of perishable food went live with its ERP, utilizing a legacy front end that processed orders from field personnel using handheld devices. The process consisted of field personnel entering data into a legacy front end that, in turn, updated the ERP's back-end system. When the front-end legacy system failed (as it was unable to handle the volume of data) the enterprise was forced to use manual data entry for orders completed by field personnel. The customer service personnel, who had no experience in entering orders, made errors when entering the data. Orders were incorrect and out of sequence, and this played havoc with the back-end warehousing operations of the business. The warehousing personnel, also new to the system, had considerable difficulty dealing with incorrect order details, returns and corrections. Incorrect deliveries were made, inventory-level information became inaccurate, and the entire episode resulted in a significant write-off.

At the core of the Oracle EBS system is a single relational database. This database uses complex technology to ensure that it can feed the system the necessary information to complete all business processes. The complexity of the database and the amount of information that is fed in and extracted from it require careful controls to be instituted.

A Single Point of Failure: Key Controls

Business continuity management plans need to be revised, taking into consideration the ERP as a single point of failure. Four characteristics of ERPs that may affect business continuity planning (BCP) are:

- The large number of modules that cover a broad range of the enterprise's business processes
- A large, integrated database
- The physical and logical intertwining of all modules and data, and the possible need for them to be recovered at the same time
- Access to various modules by suppliers and other third parties

Because of these characteristics, rapid recovery may necessitate a complete rethink and redesign of the enterprise's BCP arrangements. An online, real-time system also needs an online, real-time business environment that can effectively monitor and deal with exceptions before they turn into significant problems and impact other areas. System maintenance and version control are also important in terms of maximizing system availability and integrity.

ERP systems in general, and Oracle EBS in particular, have features to technically configure and appropriately monitor the integrity of the system, monitor system performance, and initiate corrective actions, as needed, in a real-time mode that can minimize the probability of a single point of failure.

Distributed Computing Experience: Risks

Although it is sometimes overlooked, the IT architecture may need to be overhauled totally with the implementation of ERP. New skills are required to manage and maintain this environment, and the impact of this change is often underestimated.

Figure 3.1 illustrates how complex an ERP technical environment can become. This environment is indicative of the environment on which Oracle EBS and legacy applications can be run (a client-server computer architecture). Depending upon the IT architecture used in the implementation, the audit may be centralized or decentralized. Extra care needs to be taken in scoping the first-year audit of enterprises that have implemented ERP systems. In this type of environment, there is often a combination of centralized accounting controls and decentralized operationing controls.

Figure 3.1—Decentralized ERP Environment

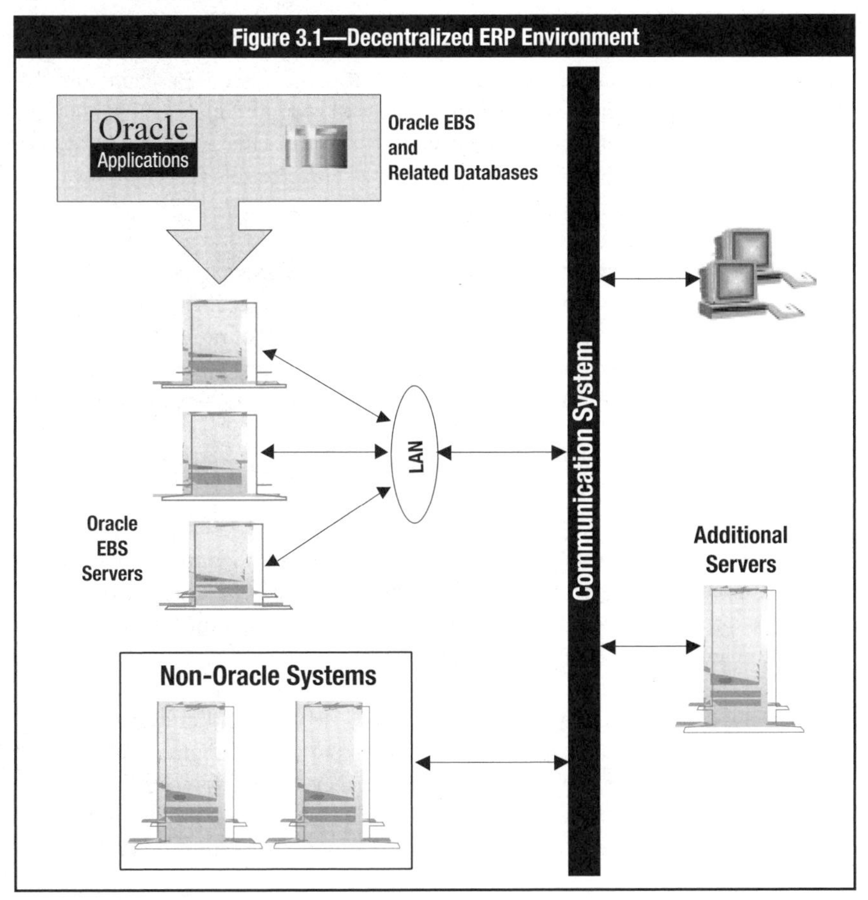

Distributed Computing Experience: Key Controls
The IT infrastructure requires the same planning as the business processes. IT staff may require training to develop new skills. These areas are often underestimated in the initial planning for an ERP implementation. IT staff may become more marketable following training in the new environment, so it is advisable to consider retention and succession plans.

System Access: Risks
By bringing a number of the enterprise's business processes together into one enterprisewide application, users potentially have access to additional information and processing functions. Recent releases of certain ERP systems are designed to allow wireless or remote access for field and sales staff and, if necessary, customers and suppliers. This level of access—directly to the system from remote locations—allows the system to be kept up to date in real time. Yet, increased remote access may create an environment in which the system is far more susceptible to hacking or other malicious tampering, which may increase the likelihood of incorrect data being introduced into the system.

System Access: Key Controls
Oracle EBS contains a number of security parameters covering passwords, intruder lockout, superuser access, etc., that, when set appropriately, serve to secure the system. Other ERP systems have varying degrees of security functionality, and some require add-on packages to adequately secure them. User access to the system should be designed and built in accordance with the enterprise's security policy or needs. Some of the factors to be considered include:
- Segregation of access/duties
- Access to only the transactions or objects required by users to perform the job or process role assigned to them (the "least privilege" concept)
- Access based on the risk assessment of the consequences of providing the additional access vs. the cost of implementing tighter security (e.g., access to view all plants vs. maintaining separate security profiles for users in each plant)

Security is covered in greater detail in chapter 4.

Data Quality: Risks
As an ERP system may rely on a single, central database, the integrity of the data within the system is paramount. Data fed from legacy systems may be inaccurate, incomplete or duplicated, resulting in operational difficulties in a more automated and integrated environment.

For example, a higher education institution converted its supplier master file, complete with fax numbers, for each supplier. In the new ERP environment, the institution moved to online faxing of purchase orders on approval by the appropriate delegated authority. As some of the fax numbers were out of date, this resulted in a number of misdirected or lost faxes. The institution had also implemented the ERP solution with third-party middleware and was unable to receive fax completion or OK messages without logging out of the ERP system and logging in again.

Data can also be locked away in the complex data structures of ERP systems. As e-business opportunities increase and prevail, it will become increasingly important to be able to unlock the data within an enterprise's systems. Data quality is becoming more important in the e-enabled ERP environment, as external parties access invoice and financial information via the Internet.

Data Quality: Key Controls

All data should be effectively cleansed before being loaded into the enterprise's ERP system. Cleansed information should be secured while awaiting conversion to the ERP environment. Control techniques, such as control totals (often embodied in ERP conversion and load utilities) and data editing criteria, should be employed as appropriate. Test conversions should be performed and financial reports should be reconciled between the two systems during the test and final conversions to confirm the completeness and accuracy of the data conversion. Data conversion is an area of key importance for an enterprise and its auditors.

Program Interfaces: Risks

While ERPs enable many different types of functions to be completed, some enterprises have requirements that are not met by an enterprise application program, or have requirements to transfer information with suppliers, customers and financial institutions. As a result, program interfaces are established to transfer transactional information among these systems. If interfaces are not controlled effectively, there is a risk of inaccurate, incomplete, unauthorized or untimely information being fed into the ERP system or extracted from it.

Oracle's Fusion Middleware Product is based on an SOA, which provides methods for system development and integration, enabling interoperability of services between Oracle EBS and other applications. SOA provides a consistent framework for integrating heterogeneous systems, thereby reducing risks instead of increasing them.

Program Interfaces: Key Controls

Controls over program interfaces are similar to those operating over data conversions (e.g., control totals, data editing criteria and periodic reconciliations), except they may be performed in an interactive manner rather than in batch mode. Further, the timing of the program interface can often be significant, particularly when there are several program interfaces that must be executed within a short processing window. As mentioned in chapter 2, version 12.1 of Oracle EBS allows continual monitoring of the integrity of data, including data being transferred through program interfaces.

The Importance of Establishing a Control Framework

A control framework for an ERP environment can provide a robust management tool and methodology for ascertaining the risks associated with an ERP environment and a standard for defining the established controls. This can be achieved by:

- Establishing an enterprise control framework
- Defining the control framework for an ERP environment

Establishing an Enterprise Control Framework

The Committee of Sponsoring Organizations of the Treadway Commission (COSO) developed a model for evaluating internal controls, with the objective of helping management improve the enterprise's internal control systems and providing a common understanding of internal control among interested parties. This model has been adopted as the generally accepted framework for internal control and is widely recognized as the definitive standard against which enterprises measure the effectiveness of their systems of internal control. The framework defines internal control as:

> *...a process, effected by an entity's board of directors, management and other personnel, designed to provide reasonable assurance of the achievement of objectives in the following categories:*
> - *Effectiveness and efficiency of operations*
> - *Reliability of financial reporting*
> - *Compliance with applicable laws and regulations*[4]

The framework defines five interrelated components of internal control:

- Control environment
- Risk assessment
- Control activities
- Information and communication
- Monitoring

The COSO framework is an effective starting ground for defining an enterprise's internal control framework and methodology with regard to an ERP environment. It can then be used for defining the elements of control governance for an ERP environment.

Defining the Control Framework for an ERP Environment

The implementation of an ERP system can introduce new risks and alter an enterprise's risk profile. As a result, enterprises need to redefine their approach toward risk management and control assessment to cater to the differences in risk encountered in an ERP environment and to achieve complete coverage of

[4] COSO, *Internal Control—Integrated Framework*, USA, 1992

the associated management controls. An ERP control framework is required to facilitate the assessment of risk and the completeness of controls. There are several frameworks that can be adopted to model the ERP control environment. One that is consistent with the ERP business process-driven approach and works well in practice is outlined in **figure 3.2**. The control framework consists of the following five areas:

- **Business process controls**—Include automated controls (e.g., online approval, three-way matching of purchase order amounts, goods receipt quantities, invoice particulars) and manual controls (e.g., reconciliations, manual approvals, review of exception reports) within the reengineered business processes. Business process controls are most cost effective when incorporated from the beginning of the project and throughout the design and development phase. Retrofitting controls after the implementation is often costly.
- **Application security**—Includes maintenance of user profiles that provide access to application functionality and system services. It includes user, system and security administration procedures, and it incorporates the setting of security parameters (e.g., password lengths) and the granting and removing of user access to Oracle EBS.
- **Program interface and data conversion controls**—Controls that need to be considered within the framework to address the risks associated with converting or interfacing data from legacy or external systems
- **Technology infrastructure**—Includes controls surrounding the technology platform on which the application resides. The technology infrastructure consists of the servers, operating systems, databases and network layers.
- **Project management**—Specifically the aspect of the control framework relating to change management and project disciplines discussed earlier

Figure 3.2—ERP Control Framework

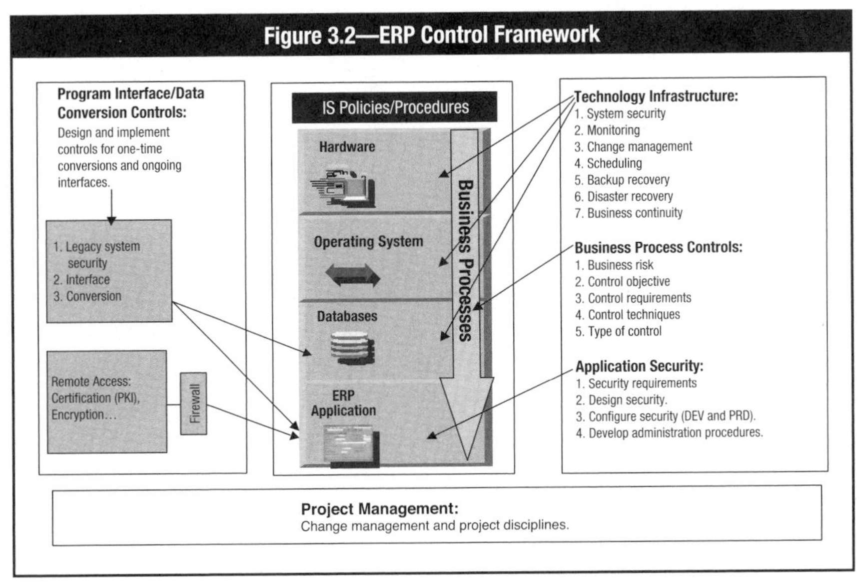

Summary

This chapter outlined risk areas and controls associated with the implementation of ERP systems. Risks related to the areas of change management, project costs, single point of failure and data conversions. Associated key controls include executive sponsorship, training, backup and recovery, and data cleansing and control totals. This leads to the conclusion that enterprises need to redefine their approach to risk management in an ERP environment, and underscores the importance of establishing a control framework to facilitate the assessment of risks and the completeness of controls.

4. ERP Audit Approach

An audit approach should be developed to address the issues involved in implementing and maintaining an ERP system. This chapter introduces a risk-based ERP audit approach and provides an understanding of the structure of Oracle EBS. Security concepts are introduced; however, auditing of security in Oracle EBS will be covered in greater detail in chapters 9 and 10.

Audit Impacts Arising From the Implementation of ERP

In an enterprise that has implemented or is in the process of implementing an ERP system, the nature and manner of auditing may need to change dramatically. An ERP implementation results in increased risk and potentially significant impacts on the internal control environment. As such, the following key principles should be considered when conducting a postimplementation audit of an ERP environment:

- Management of the change
- Audit methodology
- Role of the auditor
- Audit involvement in the project
- Audit responsibilities

Management of the Change

Audit needs to manage the change to an ERP environment by focusing on two key aspects:

- **Staffing**—The complexity of the environment usually requires a staffing model with a higher ratio of IT auditors. Traditional financial and operational auditors must transform to become integrated auditors who understand the nature of the ERP system and dependencies among different subprocesses. The audits of complex and technical areas may need to be supplemented by skilled and experienced resources.
- **Training**—A detailed knowledge of an ERP environment is necessary to effectively understand and audit security, control and implementation issues over application areas and the technical environment. Extensive training is necessary to adequately understand the new environment. Auditors may need to learn security and control implementation methodology. They may need to attend ERP training, join ERP user groups and learn to use new tools to effectively audit an ERP system.

Audit Methodology

Significant reengineering of the audit approach is needed to adjust to the new ERP environment. The enterprise's concept of its audit universe may need to change to effectively audit the new system. A risk assessment should be performed, and the audit approach should be modified accordingly. Integrated audits covering business processes and security aspects are necessary in the ERP environment. SoD and security management come together in risk management of the new reengineered business processes. Where new processes are implemented and duties are separated, any perceived lack of control is compensated for by automated controls, such as menu-driven access controls and responsibilities, release strategies, master data validation, and tolerance levels. Further, effective audit practices in the ERP environment require new automated diagnostic tools. ERPs have powerful and complex security arrangements, and testing security is not just a matter of looking up the security matrix of users to functions. Functions and responsibilities may be used, and unraveling these to find out who has access to what often requires the use of automated diagnostic tools available from third parties. Management needs to consider the impact of auditing practices and tools on the overall operational performance of the ERP system. A balance must be struck between performance and usability (a potentially negative impact from the use of auditing tools and system logging functionality) and ensuring that appropriate security measures are in place to detect unauthorized activities.

The Oracle EBS software is flexible yet complex, and it can be customized to fit each enterprise's environment. The configuration of business processes, degree of customization, scope of the implementation and version of Oracle EBS software all contribute to the uniqueness of the environment. Therefore, it is not possible to design one standard audit program that will work in every Oracle EBS environment.

Role of the Auditor

There are a number of opportunities where audit can and should contribute to the enterprise's ERP implementation. The role or roles chosen will depend on the circumstances and the enterprise's audit charter/strategy. The different levels of involvement that audit may have in the project may be characterized as follows:

- Integrated approach
- Preimplementation review
- Postimplementation review
- Quality assurance

The IT auditor adopts the role of an IT risk management consultant when adopting the integrated approach or conducting a preimplementation review.

Integrated Approach

Enterprises utilizing the integrated method focus on the design and implementation of controls for the new systems. This approach requires audit to be involved from the earliest stage of the project, assisting the project team in the design and building of the controls. The approach considers project risks and business process risk assessments, and requires auditors to perform testing to ensure that controls have been implemented properly and benefits have been pursued.

The reason for an integrated approach is to assess the impact of the development of the Oracle EBS on enterprises' business controls throughout the project life cycle. This means that when processes are mapped to activities, business risks should be assessed and analyzed, and mitigating business controls should be developed. These controls are to be implemented in a structured and balanced way. Thus, by the end of the implementation project, a situation has been reached that ensures that an effective and efficient mix of business controls is in place. These business controls should ensure the confidentiality, integrity and availability of data and data processing, thus enabling the enterprise's management to control its business and IT processes. Furthermore, the enterprise's external financial auditor will be able to place reliance on the controls and perform the audit in an effective and efficient way.

Preimplementation Review

The preimplementation review approach allows the auditor to conduct a timely review of the control design and implementation plan before going live. It usually involves a review of the:

- Business case
- Project risk management and monitoring control design
- Application security design
- Business process risk assessment and control design
- Data conversion program interface controls
- Adequacy of system testing
- Transition and system migration controls
- Change management controls

The preimplementation review may also include project management aspects, such as a review of:

- Performance measurement criteria
- User readiness to go live (e.g., adequacy of training provided, quality of user operating procedures)

A preimplementation review is highly useful because issues generally can be addressed before the system is put in place. This usually is far more cost-effective than reconfiguring the system after it is operational.

Postimplementation Review

Another role that the auditor can undertake in the project is to conduct a postimplementation review. The focus of the auditor is to assess whether risks have been addressed with appropriate controls in the new system after its implementation. This type of review considers:

- Control assessment of business processes
- Application security assessment and checks for SoD conflicts
- Program interface controls
- Achievement of project objectives and the business case

- User satisfaction and outstanding issues
- Change management processes

Audits in the first year after implementation may be complicated, particularly if the auditor has no prior involvement. This is because the enterprise is in a state of significant change, with the enterprise conducting business in a markedly new way with the implementation and acceptance of an ERP system. Further, it may be challenging to get all of the information that is needed to conduct an audit because:

- Critical information may not be known.
- Resources may not be available.
- There may be processing problems if some things were not considered during implementation.
- The system may have some bugs that could affect data integrity.

The audit process may be further complicated due to limited in-house knowledge or experience in conducting ERP audits. In particular, audit tools and methodologies may not be sufficiently mature or robust for the new ERP environment.

Change management is a key consideration, particularly during the early stages after the system is put into operation, when many changes may still be proposed (and made) to the implemented system. Adequate and structured change management processes should be in place to support the authorized and appropriate development, testing and migration of changes to the production environment. These processes should include the existence of adequate policies, procedures and control mechanisms surrounding the implementation of changes, such as:

- The configuration of separate application instances for development, testing and production
- The restriction of access to the separate application instances (e.g., access to development, testing and production) to maintain SoD
- The restriction of access to production instances to authorized personnel
- Formal change requests with justification, impact analysis and backout plans
- Appropriate approval
- Stakeholder involvement in the design of test plans
- Program, integration, stress and user acceptance testing
- The approval of testing results and migration to production

A postimplementation review should ensure that such processes and mechanisms have been followed. Furthermore, it is important to make policies and procedures available covering all aspects of administration of the system to help maintain the ongoing integrity of the system.

Quality Assurance

A quality assurance audit requires the auditor's participation throughout the life of the project and, as such, is the most comprehensive approach. It focuses on the overall quality of the business process reengineering program and considers specific deliverables at each key project milestone. Quality assurance may also focus on the project itself, timely delivery and orderly risk management. In most cases, quality assurance reviews are performed by experts in the field, or even third-party vendors for more technical quality assurance reviews.

Audit Involvement in the Project

While the approach that is adopted will vary according to the enterprise implementing the ERP system, the most effective approach usually involves auditors having an extensive and active involvement during implementation of the project. Involvement throughout implementation allows auditors to contribute to the establishment of the most effective control environment possible, as controls are built in during configuration along with other changes. If this approach is adopted, an external or independent party may need to be called on to perform any postimplementation review.

New security, audit and control tools may need to be developed/acquired to facilitate the effective implementation, operation and review of the control environment. Whatever the extent of audit involvement, it is worth reiterating that it is far more cost-effective to address issues during implementation than to reconfigure the system after implementation. It is also worth noting that audit's role remains ongoing after implementation, during the stabilization phase and continuing into the review of future enhancements.

Audit Responsibilities

Each environment needs to be evaluated on an individual basis from a risk perspective. A combined team will need to decide on the extent of involvement from the financial, operational and IT audit groups. In addition, subject matter specialists, such as tax personnel, may be needed to supplement the team. There are no hard-and-fast rules to split roles and responsibilities among the internal audit groups. An evaluation needs to be made as to how the roles and responsibilities should be defined. The important issues are that users and management should have:

- A seamless and efficient audit
- A well-integrated and knowledgeable team

Recommended Oracle EBS Audit Framework

In chapter 3, the need for an overall control framework to facilitate the assessment of risk and the completeness of management controls was identified. The audit framework in **figure 4.1** shows the key components of an ERP environment from an audit perspective. The first component of this audit framework is business processes or cycles. As described earlier in this guide, Oracle EBS software is an integrated application system; therefore, it is important to assess the end-to-end business processes and application infrastructure when performing risk and control assessments. The entire system of controls needs to be considered. A risk at the beginning of the process may be addressed by a key control at another point in the process or a series of compensating controls throughout the process.

Figure 4.1—Business Processes, Data, Application Infrastructure and Technology Infrastructure

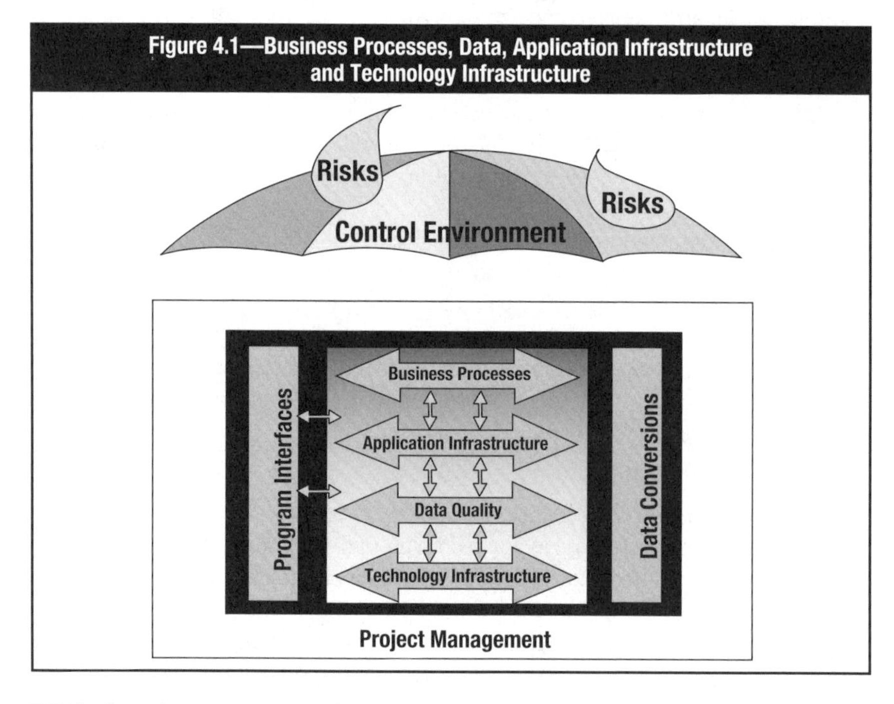

While the primary concern when auditing Oracle EBS software should be with the business process, the auditor needs to know what system functionality needs to be audited. For that reason, it is important for the auditor to obtain an understanding of the relationship between audit business processes or cycles and the Oracle EBS modules. It is important to note that the definition of business cycles may differ among enterprises. For example, an entity may combine the expenditure and inventory cycles into a larger cycle called "purchase to pay." Note, though, that the expenditure cycle links to other cycles, such as asset management; therefore, it may be preferable to assess it separately.

The revenue cycle is sometimes referred to as "quote to cash." Service organizations, such as educational institutions, may have a derivative of this cycle and refer to it as "enrollment to cash," and instead of an inventory cycle, it may have a student education cycle involving admission to graduation. Organizations dealing with large projects may have a project accounting cycle that combines the revenue and expenditure cycles and tracks the revenue and expenditure for particular projects to completion.

Linking audit cycles to the Oracle EBS modules is shown in **figure 4.2.** The financial accounting cycle (involving setting up the COA, processing journal entries and preparing the financial statements) links to the Financial Management module. The operational business cycles of revenue, expenditure and inventory link to both the Financial Management and the Asset Lifecycle Management (ALM)/CRM/Procurement/Product Lifecycle Management (PLM)/Supply Chain Management (SCM)/Manufacturing (MAN) modules, as does the fixed asset cycle (involving the maintenance of the fixed asset register, depreciating assets, managing fixed assets, and acquiring and disposing of fixed assets). The personnel/payroll cycle links to both the Financial Management and the Human Capital Management modules. The Cross-Applications and the Industry Solutions module groups do not relate to individual audit business processes; rather, they function across all the business processes.

Figure 4.2—Linking Audit Cycles to Oracle EBS Modules

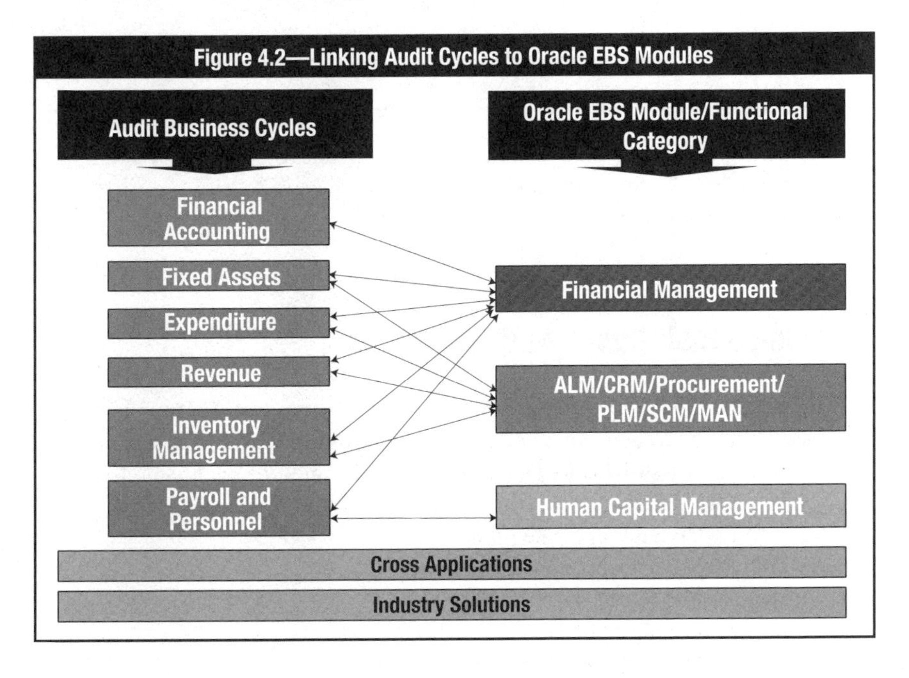

Figure 4.3 illustrates the relationships among the core business cycles. Sales are made in the revenue cycle, generating a demand for the production of finished goods from the inventory cycle. Production of goods necessitates the purchase of raw materials through the expenditure cycle. All of this activity is recorded online and in real time in the financial accounting cycle.

Figure 4.3—Core Business Cycles

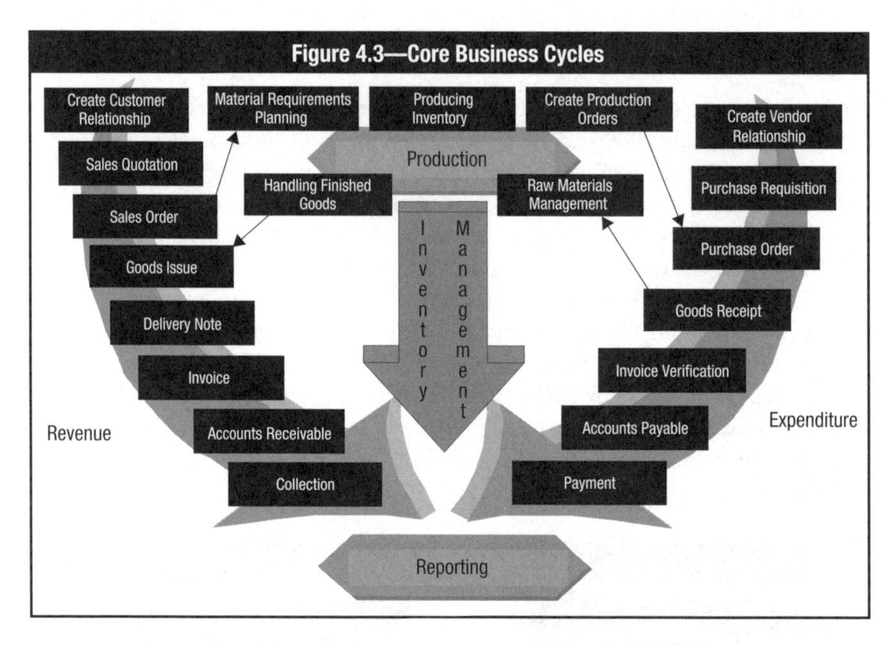

The following chapters in this technical reference guide describe how to audit the following core business cycles:

- Financial accounting (GL)
- Expenditure

Adopting a Risk-based Audit Approach

Having divided the control framework into components and subcomponents, a risk-based audit approach can be undertaken. (This guide assumes the adoption of a contemporary risk-based audit approach for the audit of the Oracle EBS environment.) This may involve:

- Gaining an understanding of the processes
- Identifying the significant risks
- Determining key controls
- Assessing the design of controls
- Assessing the operating effectiveness of controls

Techniques for gaining an understanding of the business process or function of the component or subcomponent under review include:
- Interviewing key personnel (i.e., implementers, users, business process owners, others knowledgeable in the area)
- Reviewing key documents and reports
- Observing personnel in the performance of their duties
- Developing process flow diagrams and descriptions

Understanding the process will provide the auditor with a framework for assessing the risks related to the process. Significant risks are risks that would negatively impact the achievement of the enterprise's objectives. Some techniques that may be employed include:
- Using a checklist approach (Appendix 6 of this guide provides ICQs for risks and controls in the areas of core business cycles and security administration covered in chapters 6, 8 and 10.)
- Using history or past experience (The risks included in this guide are based largely on past experience.)
- Determining objectives of implementation and assessing potential or existing barriers
- Analyzing the requirements of legislation, policy or procedures
- Adopting and applying the IT Assurance Framework (ITAF) and COBIT
- Utilizing a systems audit approach, considering inputs, processing or calculations, and outputs
- Adopting a quality management perspective
- Performing a strategic analysis (e.g., a strengths, weaknesses, opportunities and threats [SWOT] analysis or applying portfolio theory analysis)
- Questioning assumptions being made
- Assessing the needs of various stakeholders and the extent to which they are being met
- Evaluating hot spots (e.g., looking at the best and worst instances/stratified populations, key person dependencies, security issues, going concerns or business continuity issues, integration or conversion difficulties, valuations, accountabilities and reconciliations)

Having identified the significant risks, key controls need to be identified. Determining the key controls involves:
- Understanding the control culture of the enterprise (e.g., a "just-enough" control philosophy is employed consistent with the pursuit of an overall cost leadership position in the industry, or a "belts-and-braces" philosophy is employed consistent with the pursuit of a position of differentiation based on image, product, distribution or service characteristics that need to be protected)
- Exercising judgment to determine the key controls in the process or function and whether the control structure is adequate (Any weaknesses in the control structure need to be reported to executive management and resolved.)

There are two basic types of controls in an ERP environment:

- People-based controls (management controls, such as report monitoring and manual reconciliations or approvals)
  - Purely manual controls
  - IT-dependent manual controls
- System-based or application controls
  - Inherent or embedded controls (relying on controls built into the operation of the system, e.g., edit and validation routines)
  - Configurable controls (utilizing customizing options at installation or at a later time to control and direct processing operations in the system, e.g., certain password controls preventing the use of the last five passwords, passwords containing any three-character string contained in the user ID and passwords beginning with three identical characters)
  - Logical access controls (restricting access to computer system functions based on the ERP and technical infrastructure security authorization techniques)

The next step is to evaluate whether the system- and people-based controls and procedures identified are adequate to manage the related risks. These key controls need to be tested to confirm that they are operating effectively and as intended by management. The control testing techniques described in this guide focus on configurable controls and logical access security because:

- Manual controls are not unique to an ERP environment and testing techniques have already been covered in other forums and literature
- Automated controls (e.g., inherent, configuration settings and logical access security) tend to be more pervasive in an ERP environment
- Inherent controls are essentially built into the Oracle EBS software and, therefore, do not require control testing

Oracle EBS Configuration Concept and Testing Configurable Controls

As with other application packages, configuration settings allow enterprises greater flexibility to customize and direct the operation and processing of Oracle EBS software in their environments. The Oracle EBS application programs read the configuration settings, and processing is directed accordingly. Configuration settings are stored in tables in the Oracle EBS software. They may be field values in a table (e.g., security parameters), lookup codes (quick codes) or qualifying conditions.

Lookup codes (or quick codes) are internal names of values that are defined in a lookup type. They are used through Oracle EBS to define lists of values to provide valid values for certain items of information. Users do not see the codes since they are used internally; instead, users see the list of values.

Qualifying conditions are used through Oracle EBS to ensure that specific conditions have been met prior to allowing the transaction to be processed. An example of a qualifying condition includes the qualifier conditions in Oracle Pricing. These conditions determine whether, for example, a price qualifies for a discount.

Testing of configurable controls should include testing the access to the configuration setup, which may be performed in three ways:

- Checking configuration, including:
  - Configuration screen online
  - Configuration report
  - Tools
  - Database table
- Reperforming a control, including:
  - Sitting next to the user
  - Inputting certain data in the system
  - Using predefined scenarios and data in a testing environment
- Data analysis, including:
  - All transactions
  - Identification of specific exceptions
  - Validation of complex, configurable calculations
  - High-volume, computer-generated transactions

Oracle EBS Security Concepts

This section provides an overview of the Oracle EBS security concepts, introducing and explaining key Oracle EBS security functionality. Security is an integral part of the control techniques and procedures in all business areas. It controls the user's access to Oracle EBS.

With Oracle EBS R12, the Core Security area of Oracle User Management now includes an RBAC model that builds on the existing Function Security and Data Security models. A new set of Administrative Features that build on Core Security are also introduced in this release. Administrative Features include Delegated Administration, Provisioning Services, and Self Service and Approvals. **Figure 4.4** illustrates the Oracle User Management layers. Each of these layers will be discussed in the following sections.

Figure 4.4—Oracle User Management Layers

Administrative Features	Self Service and Approvals
	Provisioning Services
	Delegated Administration
Core Security	Role-based Access Control
	Data Security
	Function Security

Function Security

Function Security is the base layer of access control in Oracle EBS. It restricts user access to individual menus and menu options within the system, but does not restrict access to the data contained within those menus.

Data Security

Data Security is the next layer of access control. It provides access control on the data that a user can access and the actions a user can perform on those data. Access is restricted to individual data that are displayed on the screen once the user has selected a menu or menu option.

It is important to note that data security is only compatible for applications that have been written to leverage the Data Security Framework.

RBAC

RBAC is the next layer that builds upon Function Security and Data Security. The RBAC model augments the existing access control model in Oracle EBS by providing additional methods to organize data security policies and existing function security (via roles). Security privileges in Oracle EBS have, up to this point, been managed on an individual user basis, with different types of privileges assigned to each user directly. By leveraging the RBAC model, users will no longer need to be directly assigned the lower-level permissions and responsibilities since these can be implicitly inherited based on the roles assigned to the user. Roles can now be defined to consolidate responsibilities and other roles through role inheritance, as well as lower-level permissions (functions) and data security policies.

As part of the RBAC model, administrators are able to create role categories to bundle roles and responsibilities to make the process of searching for them easier. For example, all sales and marketing related roles could be included in the Sales and Marketing category.

Roles can also be included in role inheritance hierarchies that can contain multiple subordinate roles and superior roles. With role inheritance hierarchies, a superior role inherits all of the properties of its subordinate role, as well as any of that role's own subordinate roles.

Oracle's RBAC implementation within EBS does not replace previous security models; rather, it is an additional layer that should be integrated carefully. Implementations may continue working with the existing security model based on responsibilities, forms and functional security, where other implementations may move to a full role model or hybrid approach. In any case, access rights and functional security setup should be reviewed carefully.

Some applications, e.g., Oracle Learning Management (OLM), work only with roles and the RBAC model. This application includes a predefined set of roles and requires the UMX application to be configured to manage roles, grants, permissions, etc. On the other side, all non-RBAC modules will continue working within an RBAC model. Existing responsibilities could be "embedded" inside roles and managed through the UMX Self Service user interface. This integration of the two layers will be further detailed in chapter 9.

Delegated Administration

Delegated Administration is a privilege model that builds on the RBAC system. It provides enterprises with the ability to assign the required access rights for managing roles and user accounts. Instead of relying on a central administrator to manage all its users, an enterprise can create local administrators and grant them sufficient privileges to manage a specific subset of the enterprise's users and roles. Delegation policies are defined as data security policies. The set of data policies that are defined as part of the Delegated Administration are known as Administration Privileges.

Administration Privileges determine the users, roles and organization information that delegated administrators (local administrators) can manage. Although each privilege is granted separately, the three work together to provide the complete set of abilities for the delegated administrator:

- **User Administration Privileges**—Enable the local administrator to determine the users and people able to be managed. Local administrators can be granted different privileges for different subsets of users.
- **Role Administration Privileges**—Define the roles that local administrators can directly assign to and remove from the set of users they manage.
- **Organization Administration Privileges**—Define the external organizations that a local administrator can view in Oracle User Management.

Provisioning Services

Provisioning Services are modeled as registration processes that enable end users to perform some of their own registration tasks, such as requesting new accounts or additional access to the system. They also provide administrators with a more efficient method of creating new user accounts, as well as assigning roles. When a user completes registration using the registration process, the system captures the required information from the user, and subsequently assigns that person a new user account, role or both. Oracle User Management includes three types of registration processes:

- **Self Service Account Requests**—Provides a method for individuals to request a new user account
- **Requests for Additional Access**—Users can request additional access through the Oracle User Management Access Request Tool (ART), available in the Global Preferences menu

- **Account Creation by Administrators**—Administrators benefit from registration processes that have been designed to streamline the process of creating and maintaining user access. Each account creation registration process can be made available to selected administrators.

Self Service and Approvals

Once registration processes have been configured as required, individuals can subsequently perform Self Service registration tasks, such as obtaining new user accounts or requesting additional access to the system. Organizations can also use the Oracle Approvals Management Engine (AME) to create customized approval routing for such requests. For example, an organization may enable users to request a particularly sensitive role. However, prior to the user being granted access to the role, the organization can require two senior staff members to approve the request.

AME has implemented the RBAC model from AME.B onward. AME has five seeded roles:

- Approvals Management Administrator
- Approvals Management Analyst
- Approvals Management System Viewer
- Approvals Management System Administrator
- Approvals Management Process Owner

Approval rules within the registration processes should be reviewed and considered as sensitive functions.

Multiple Organizational/Ledger Sets Functions

Multiple Organizational (Multi-Org) is a function within Oracle EBS that allows enterprises to define organizations within their structure and the relationships among these organizations. This function:

- Affects how transactions flow through the different organizations and how the organizations interact with each other
- Allows greater flexibility in how transactions are processed in different organizations
- Supports three primary organization models:
  - Human resources (HR)
  - Assets
  - Accounting/distribution/materials management
- Segregates and secures access to data based on the organization's setup

Ledger Sets is the method by which Oracle GL organizes accounting data. A Ledger Set shares a common functional currency, COA and Accounting Calendar.

Multi-Org is an Oracle EBS feature that allows diverse businesses with significantly different operating requirements to exist in a single Oracle EBS instance. This is achieved by allowing separation of the way transactions are processed and data are stored according to operating units. Within the context of Multi-Org, Ledger Sets is a type of organization and defines how financial transactional data are collected and stored. Ledger Sets is defined within the GL module. It is possible to create multiple Ledger Sets in a single organization environment. Also, multiple inventory organizations can be created in a single organization setup. However, data can be secured across multiple operating units only in a Multi-Org environment.

The components of Multi-Org are:

- Organization types based on the following hierarchical structure:
  - Business group
  - Government Reporting Legal Entity (GRLE)
  - Legal entity
  - Ledgers/Ledger Sets
  - Operating unit
  - Inventory organizations
  - HR organizations
- Organization classifications:
  - Classifications define what type of a role an organization performs.
  - Organizations are assigned classifications to define relationships among organization types.
  - One organization can be assigned any combination of legal entity, operating unit and/or inventory organizations.

Which organization models apply depends upon the Oracle EBS modules installed. **Figure 4.5** shows an example of an organizational model for the accounting/distribution/materials management model.

Figure 4.5—Example Organizational Model

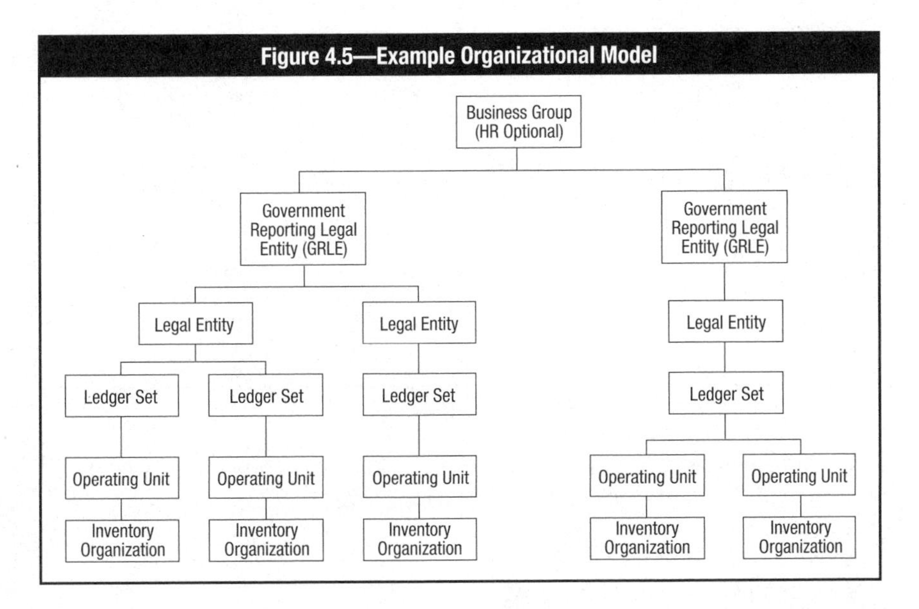

Oracle User Management enables organizations to define administrative functions and manage users based on specific requirements, such as job role or geographic location.

The key features of Oracle User Management can be divided into two main areas: Core Security and Administrative Features.

Security Audit Steps

Figure 4.6 show a representation of the security audit steps and the reports used in each step.

Figure 4.6—Oracle EBS Security Audit Steps

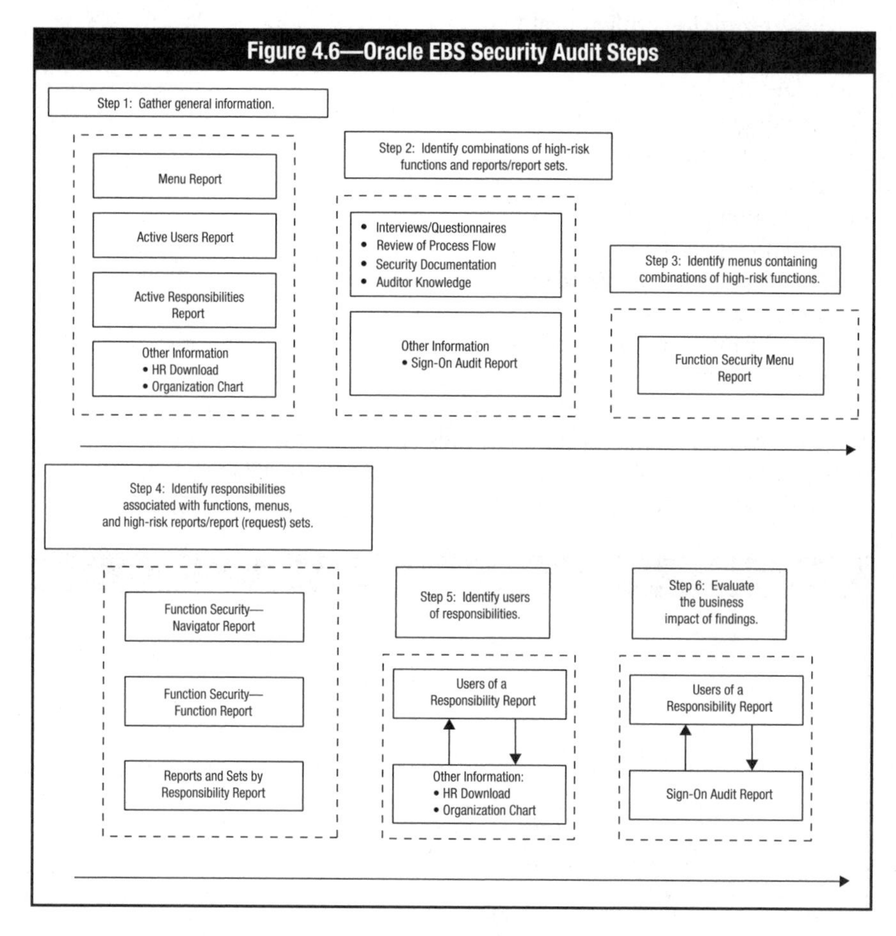

Step 1. Gather General Information

The following reports should be run and printed for the period under review:

- Menu report (not the Function Security Menu report)
- Active Users report
- Active Responsibilities report
- SOX Compliance RBAC Reports

These reports provide a listing of all the menus in use and their structures and all the active users on the system, responsibilities and roles assigned. This presents the auditor with an overall picture of the system (i.e., the menus being used, the number of users and their responsibilities).

It may also be useful to obtain an organization chart of employees within the department under review. A download of employees from payroll may also prove useful to assess the appropriateness of the number of users and their job functions within the enterprise.

Step 2. Identify Combinations of High-risk Functions and Reports/Report Sets
A list of the target high-risk functions and Reports/Report (Request) Sets should be compiled. This can be achieved through a combination of interviews or questionnaires with business management; a review of process flow and security documentation; and the auditor's knowledge of the business, Oracle EBS software and related risks.

The auditor can determine what forms (functions) key users have executed within the past three months using the Sign-On Audit Forms report (refer to chapter 9, Oracle E-Business Suite Security), if the Sign-On: Audit-level Profile Option has been set. This frequently provides insight into the real form (function) usage of key business roles within the enterprise. It may also uncover Oracle EBS or customized forms (functions) with which the auditor is not familiar or which possibly have not been uncovered through previous inquiries. It should be noted that this can be used only as a guide because the audit data may be purged by the system administrator at any time.

In R11*i* there is a profile option called Concurrent: Report Access Level that defines who can access the output of concurrent requests, which could be set to User or Responsibility. In R12, this was replaced by RBAC. The UMX RBAC is used to control who can view request output files.

It is also important to consider early in the audit specific ERP risks that are not limited to the production system. Additional areas of risks that should be considered by the auditor include:

- **Data extracts and interfaces with legacy systems, business intelligence or reporting applications, etc.**—Extracting data from production environments may lead to loss of confidentiality or integrity (SQL import). Some applications (e.g., Web ADI-BNE) allow import-export from a client workstation to EBS through Microsoft Excel files. Although these functions are convenient, they could add significant risk if not implemented and monitored adequately.
- **Forms customizations**—Can be a significant risk. All added HTML pages, reports, concurrent programs and forms should be reviewed to determine whether there is a high-risk custom function.

- **Infrastructure management allowing environment cloning and copying**—Usually part of the high privileged functions. The auditor should evaluate the process to perform such tasks and who is authorized to do so. As the data in the environment could be highly sensitive, data integrity, confidentiality and availability need to be considered as part of the copying/cloning process. Cloning environments, which contain sensitive data to create testing or training environments, could pose a significant risk.

Step 3. Identify Menus Containing Combinations of High-risk Functions
The Function Security Menu report should be run and printed for the period under review to identify menus containing target high-risk functions. The report lists all of the functions associated with menus. All of the target high-risk functions should be identified and the menus with which they are associated flagged for the next testing step.

Step 4. Identify Responsibilities Associated With Functions, Menus (Containing High-risk Functions) and High-risk Reports/Report Sets
The following reports should be run and printed for the period under review:
- **Function Security Navigator report**—Lists all of the responsibilities and their menus. For the menus identified in step 2 as containing the target high-risk functions, the responsibilities should be flagged for the next testing step.
- **Function Security Function report**—Lists all of the functions associated with responsibilities. All of the target high-risk functions should be identified and the responsibilities with which they are associated flagged for the next testing step.
- **Reports and Sets by Responsibility report**—Lists the Report Sets and concurrent programs that responsibilities can access. All of the target high-risk Reports/Report Sets should be identified, and the responsibilities with which they are associated should be flagged for the next testing step.

The auditor should, at this stage, have a list of responsibilities to review that provide access to the target high-risk functions. The list may contain duplicate responsibilities identified via the different reports produced in step 3. Duplicate responsibilities should be deleted.

Another method that may be employed to identify the target high-risk functions and associated responsibilities is security documentation. If application security has been sufficiently documented, the project team may have captured all of the relevant functions and responsibilities in the course of the security implementation. When relying on security documentation, the auditor should consider corroborating with at least two members of IT/security management to ensure that the documentation has been maintained accurately and is up to date. Results should also be confirmed with management.

Assessing access based on the responsibilities created by the organization is also unlikely to yield accurate results since they may have been created with additional functions that provide excessive access or, through the combination of responsibilities, users may have inadvertently been provided with access to functions contrary to management's intentions.

The auditor needs to exercise judgment in the situation under review to determine the appropriate functions for testing.

Step 5. Identify Users of Responsibilities (Containing Combinations of High-risk Functions)

Run and print the Users of a Responsibility report for the period under review. It is recommended that the responsibilities that can access the target high-risk functions be specified in the report parameters before the report is run so that only relevant responsibilities are reported. Then, the report will list only the users who have access to the responsibilities specified. The identified users should be compared to their job functions, obtained from the HR download or organizational chart, and the appropriateness of their access should be assessed in addition to achieving a satisfactory SoD.

Step 6. Evaluate the Business Impact of Findings

Compare the Sign-On Audit report run in step 1 to the users listing obtained in step 5. This allows the auditor to determine the percentage of users who can execute the function, compared to those who have actually made use of the ability. This can be a powerful argument in illustrating the need for tighter security.

Testing SoD/Excessive Access

Traditional systems of internal control have relied on assigning certain responsibilities to different individuals or segregating incompatible functions. SoD is intended to prevent one person from having access to assets and responsibility for maintaining the accountability of such assets. For instance, in an inventory management system, different individuals typically are responsible for duties such as:

- Initiating or requesting a purchase
- Placing and inputting POs
- Receiving goods
- Ensuring custody of inventories
- Maintaining inventory records and/or authorizing adjustments to costs or quantities, including authorizing disposal or scrapping
- Making changes to inventory master files
- Performing independent inventory counts
- Following up on inventory count discrepancies
- Authorizing production requests and/or materials transfers
- Receiving/transferring goods into/from manufacturing
- Shipping goods

Further, physical counts of inventory should be performed by someone independent of the custody of inventory and with no access to inventory records. An individual who is independent of the custody and recording of inventory should also follow up on discrepancies noted in the comparison of the counts to inventory records. If one individual has responsibility for more than one of these functions, that individual could misappropriate assets and conceal the misappropriation. For example, if one individual has the ability to process sales orders and access the inventory management master files, that person can modify product selling prices and process unauthorized sales. Legacy system environments necessitated and facilitated the SoD because of the predominantly manual control framework fragmentation of legacy systems also facilitated the SoD since purchasing systems, warehousing systems and GL systems were all separate.

However, this traditional notion of SoD needs to be refined in a fully automated Oracle EBS environment. ERP systems have shifted the emphasis to user empowerment, enabling users to have access across business functions or, alternatively, to handle physical assets and record their movements directly into the computing and accounting system. As shown earlier in this guide, along with this transition, controls have moved up front in the process and have become increasingly automated with online release, automated matching of transactions, increased integration, etc. The notion of the good business practice of the SoD control needs to be developed to include a risk management perspective and a trade-off or balance between functional access and security in the new ERP environment. The ERP environment should be assessed from a risk and control perspective. Key control steps should be modeled for each business process in the organization, and appropriate trade-offs between empowerment and the need to minimize the risk of fraud or unauthorized transactions should be made.

In the model of the purchase to pay cycle, shown in **figure 4.7**, a goods receipt is considered in the same way as an invoice approval since it is the entry of a goods receipt that in many cases completes the matching process and gives rise to an authorized payment, if pay-on-receipt has been enabled. Any user who has access to a single step in the process may be considered to present a low risk from an SoD perspective (e.g., the individual must collude to effect a fraudulent or unauthorized transaction). Any user who has access to two steps in the control model may be considered a medium risk from an SoD perspective (i.e., the individual must influence at least one other person to effect a fraudulent or unauthorized transaction). Any user who has access to three steps in the control model may be considered a high risk from an SoD perspective (i.e., the individual can most likely effect a fraudulent or unauthorized transaction that may go undetected, at least for some time).

Figure 4.7—Purchase to Pay

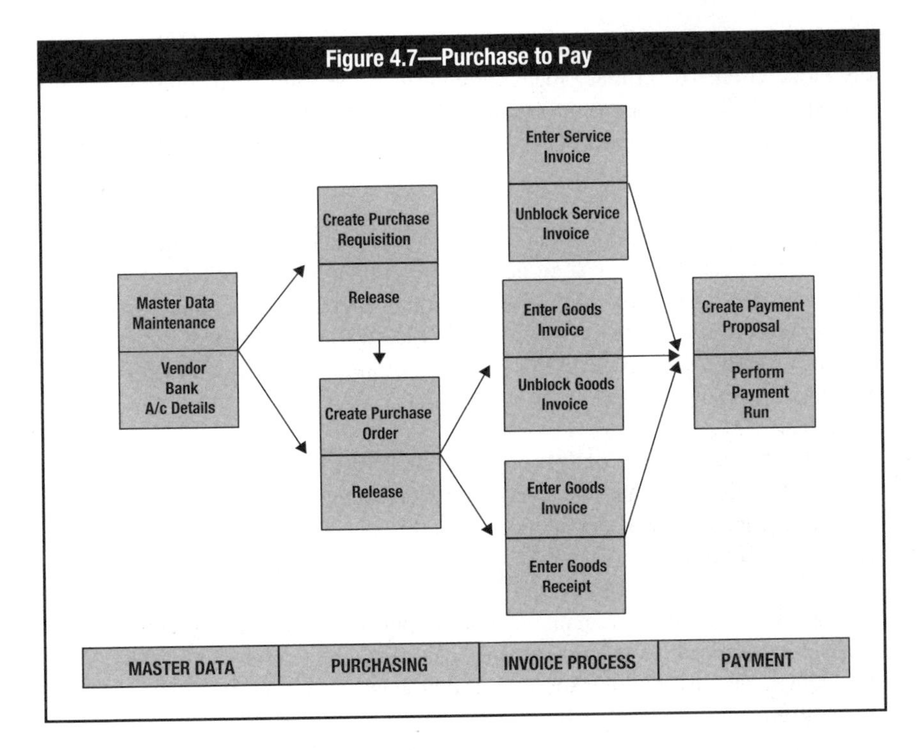

Two examples of a high-risk situation are:

- Users with access to maintain master data and the ability to enter a service invoice and match it with a PO or override system hold can make a payment to themselves.
- Users who can enter and release a PO and perform a goods receipt have the potential to misappropriate goods.

Note: SoD risks can be decreased through compensating controls such as a requirement that supporting documents be reviewed manually and approved prior to payment.

Specific techniques for testing automated controls (i.e., access security and configurable controls) in Oracle EBS software are described in chapters 6, 8 and 10, which focus on testing the core financial business cycles. Depending on the level of risk and the amount of testing required, audit testing rotation plans may be established for each of the components or subcomponents of the audit framework. A rotation plan would allow audit areas to be rotated, meaning that not every component would be tested on an annual basis but over an audit period, e.g., three years.

Risk and Control Documentation

Risks and controls vary among organizations, depending on external and internal factors and the configuration of Oracle EBS. This guide does not attempt to comprehensively address legal and regulatory compliance risks for any jurisdiction. The primary focus for controls is on automated controls that are unique to an ERP system or the Oracle EBS environment. Auditors should be familiar with manual controls (e.g., reconciliations), manual approvals and dual signatories.

Specific assurance techniques vary from enterprise to enterprise, depending on the extent of customization, specific configuration options selected and whether automated diagnostics are used. The techniques, therefore, do not necessarily include prescriptive lists of specific Oracle EBS testing techniques, but rather provide a sample tool set for use or reference as appropriate in the user's organizational environment.

Note: Customize and customization are general terms used in this guide to refer to the tailoring of an Oracle EBS Suite environment to suit the customer's specific needs, by way of configuration, modification or additional development of tables or programs.

Because there may be more than one control per risk, a numbering sequence for risks, controls and testing techniques has been adopted throughout each of the chapters dealing with the auditing of core business cycles, as shown in **figure 4.8**.

Figure 4.8—Numbering Sequence for Risks, Controls and Testing Techniques

Number	Description
Risks	
1.1	Risk number 1 for the first subprocess
1.2	Risk number 2 for the first subprocess
2.1	Risk number 1 for the second subprocess
Controls	
1.1.1	Control number 1 for risk number 1 of the first subprocess
1.1.2	Control number 2 for risk number 1 of the first subprocess
1.2.1	Control number 1 for risk number 2 of the first subprocess
2.1.1	Control number 1 for risk number 1 of the second subprocess
Testing Techniques	
1.1.1	Testing technique for control number 1 for risk number 1 of the first subprocess
1.1.2	Testing technique for control number 2 for risk number 1 of the first subprocess
1.2.1	Testing technique for control number 1 for risk number 2 of the first subprocess
2.1.1	Testing technique for control number 1 for risk number 1 of the second subprocess

In each of the business cycles:
• The functionality of the Oracle EBS business process and its subprocesses, from a controls or operational audit perspective, are described.
• Specific risks are identified.
• Potential automated controls are outlined.
• Sample assurance techniques are suggested.

In this way, the relationship among risks, controls and testing techniques for subprocesses (such as Master Data Maintenance, sales order processing and invoicing) should be evident as the reader progresses through each of the chapters dealing with auditing the core business cycles.

Audit Framework and COBIT

When control issues are identified, the auditor should attempt to uncover and report to management the causes of the problem, together with recommendations. In this respect, COBIT 4.1 helps meet the multiple needs of management by bridging the gaps among business risks, control needs and technical issues. It provides good practices across a domain and process framework, and presents activities in a manageable and logical structure. COBIT's good practices provide a measure to judge against when things do go wrong and can assist in identifying problem causes.

The COBIT framework consists of the following four domains:
• Plan and Organize (PO)
• Acquire and Implement (AI)
• Deliver and Support (DS)
• Monitor and Evaluate (ME)

Some of the processes contained within these domains that are of particular relevance when assessing ERP systems include:
• AI6 Manage changes.
• DS5 Ensure systems security.
• DS9 Manage the configuration.
• DS11 Manage data.

Summary

This chapter looked at the audit impacts arising from the implementation of an ERP system (e.g., changes to audit methodologies) and recommended an Oracle EBS audit framework (including, for example, business processes and application security). It detailed the steps involved in taking a risk-based audit approach to ERP, including a description of key concepts (e.g., authorization and configuration) and methods of testing configurable controls, access security and SoD. Finally, the relationship between the recommended Oracle EBS audit framework and COBIT, including COBIT's usefulness in identifying the causes of control breakdowns, was discussed.

Page intentionally left blank

5. Oracle E-Business Suite—Financial Accounting Business Cycle

Introduction

The Oracle EBS Financial Accounting module (GL) has an integrated architecture that facilitates data integrity, auditability and control. The GL contains many of the same features and functions of other GL packages. The GL simultaneously maintains all financial balances—such as actual, budget, summary, foreign currency and statistical—in a single ledger. The posting process allows the automatic synchronizing of journals, credit balances, debit balances and summary balances (if configured).

The GL uses a graphical user interface (GUI) with an intuitive feel. **Figure 5.1** shows the GL screen.

Figure 5.1—GL Screen

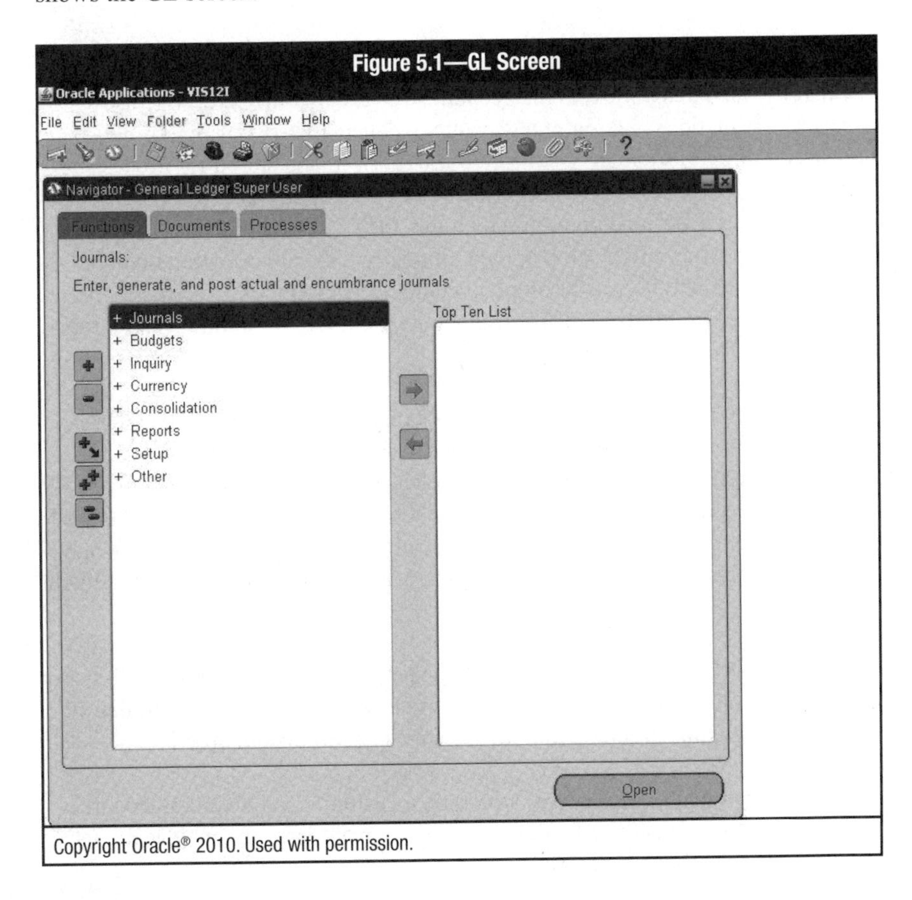

The Oracle EBS GL contains a number of features and functions for GL processing, some of which are described in this chapter.

The financial accounting cycle has three main subprocesses:
• Master Data Maintenance (Chart of Accounts)
• Journal Processing
• Reconciliation and Financial Reporting

Specific control features and functions of Oracle EBS are described for activities within these major subprocesses.

Master Data Maintenance (Chart of Accounts)

Maintaining the GL master data includes all of the activities related to designing, implementing and managing the accounting framework, which is comprised of a GL Chart of Accounts, coding structure, etc.

Chart of Accounts Structure

A Chart of Accounts (COA) is the account structure that the enterprise uses to record transactions and maintain account balances. The COA determines how accounting information is collected, categorized and stored for processing and reporting purposes. The COA structure (e.g., company-division-cost center-account) is defined upon initial Oracle EBS GL setup. Changes to the COA structure after initial setup are operationally difficult and often involve data conversion activities or a complete reengineering of the COA structure. Therefore, it is critical that sufficient thought be put into defining this structure to meet future reporting needs.

The GL is assigned to a financial reporting entity that uses a particular COA, functional currency and Accounting Calendar. For a globally dispersed enterprise that shares common features, such as a common COA, functional currency or Accounting Calendar across its subentities, Ledger Sets can be used to consolidate operations across these subentities. Ledgers and Ledger Sets have replaced the concept of Set of Books, which was used in versions prior to Oracle EBS R12.

Oracle GL provides the functionality to manage financial information within any enterprise structure. Multiple companies can be maintained within a similar or different accounting structure, and their results can be consolidated.

With a multicompany (Multi-Org) structure, a customer can maintain ledgers, budgets and control security for each company; create intercompany accounts; and produce consolidated reports.

The GL has three main subledgers: Accounts Receivable, Accounts Payable and Cash Management. Oracle EBS Accounts Payable and Accounts Receivable cannot process multiple ledger sets unless Multi-Org is enabled. Therefore, it is important to incorporate all reporting entities into one COA.

Foreign and domestic locations can share one COA, as long as the functional currency and Accounting Calendar remain the same. The foreign journal entries are converted to domestic currency at the time of posting. To translate balances, the GL's translation functionality may be used. Currency conversion rate types are defined and assigned for the automatic conversion paths as follows:
• Define or enable currencies: Setup➜Currencies➜Define
• Define conversion rate types: Setup➜Currencies➜Rates➜Types
• Define daily rates: Setup➜Currencies➜Rates➜Daily
• Define historic rates: Setup➜Currencies➜Rates➜Historical

Each Ledger Set is assigned an Accounting Calendar, functional currency and COA and has an associated Period Type. When a Ledger/Ledger Set is assigned a calendar, transactions can only be posted to the Ledger/Ledger Set for the appropriate Period Type. The Period Type normally refers to the Open Periods as defined by the Accounting Calendar. The Ledger/Ledger Set accesses only the appropriate Period Type.

Accounting Calendar

The Accounting Calendar defines the accounting periods and fiscal years in Oracle GL. Transactions can only be posted in an Open Period. Several periods can be open at one time, to allow for prior period transactions and future period transactions to be posted. Periods should be closed once no more transactions are required to be posted into that particular period, to prevent accidental entries into the period.

The following paths are for defining:
• Accounting Calendars: Setup➜Financials➜Calendar➜Accounting
• Period Types: Setup➜Financials➜Calendar➜Types

Multiple Reporting Currencies

Multiple Reporting Currencies (MRC) is used to maintain transactions and account balances in multiple currencies. MRC is needed specifically for use by enterprises that must regularly and routinely report their financial results in multiple currencies. MRC is not to be used as a replacement for the GL's translation functionality, which is used to translate amounts from another currency to the functional currency at the account-balance level. MRC is used to convert amounts from a functional currency to another currency at a transactional level. If the MRC feature is used, reporting currencies need to be defined before the reporting Ledger/Ledger Set is defined.

The following path is for defining Ledger Sets: Setup➜Financials➜Ledger Sets.

Account Numbers

Key Flexfields are used to uniquely identify information such as GL accounts, inventory items, fixed assets and other entities that are important for enterprises to access. Key Flexfields consist of segments (e.g., company, department, account). Each segment has a name, meaning and value that have been specified by the enterprise. By using Key Flexfields, the enterprise can build customized fields for entering and displaying information relating to its business, such as part numbers and account numbers.

An enterprise will use codes to identify the entities' information (e.g., part numbers, account numbers). These codes should be developed by the implementation team to customize all Key Flexfields to meaningful code segments to describe each Key Flexfield entity. The entrprise should decide the following for each Key Flexfield during the implementation phase:

- How many segments an entity has
- What each segment means
- What values each segment can have
- What each segment value means
- Rules that govern valid combinations of segment values (Cross-Validation Rules)
- A definition of dependencies among the segments

A combination of segment values (or Key Flexfield code combination) uniquely describes an entity stored in a Key Flexfield. Therefore, a change made to one segment in a Key Flexfield changes the combination of segment values. It is also possible to create a combination of segment values for a Key Flexfield without the use of a combination form. This can be done by allowing the creation of a valid combination, as the user enters the values in the Key Flexfield for the first time, directly into a Flexfield pop-up window. It is important to note that any Cross-Validation Rules must be satisfied with the new combination.

GL account numbers are called Accounting Flexfield combinations in Oracle EBS terminology. The Flexfields must be defined during initial Oracle EBS setup based on company-specific COA requirements, e.g., one enterprise may wish to customize the Accounting Flexfield to have only three segments (company, department and account), whereas another enterprise may choose to have six segments (company, cost center, account, product, product line and subaccount). New account combinations are designated as posting or summary-level accounts. Each Key Flexfield must be defined in the Define Key Flexfields screen as well as in the Value table. To define the Key Flexfield segments and values, follow these paths:

- Key Flexfield segments: Setup➔Financials➔Flexfields➔Key➔Segments
- Key Flexfield values: Setup➔Financials➔Flexfields➔Key➔Values

Dynamic Insertion and Cross-Validation Rules

COA maintenance in the Oracle EBS GL includes two unique functions: Dynamic Insertion and Cross-Validation Rules. If the Dynamic Insertion flag is set to Yes in the Oracle EBS GL module, new account combinations are automatically created as they are entered in transactions. An example follows.

An AP clerk receives a voucher from the assistant controller. Unknown to the clerk, the voucher has been coded to a new expense account combination. The clerk enters the voucher into Oracle EBS Accounts Payable, and it passes all online edits. If the Dynamic Insertion flag had been set to No, the clerk would have received an error preventing him/her from entering the voucher until the account combination was set up in Oracle EBS GL.

Dynamic Insertion is a valuable tool that can reduce manual efforts required in setting up accounts manually. It can also improve system efficiency because GL accounts are added only if needed. Oracle EBS account numbers can never be deleted once set up (users can inactivate old accounts). Therefore, it is very important to set up only those accounts that are needed. Reports run more quickly with a shorter COA because fewer data have to be queried. In addition, month-end processes run more quickly when the COA is shorter because fewer beginning period balances need to be updated.

With Dynamic Insertion, invalid account combinations can be created (e.g., an invalid division or cost center) because no validation against existing accounts occurs. If new account combinations are outside existing ranges in the COA, they may not be picked up in the financial statements.

With the Dynamic Insertion function, Cross-Validation Rules should also be used. Cross-Validation Rules within Oracle EBS GL define the valid combinations of account numbers that can be created. For example, Cross-Validation Rules can prevent revenue entries to the payroll cost center. Another example is Cross-Validation Rules that prevent entries to other existing divisions.

Cross-Validation Rules are somewhat time consuming to define and set up in Oracle EBS GL. In addition, they must be maintained continually as the COA evolves, but are one of the most important preventive controls in the Oracle EBS setup for ensuring integrity of the accounting data.

System Efficiency

For system efficiency and control purposes, it is usually better to leave only the current accounting month open. Month-end processes will run faster because fewer beginning balances need to be updated, and postings to incorrect months will be prevented. Changes required to previous months should follow an appropriate change management and approval process to ensure accountability for changes.

Journal Processing

Journal Processing is used in GL accounting to facilitate entering, maintaining and reporting on actual accounting information for an enterprise. Key Journal Processing concepts include:
• Journal sources
• Journal definition
• Entering/importing/generating journals
• Special journals
• Posting journals
• Balancing journals

Journal Sources
Journal sources identify the origin of journal entries and specify whether detailed reference information should be imported along with the summary information imported from the feeder systems. Journal sources can also be frozen, preventing users from making changes to any journal posted to the GL from that source. If Journal Approval is enabled, higher management levels can be required to approve journals with a specific journal source before the journal can be posted.

Journal Definition
Oracle EBS GL allows the user to predefine allocation and recurring journal entries to help ensure consistency in processing month to month. It is important to restrict access to the journal definition to maintain the integrity of the allocation and recurring journal entries.

Entering/Importing/Generating Journals
Journal entries can be processed for posting in the GL in three different formats: manual, imported or generated. Manual journal entries or journal entry batches are keyed to the ledger through the Oracle EBS GL module. Journal entries from outside of the GL are processed through the Import function, which allows for the correction of journal data imported into the ledger. Reversing, recurring and allocation journals are generated in the Oracle EBS GL module.

Since journal entries update the information used in financial statements, it is important to restrict access to Journal Processing and to ensure that only valid transactions are reflected in the system.

Special Journals
Recurring journal vouchers should be used to prevent keying errors for entries that are similar each month (e.g., depreciation or allocation entries using the same account numbers and the same or different amounts each month).

Reversing journals should be used for all accrual entries. These journals immediately appear on the Unposted Journals Report during the following month. They reduce the chance of accruals not being reversed or of keying errors upon reversal.

Intercompany journals are also a special type of journal available to record transactions between two subsidiaries of an enterprise. The Global Intercompany Self Service application interface in Oracle EBS allows users to set up intercompany journals, enter intercompany transactions, import transactions from external sources and generate intercompany reports.

Journal Approval

The Oracle EBS GL Approval process obtains the necessary management approvals for manual journal batches utilizing the Oracle EBS Workflow Builder. If this is configured in Oracle EBS, the process validates the journal batch, determines wheher approval is required, submits the batch to approvers (if required) and notifies appropriate individuals of the approval results.

It is important to restrict access to Journal Approval through the Oracle Workflow Builder to maintain the integrity of the journal entries. Journal Approval processes include:

- **GL initialization and validation process**—Initializes and validates the journal batch, and determines whether the batch requires approval
- **GL preparer approval process**—Determines whether the preparer is authorized to approve his/her own journal batch. If so, the batch is approved, the approver name is set, and notifications are sent. The Profile Option: Journals: Allow Preparer Approval under the System Administrator responsibility should be configured to control the GL journal preparer approval process.
- **GL approval process**—Finds an appropriate approver, seeks journal batch approval, and sends notifications of approval or rejection
- **GL request approval process**—Seeks journal batch approval from the selected approver
- **GL no approver response process**—Provides handling options and actions to take when the approving manager has not responded to a journal batch approval request. This includes resending the request until a certain limit is reached, then providing the preparer with the option to resend the approval request to the approver or to send the request to the approver's manager.

Posting Journals

Journal entries can be posted to the GL from two different methods: the Post Journals form and as an Autopost request. The Autopost functionality can be initiated manually or scheduled to run as a concurrent request and automatically posts the journal entries that meet the criteria for journal source, journal category and effective date combinations, as defined by the user.

Since journal entries update the information used in financial statements, it is important to restrict access to journal posting and ensure that only valid transactions are reflected in the system. Journals should be subject to review and approved when they meet the enterprise's workflow criteria.

Balancing Journals

There are two types of journal balancing techniques:

- Optional Suspense Posting feature
- Control Total feature

Optional Suspense Posting Feature

Using the Optional Suspense Posting feature in Oracle EBS GL, management can force all journal vouchers (including those imported from all other applications) to balance to a suspense account during the posting process. Oracle EBS allows for the use of Suspense Posting of journal entries, if appropriately configured.

If Suspense Posting is enabled, a journal entry does not need to balance to post because the offset amount will post to the designated suspense account. A company may want to enable Suspense Posting for a number of reasons, but primarily to expedite the flow of information into the financial books. If someone in the field does not have all of the accounting information necessary to process a transaction at hand, Suspense Posting allows the transaction to flow through with the available information. Someone in the accounting department is then tasked with clearing the suspense accounts and reclassifying the expenses to the proper accounts.

If Suspense Posting is not enabled, the standard delivered Oracle EBS functionality will allow an out-of-balance journal entry to be created and saved, but will not allow the journal entry to post to the ledger. Another option, therefore, is to leave the Optional Suspense Posting flag set as No, stopping all journal vouchers from posting until they are corrected manually.

Control Total Feature

The Oracle EBS Control Total feature should be used for all journal vouchers keyed directly into the Oracle EBS GL. This feature performs two checks before allowing these entries to be posted to help ensure that the total debits entered equal the total:

- Credits entered
- Debits expected to be entered (keyed initially in the header)

Reconciliation and Financial Reporting

The purpose of closing activities is to finalize financial data from systems. The data are accumulated into logical accounts that indicate the nature of transactions to close and balance the financial records at period end. This subprocess also makes financial data available for reporting and analysis.

Closing the accounting periods on the system is performed to:

- Control journal entry
- Control journal posting
- Compute period and year-end actual and budget account balances for reporting

Figure 5.2—Closing a Period

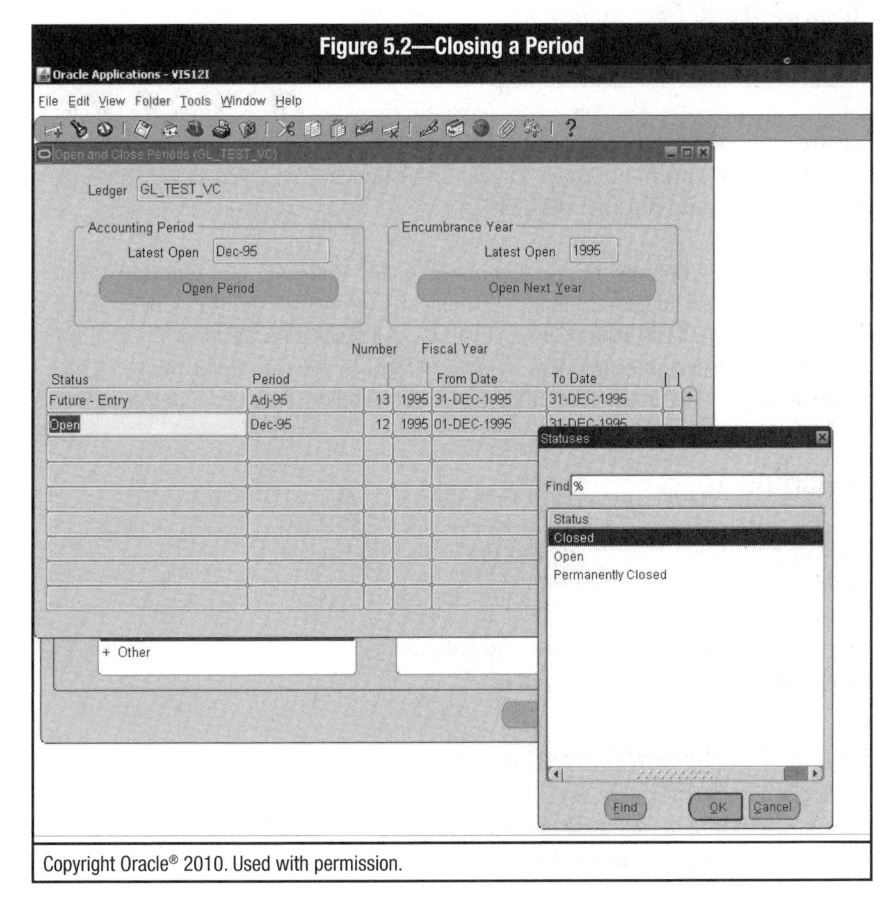

To close an accounting period, as shown in **figure 5.2**:

- Navigate to the Open and Close Periods window (Setup/Open/Close). This will display all of the accounting periods defined within the Ledger Sets.
- Set the open period to Close.

- Enter a new status for the period:
  - Enter Closed to prevent entering or posting journals to that period. A closed period can be reopened at any time.
  - Enter Permanently Closed to prevent entering or posting journals to that period. A permanently closed period cannot be reopened.
- Save the change.

Financial reporting includes preparing internal and external financial reports that may be used for management decision making, financial analysis, regulatory compliance, etc.

Reviewing and Reconciliation

Oracle EBS may not ensure that the GL is in balance. At month end, the GL should be balanced manually to help ensure that all new account numbers were set up with the correct account type (e.g., an asset could be set up in the ledger incorrectly as an expense account). The total of revenues less expenses and tax should equal net profit after tax. The ending balance of assets less liabilities should equal equity plus net profit after tax.

Oracle EBS may not ensure that all entries posted to the GL are reported on the Oracle EBS financial statements. All financial statements should be reconciled to the GL (assets, liabilities, equity revenue, expenses and net income) at month end, to help ensure completeness. Oracle EBS does provide reporting tools that could assist with detecting reconciliation issues and performing integrity checks.

Reporting Tools

The GL provides a number of reporting capabilities:
- Financial Statement Generator
- Online inquiries
- GL standard accounting reports and listings
- Enterprise Planning and Budgeting

Financial Statement Generator

Financial Statement Generator (FSG) is the built-in GL reporting tool. All FSG reports, including the balance sheet and income statement, must be defined upon initial Oracle EBS GL setup. This is because the reports are based on company-specific COA definitions.

It is very important to maintain the FSG reports consistent with the maintenance of the COA. FSG report definitions should be reviewed each time that new account numbers are added to the ledger to help ensure that the accounts will be included and reported in the correct place on the financial statements. Wide account number ranges are used in the FSG report definitions as much as possible so new account numbers can be added easily without requiring changes to the financial statements.

FSG reports are available online immediately and can be printed easily. Unfortunately, FSG reports may be difficult to secure effectively. The FSG program can be assigned to the report security group in a responsibility. However, users who have access to FSG through Oracle Financials EBS security can print any/all reports for the Ledger/Ledger Sets or COA. Reports can be run using the Submit Requests window, where Required Request (Single Report) is selected. FSG report sets cannot be run from the standard request submission. **Figure 5.3** shows the FSG screen (Reports➔Request➔Financial). The desired report is selected from the list.

Figure 5.3—Financial Statement Generator

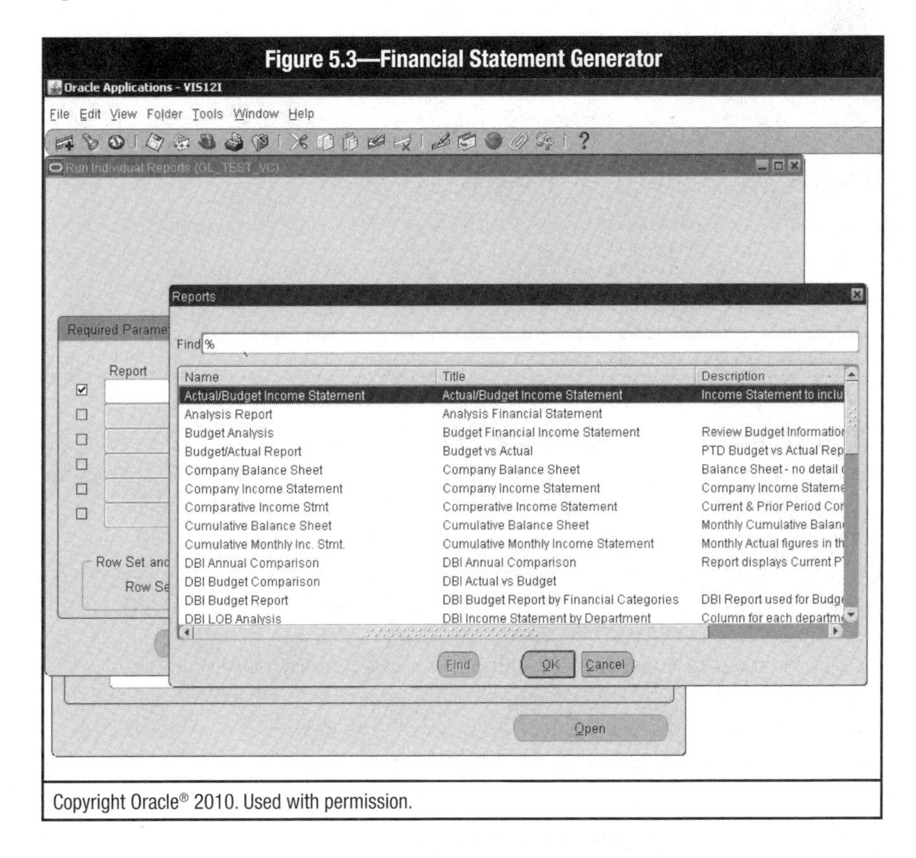

Online Inquiries

Online inquiries can be used to search for detailed information quickly. For example, an online inquiry can be performed on an account balance or journal entry. Financial statements, accounting reports and listings can also be reviewed online. **Figure 5.4** shows the Online Inquiries screen. Account inquiry criteria are specified as necessary, including access restrictions to online inquiries.

Figure 5.4—Online Inquiries

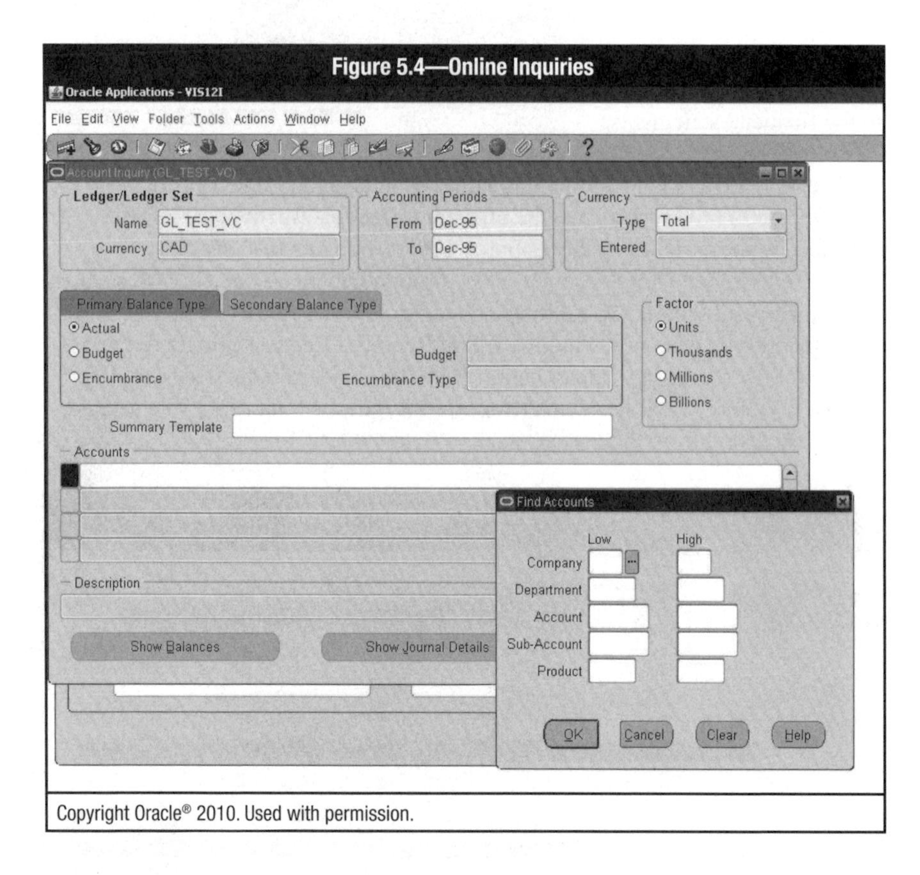

GL Standard Accounting Reports and Listings

Standard reports include Trial Balances, Journals reports, GLs and Account Analysis reports. Listings to review key nonstandard information include COA listings, Row Sets, Column Sets, Content Sets, report hierarchies, consolidation definitions and recurring journals. Runtime options for detail or summary information, sort sequence, and the selection of data can be set for the desired reports.

To run a standard report, use the Submit Requests window, and select the required request or Request Set. **Figure 5.5** shows the Standard Reports and Listings screen (Reports→Request→Standard). The desired report is selected from the list to be submitted, as per a normal request.

Enterprise Planning and Budgeting

Enterprise Planning and Budgeting (EPB) is the replacement for the Oracle Financial Analyzer (OFA) and Oracle Sales Analyzer tools. The Oracle EBS GL is integrated with EPB. EPB enables enterprise-level organizations to control their planning cycles with tailored EPB business processes. Each business process can be defined according to business rules and allows complex rules,

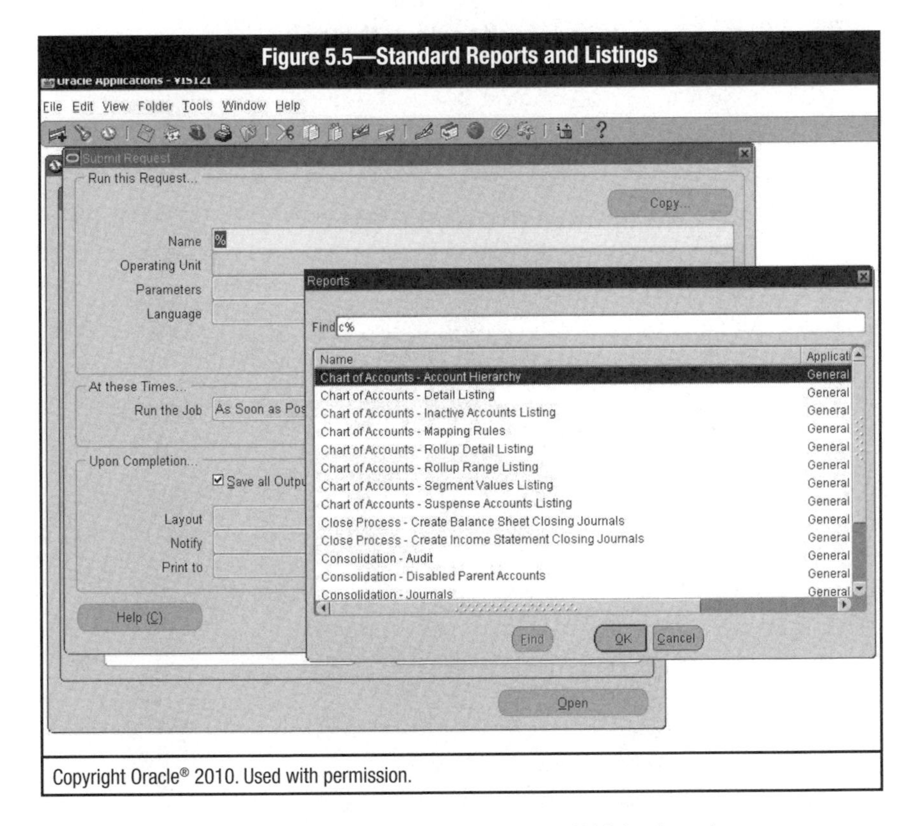

Figure 5.5—Standard Reports and Listings

calculations, tasks and workflow controls. Responsibilities based on users and hierarchies allow control to be maintained over reporting and budgeting processes and for controlling business process required actions and approvals. The calculation engine for EPB is based on the online analytical processing (OLAP) technology and provides powerful modeling and analysis capabilities. Links with the Oracle GL simplify data acquisition and maintenance tasks, allowing greater emphasis to be placed on developing and monitoring budgeting processes. EPB can also be used as a standalone product for customers who do not use Oracle EBS.

Summary

This chapter provided an overall understanding of the operation of the Financial Accounting business cycle in an Oracle ERP environment. It summarized important processes from an audit perspective—including Master Data Maintenance, Journal Processing, and Reconciliations and Financial Reporting.

Page intentionally left blank

6. Auditing Oracle E-Business Suite—Financial Accounting Business Cycle

In chapter 5, the main subprocesses of the Financial Accounting business cycle were outlined. This chapter looks at the significant risks, key controls and techniques to test the controls for each of the following subprocesses:
- Master Data Maintenance
- Journal Processing
- Reconciliation and Financial Reporting

Refer to **figure 4.8** for the key to the numbering sequence for risks, controls and testing techniques.

Master Data Maintenance

Master Data Maintenance: Risks

The COA defines how information will be reported. Inaccurate, incomplete, invalid or untimely creation or maintenance of master data can affect multiple areas within the enterprise.

Significant risks include:

1.1	The entry of and changes to master data may be unauthorized, invalid, incomplete, inaccurate or untimely.
1.2	Master data may not remain current and pertinent, e.g., inactive accounts remain active.

Master Data Maintenance: Key Controls

Key controls over Master Data Maintenance include:

1.1.1	Oracle EBS security should restrict the number of people with access to create/change the COA structure, GL setup and vendor master data.
1.1.2	Oracle EBS Role and Responsibility functionality should restrict to authorized personnel the ability to add, change or delete: • Flexfield segments • Flexfield values • Flexfield Security • Value Sets • Account combinations • Ledgers • Ledger Sets • Ledger accounting configurable settings

1.1.2 *(cont.)*	• Cross-Validation Rules • Archive and purge functionality • Calendars • Currencies
1.1.3	Cross-Validation Rules that define valid segment value combinations should be enabled and developed to ensure the accuracy of data entry. Oracle EBS prevents the creation of account combinations that violate the Cross-Validation Rules. Furthermore, Oracle EBS prevents input to account numbers outside of the effective dates specified in the account setup. It is important to set up Cross-Validation Rules when the Dynamic Insertion feature is activated because Dynamic Insertion allows flexibility to create *ad hoc* accounting combinations. Predefining Cross-Validation Rules will allow valid accounting combinations only during Dynamic Insertion.
1.1.4	The Oracle EBS GL should edit and validate modifications to accounts or GL criteria online during data entry.
1.1.5	Management should review key COA reports after changes are made to ensure proper setup.
1.1.6	Management should review and monitor the COA structure on a regular basis.
1.2.1	Management should review master data periodically to check their currency and ongoing pertinence.

Master Data Maintenance: Testing Techniques

Testing techniques for master data controls include:

1.1.1	Extract a list of the target high-risk functions/forms as explained in chapter 4. Review the function/form listing in 1.1.2 Master Data Maintenance: Key Controls to ascertain who has access privileges to the following functions/forms: • Account Generation processes—Select Workflow to generate Accounts • Account Hierarchy editor—Edit Account Hierarchies • Auto Allocation Workbench: GL—Define, run, schedule and monitor the GL Auto Allocation process • Concurrent Program controls—Define Concurrent Program controls • Conversion Rates Types—Define Conversion Rates Types • Currencies—Currencies form • Daily Rates—Define daily rates • Define Content Set • Define Mass Allocations

1.1.1 *(cont.)*	• Document Sequence Assignments—Sequential numbers: Document Sequence Assignments form • Document Sequences—Sequential numbers: Document Sequence form • Elimination Sets—Define Elimination Sets • Encumbrance Types—Define Encumbrance Types • Financial Item—Define Financial Item • Generate AutoAllocation—Enter parameters to generate AutoAllocation requests • Generate AutoAllocation—Run AutoAllocation request • Generate AutoAllocation—Run Mass Allocation request • Generate AutoAllocation—Schedule Mass Allocation request • Generate Eliminations • Generate Mass Allocations • GL Accounts—Define GL Accounts • Consolidation Workbench • Historical Rates—Define Historical Rates • Intercompany Accounts—Define Intercompany Accounts • Intercompany Clearing Accounts—Define Intercompany Clearing Accounts • Intercompany Transaction Types—Define Intercompany Transaction Types • Ledger Sets—Define Ledger Sets • Mass Maintenance Workbench—Define and run Mass Maintenance requests • Mass Maintenance Workbench: Prevalidate—Submit Prevalidation from Mass Maintenance Workbench • Mass Maintenance Workbench: Reversal—Submit reversal from Mass Maintenance Workbench • Mass Maintenance Workbench: Submit—Submit move/merge or mass creation from Mass Maintenance Workbench • Period Rates—Define period-end and period average rates • Period Types—Define Period Types • Profile System Values—Profile System Values form • Profile User Values—Profile User Values form • Revaluations—Process Navigator definition for revaluation purposes • Revalue Balances • Rollup Groups—Flexfield Rollup Groups: Key Mode form • Statistical Units of Measure—Define statistical units of measure • Storage Parameters—Update Storage Parameters • Subledger Import Process—Process Navigator definition for Subledger Import Process • Subsidiaries—Define Subsidiaries • Summary Accounts—Define Summary Accounts

1.1.1 *(cont.)*	• Suspense Accounts—Define Suspense Accounts • Tax Codes • Tax Codes and Rates • Tax Options—Define Tax Options • Transaction Calendar—Define Transaction Calendar • Translate Balances—Translate Balances • Translations—Process Navigator definition for Translation process
1.1.2	Extract a list of the target high-risk functions/forms. Review the function/form listing in 1.1.1 to ascertain who has access privilege to create, modify and delete the following functions: • Archive and Purge Account Balances and Journal Entries • Assign Descriptive Flexfield Rules • Assign Flexfield Security Rules • Assign Key Flexfield Security Rules • Calendars • Column Set • Cross-Validation Rules • Descriptive Flexfield Security Rules • Descriptive Flexfield Segments • Descriptive Flexfield Values • Flexfield Security Rules • Flexfield Value • Flexfield Value Sets • Key Flexfield Security Rules • Key Flexfield Segments • Key Flexfield Values • Mass Maintenance Workbench: Purge • Purge Consolidation Audit Data
1.1.3	Inquire whether the organization is utilizing the Dynamic Insertion functionality during COA maintenance. This can be verified by navigating to the Key Flexfield Segments form (Set-up➔ Financials➔ Flexfields➔Key➔Segments) and selecting the name of the application (Oracle GL) and Flexfield (Accounting Flexfield). The Allow Dynamic Inserts checkbox should be enabled if the organization is utilizing Dynamic Insertion. If this has been set, there will be no control in place to validate account combination creation. To address this, the Cross-Validate Segments checkbox should also be enabled. The Cross-Validate Segment applies Cross-Validation Rules to account combination creation. **Figure 6.1** shows the Key Flexfields Segments screen used to test whether the Allow Dynamic Inserts and Cross-Validate Segments checkboxes have been set correctly.

Figure 6.1—Cross-Validation Test

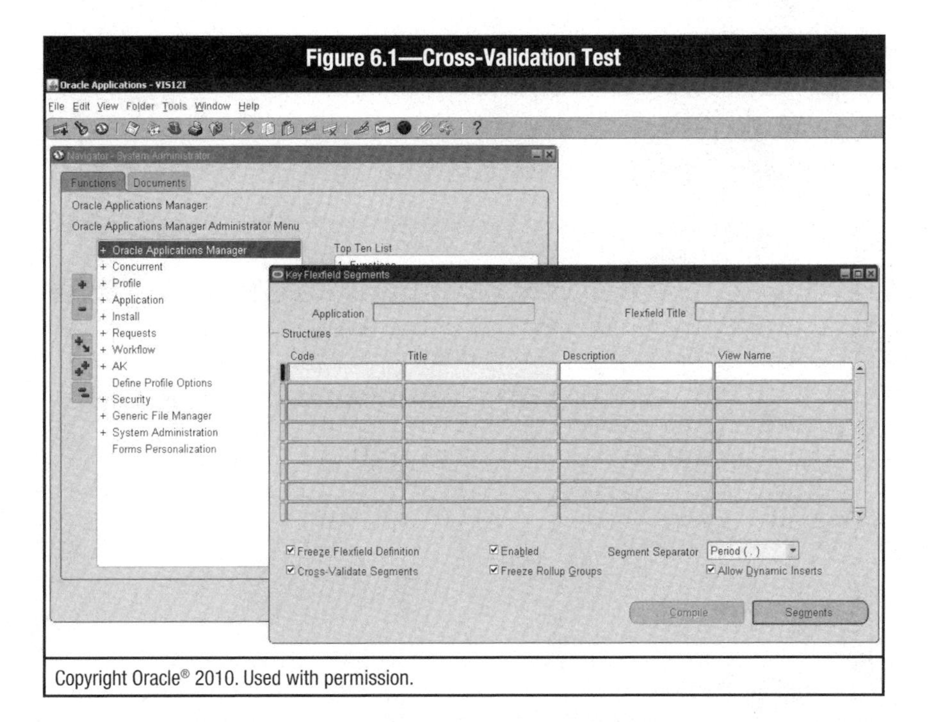

1.1.3 *(cont.)*

There are several standard Oracle EBS reports available for management to use to monitor changes to the Cross-Validation Rules, as follows:

- The Cross-Validation Rule Violation report provides a listing of all the previously created Flexfield combinations that violate Cross-Validation Rules for a given Flexfield structure. The report program can also be formatted to disable the existing combinations that violate new rules.
- The Cross-Validation Rules Listing report lists all the Cross-Validation Rules that exist for a particular Flexfield structure. This is the information defined using the Define Cross-Validation Rules form, presented in a multiple-rule format that can be reviewed and retained.

Ask management whether there is a procedure for reviewing these reports. Ascertain if they are reviewed on a regular basis and whether exceptions are identified for follow-up and resolution. Inspect physical evidence that the reports are run, approved and retained and that exceptions are resolved, as necessary. The reports can be run using the Requests window following the path Requests→Run, using the System Administrator responsibility. **Figure 6.2** shows the list of the reports previously explained in Oracle EBS.

Figure 6.2—Cross-Validation Rules Review Reports

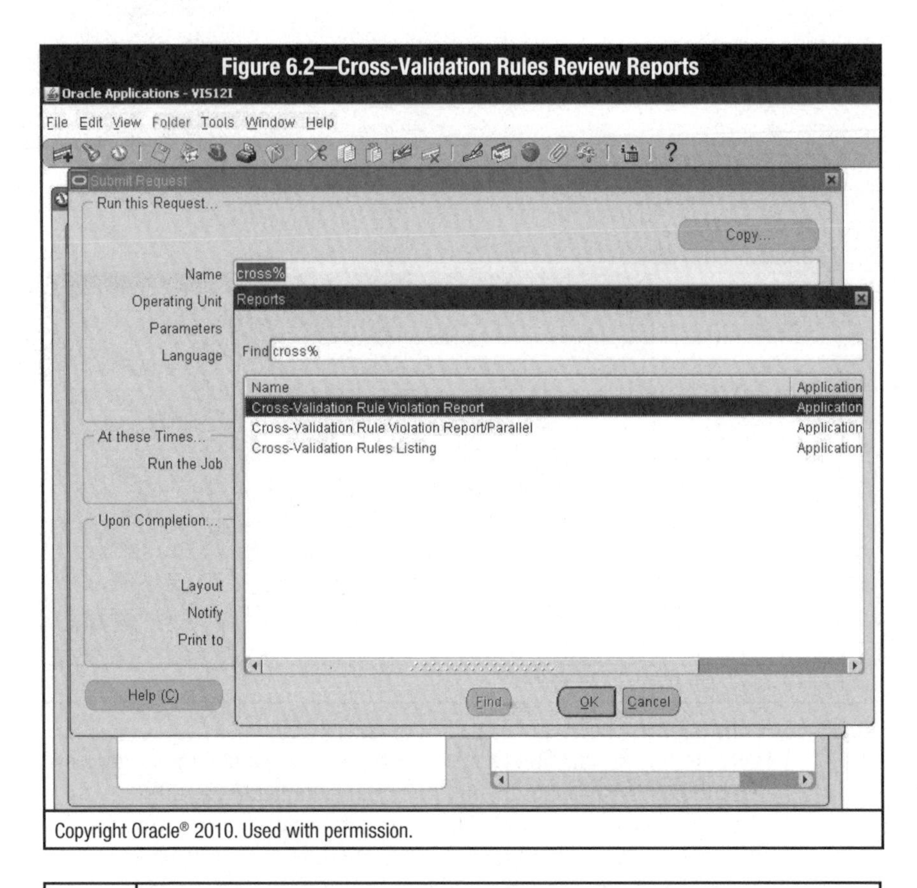

1.1.4	Oracle EBS utilizes several online edits and validation tools, including required fields (both Oracle EBS-defined and user-defined), QuickPick functionality and default field values. These tools facilitate the accurate recording of journal entries and COA updates. Transactions cannot be processed if a required field is blank; for instance, the Date field in the Journal Entry form. QuickPick functionality provides pop-up windows that contain predefined options from which the user can choose. Fields that generally require the same input can be defined to contain default values, so input is not required. Review the setting by navigating to the properties of a required field through the path Help➔Diagnostics➔Properties➔Item and confirming that the null value equals False, which defines that field as required. A password is required to view the attributes of a field. The preloaded password is “apps.” If this password has not been changed, the security of the ledger system may be compromised. Request the password to perform the test, if the organization has changed the password. **Figure 6.3** shows the properties of item Name in the GL Accounts function.

Figure 6.3—Online Edit Test

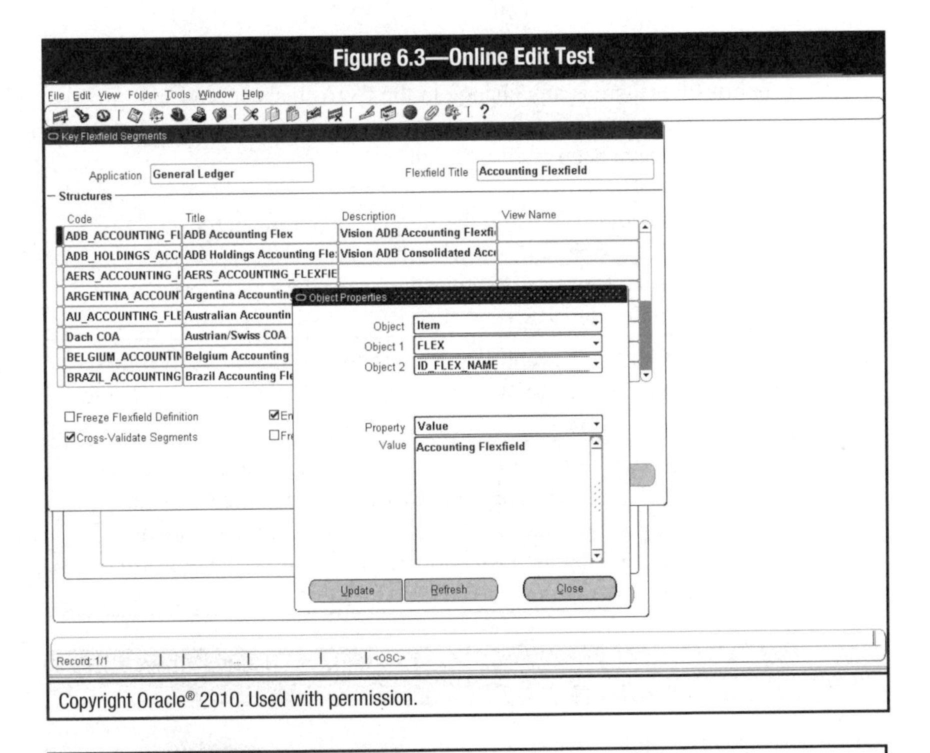

1.1.5	Using the several standard Oracle Reports available for management, monitor changes to the COA, as follows: • The Chart of Accounts—Account Hierarchy report lists the detail accounts that roll up into each Summary Account for all summary templates. • The Chart of Accounts—Detail Listing identifies both detail and Summary Accounts for the current Ledger Set, including disabled detail accounts. This report also provides the start and end dates, if any, for each account. • The Chart of Accounts—Segment Values Listing lists all segment values for a specific account segment. This listing includes information about each segment value, such as whether the segment value is enabled, whether it is a parent, the account type, and the start and end dates for the parent segment value. • The Chart of Accounts—Suspense Accounts Listing lists all of the suspense accounts for the current Ledger/Ledger Set. This listing provides the suspense account for each journal entry source and category.

1.15 *(cont.)*	Ask management whether there is a procedure for reviewing these reports after there is any change in the COA. Inspect physical evidence that the reports are run, approved and retained and that the exceptions have been resolved, as necessary. The reports can be run using the Requests window in the Reports➔Request➔Standard with the relevant GL report set (e.g., GL Superuser). **Figure 6.4** shows the list of the reports previously explained in Oracle EBS.
1.1.6	Ascertain whether the reports related to the COA, shown in **figure 6.4**, are reviewed on a regular basis and the exceptions identified for follow-up and resolution.
1.2.1	Ask management whether there is a procedural periodic review by management of master data to check whether they are up to date and for continued pertinence. Inspect evidence of the review for performance, and the identification and approval of the removal of master data.

Figure 6.4—Review Chart of Accounts Reports

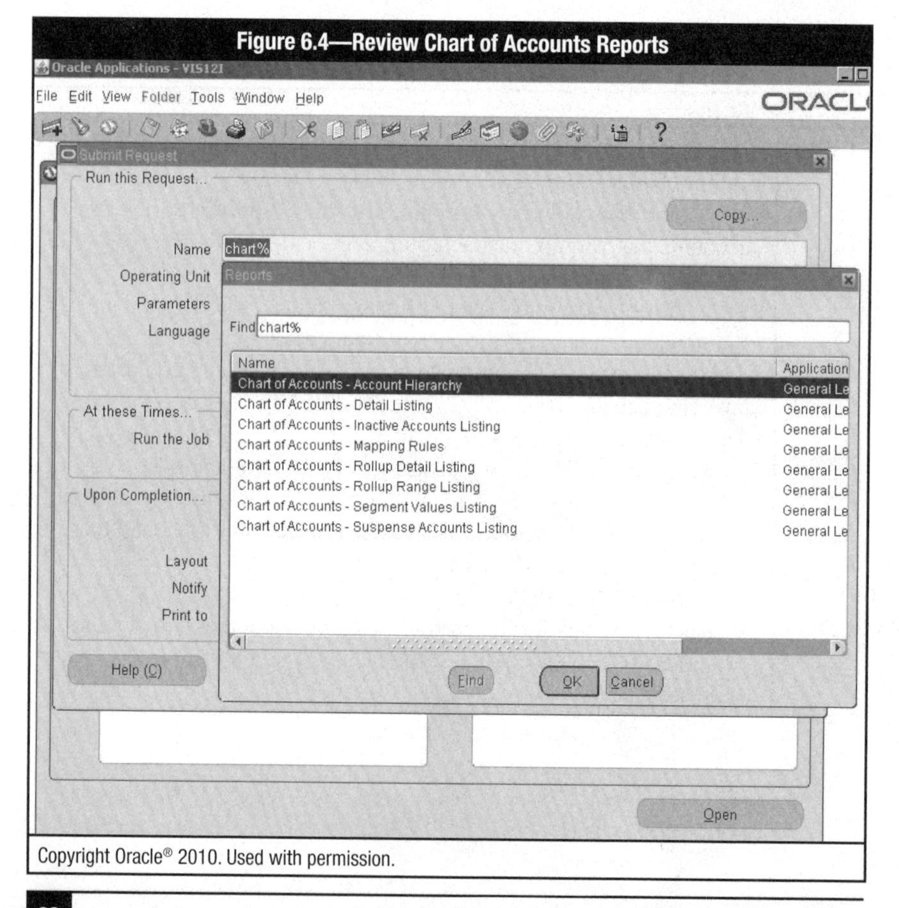

Journal Processing

Journal Processing: Risks

Controls should ensure that postings to the GL are accurate, timely and complete. Oracle EBS R12.1 functionality should be utilized to automate as many entries as possible with, for example, recurring entries and model entries. In addition, Oracle EBS utilizes automated balancing, edits and validations.

Significant risks include:

2.1	Invalid journal entries may be booked to the GL.
2.2	Journal entries may be posted more than once to the GL.
2.3	All journal entries may not be posted to the GL.
2.4	Journal entries may not be posted to the correct period.
2.5	Journal entries may be inaccurate or out of balance.

Journal Processing: Key Controls

Key controls over the processing of journals include:

2.1.1	Access to the entry, import, definition, setup and generation of journal entries in the Oracle EBS GL module should be restricted appropriately.
2.1.2	Access to the approval and posting of journal entries in the Oracle EBS GL module should be restricted appropriately. The ability to create vs. post journal vouchers should be segregated to different personnel by Oracle EBS Security. Journal Approval workflow should be set up appropriately (if applicable).
2.1.3	Oracle Cross-Validation Rules should be used to help prevent inaccurate journals; entries of invalid account combinations in all modules should be stopped at the time of data input with a hard warning message (error message).
2.1.4	Oracle EBS reversing and recurring Journal Entry features should be used to prevent omitted or inaccurate journal entries.
2.2.1	Oracle EBS should be configured to require a unique journal number, batch name or date for a journal to post to the ledger. It should not allow the posting of a journal entry with the same journal number, batch name and date as an already-posted entry.
2.3.1	Monthly journal entries should be reconciled against the monthly closing schedule to ensure that all regular monthly entries have been posted.

2.4.1	Oracle EBS should prevent posting of journal entries to closed or future periods. In the GL module, journal entries (both manual and system-generated) should be posted only to those periods with a status of Open. Journal entries should be entered and saved in periods with a status of either Open or Future Entry. Other period statuses are Closed, Permanently Closed and Never Opened. The status of periods should be changed through the Open and Close form, with the exception of Permanently Closed and Never Opened.
2.4.2	Access controls over accounting periods and the ability to open and close periods in the Oracle EBS GL module should be restricted appropriately.
2.5.1	Online edits and validation tools should be used in the data entry process.
2.5.2	Although out-of-balance journals can be entered and saved, Oracle EBS will prevent the posting of out-of-balance journal entries. It should either prevent posting of out-of-balance journal entries or force journals to balance using a suspense account. Accounts used for suspense posting of journal entries should be configured properly in Oracle EBS, and balances should be reviewed and cleared on a regular basis.
2.5.3	Oracle EBS should prevent changes to journal vouchers once they have been posted. Required changes to posted journal entries should be made through a separate, subsequent journal entry. Only authorized personnel should have access to reverse journal functions.
2.5.4	Although optional, Oracle EBS offers the use of Control Totals when manually entering journal entries to provide automatic comparisons of Control Totals against total debits entered and total debits entered against total credits entered, to ensure the accuracy of data input.

Journal Processing: Testing Techniques

Testing techniques for Journal Processing include:

2.1.1	Extract a list of the target high-risk functions/forms. Review the function/form listing in 1.1.1 to ascertain who has access privileges to the following functions: • Correct Journal Import Data • Define Recurring Journal Formula • Delete Journal Import Data • Enter Encumbrances • Enter Intercompany Transactions • Enter Journals • Process Navigator Definition for Entering Journal Process

2.1.1 *(cont.)*	• Generate AutoAllocation: Run Recurring Journal • Generate AutoAllocation: Schedule Recurring Journal • Generate Recurring Intercompany Transactions • Generate Recurring Journals • Import Journals • Define Journal Categories • Query Journals and Journal Batches • Define Journal Sources • Define Recurring Intercompany Transactions Guidelines provided in the International Standard on Auditing (ISA) 240 (paragraphs 76 to 79) regarding standards for journal entry testing should also be taken into consideration.
2.1.2	Extract a list of the target high-risk functions/forms. Review the function/form listing in 2.1.1 Journal Processing: Testing Techniques to ascertain who has access privileges to the following functions: • Define Autopost Criteria • Post in the Enter Journals or Enter Encumbrances forms • Enter, change or delete employees who make up the Journal Approval hierarchy: • Post Journals • Run Autopost Requests Confirm that SoD is in place. Navigate to the System Profiles window in the System Administrator responsibility (Profile→System), and search "jo%" in the Find field. Depending on the way in which the organization has set up its system profiles, ensure that the following profiles are restricted at the site, application, responsibility and user levels: • Journals: Allow Posting During Journal Entry should equal No. • Journals: Allow Preparer Approval should equal No. • Journals: Find Approver Method should be set to Go Up Management Chain, or Go Direct, One Stop and Then Direct (depending on the organization's requirement).
2.1.3	This control should be tested as per point 1.1.3 in Master Data Maintenance: Testing Techniques shown previously.

Figure 6.5 shows the Journals: Find Approver Method screen and the options available.

Figure 6.5—Journals: Find Approver Method

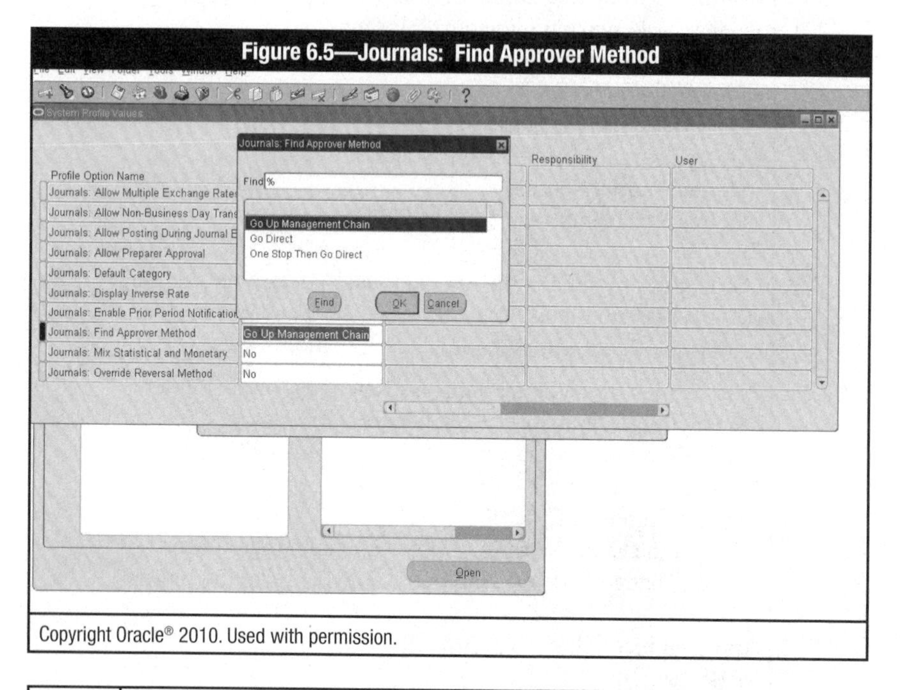

2.1.4	Through inquiry and inspection of journal entry procedures documentation, ensure that the organization uses the Oracle EBS reversing and recurring Journal Entry features to prevent omitted or inaccurate journal entries. Navigate to the Generate Recurring Journals screen (Journals➔ Generate➔Recurring) to bring up a list of all recurring journals that have been set up (**figure 6.6**), for evidence of performance. Navigate to the Generate Reversal Journals screen (Journals➔ Generate ➔Reversal) to bring up a list of all reversed journals (**figure 6.7**), for evidence of performance.

Figure 6.6—Recurring Journal Entries

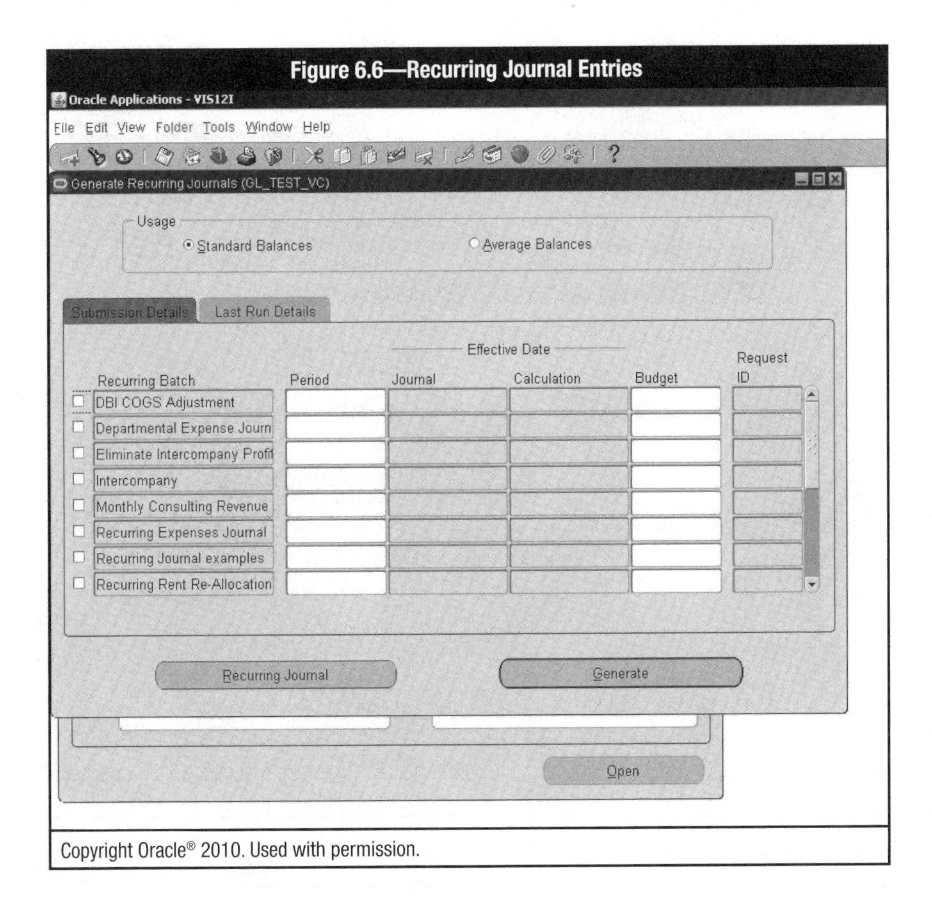

Figure 6.7—Reversed Journal Entries

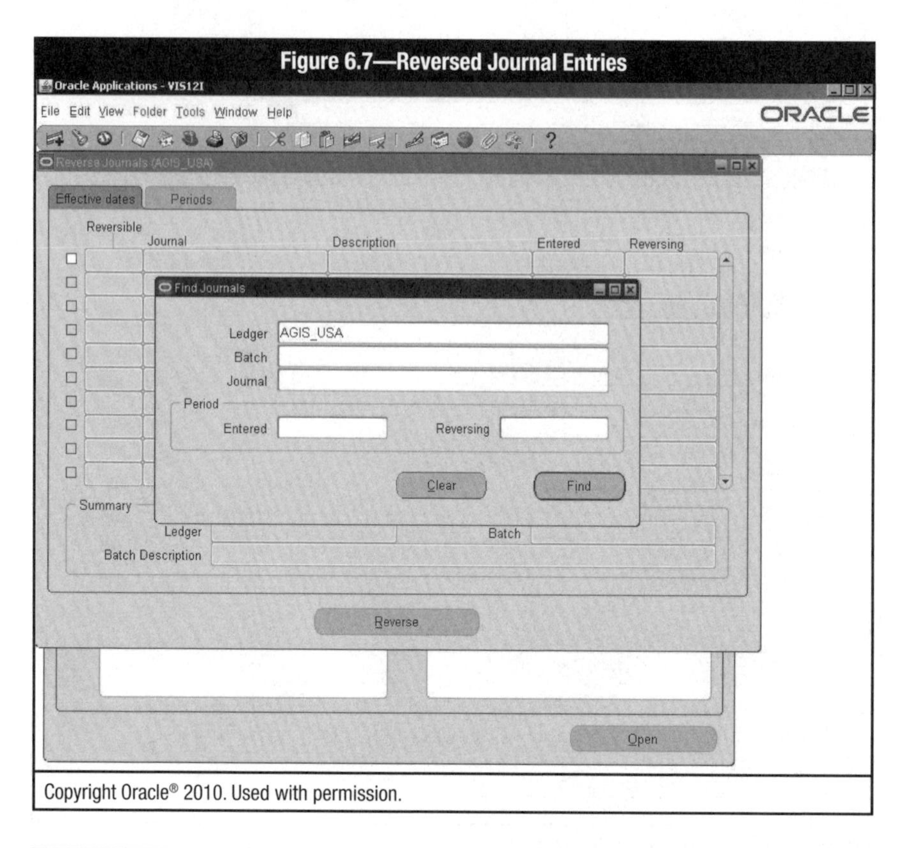

2.2.1	When processing journal entries in the Oracle EBS GL module, the standard, delivered functionality of the application requires a unique combination of journal number, batch name and date for a journal entry to be created. These three fields form the key for the data table, which is verified as unique prior to the database accepting the input. If a record already exists with the same combination of journal number, batch name and date, Oracle EBS will reject the record as a duplicate and require the user to change one of the fields before continuing with processing.
2.3.1	To monitor the progress of posted journal entries against the monthly closing schedule, use one of the following several standard Oracle Reports available for management: • Journal reports are available in three different formats. Review the journal batches and journals associated with each batch for posted journals, unposted journals or error journals.

2.3.1 *(cont.)*	• The Journals—Batch Summary report identifies posted journal batches for a particular balancing segment, currency and date range. • The Journals—Entry report lists all journal activity for a given period or range of periods, balancing segment value, currency, and range of account segment values. • The Journals—Day Book report (header description and line description) lists all posted journal entries and journal details chronologically by accounting date for a specified range of dates, journal sources and journal categories. • The Journals—Line report lists all the journals, grouped by batch, for a particular journal category, currency and balancing segment value. • The Journals—Tax report is used to verify manually entered GL tax journals. It enables the review of journal taxable lines and the tax lines they generate for posted or unposted journals. The report includes balancing segment, tax type and tax code. • The Journals—Document Number report lists all the journals in ascending order by document number. The report details include the creation date, batch date, journal name, category, posting status, posted date, currency and journal amounts. • The Journals—Entered Currency report lists all the journal batches and associated journals for posted, unposted and error journals entered in the specified currency. Ask management whether there is a procedure for reviewing these reports. Ascertain whether the reports are reviewed on a regular basis during the monthly closing cycle, depending upon the level of journal activity, to confirm that all scheduled journal entries and adjustments have been made at the appropriate time. The reports can be run using the Requests window in the Reports➔Request➔Standard with the relevant GL report set (e.g., GL Superuser). **Figure 6.8** shows the list of the Journal reports explained previously in Oracle EBS. The Daily Journal Book reports can be found in the same way, but use a "d%" search type in the Find field. **Figure 6.9** shows how the journal reports can be run. Select Item or Source and then Posted Journals, Unposted Journals or Error Journals.

Figure 6.8—List of Journal Reports

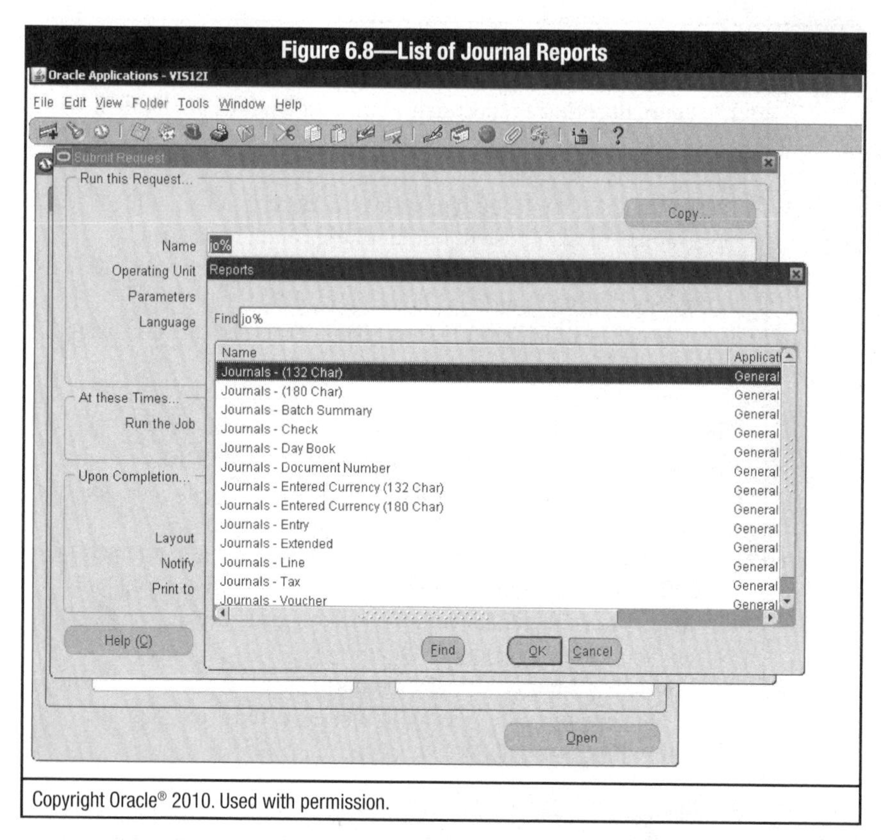

Figure 6.9—Run Journal Reports

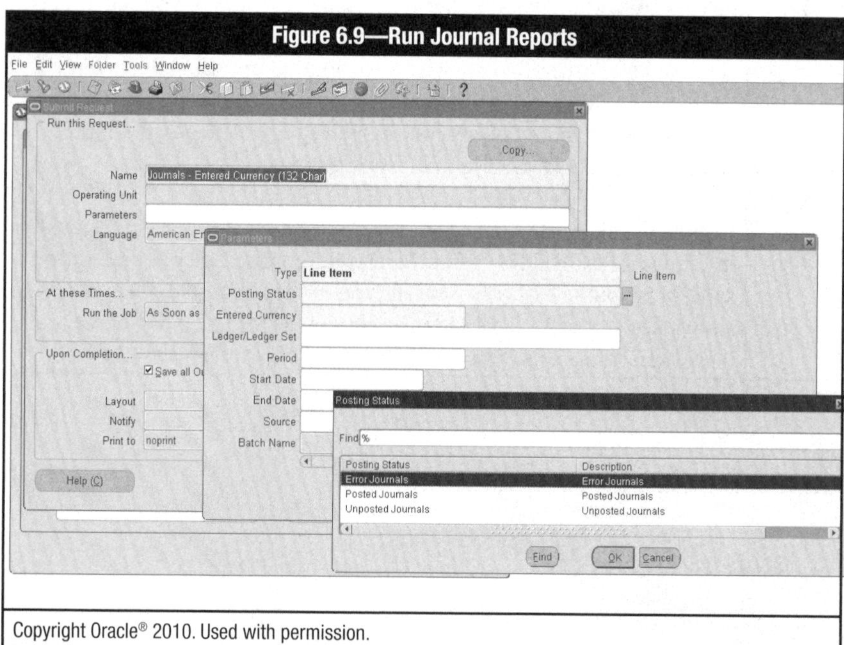

2.4.1	Go to the Setup➔Open/Close option in a responsibility with the ability to view Setup (e.g., GL Superuser). Review the status of the periods listed to determine which are: • Open • Future-Entry • Closed • Permanently closed **Figure 6.10** shows the view of the periods in the Setup Open/Close option screen.

Figure 6.10—Open and Close Periods

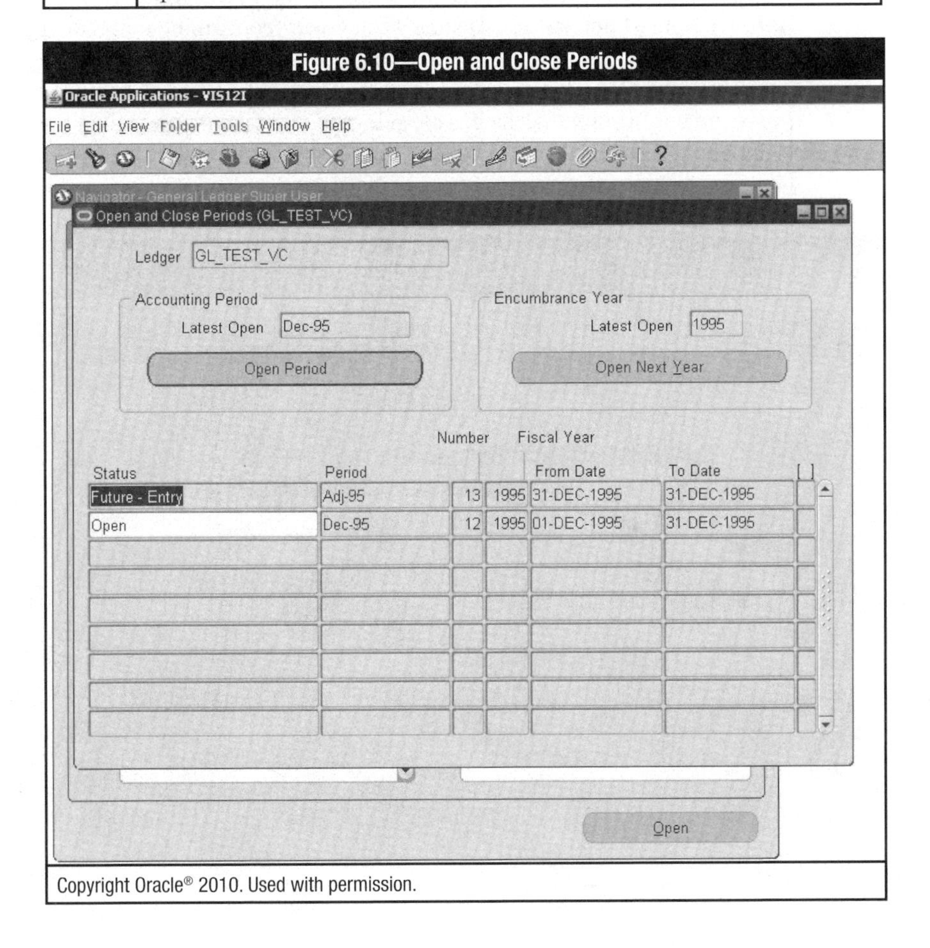

2.4.2	Extract a list of the target high-risk functions/forms. Review the function/form listing in 2.1.1 Journal Processing: Testing Techniques to ascertain who has access privileges to the following functions/forms: • Open and close periods • Process Navigator definition for period close—Parent company process • Process Navigator definition for period close—Subsidiary company process
2.5.1	Online edits and validation tools should be tested as per point 1.1.4 in Master Data Maintenance: Testing Techniques, confirming that the null value is set to False (this defines that the field value is required), viewing QuickPick lists for fields that have been identified as having lists and confirming the choices are relevant, and viewing defaults. **Figure 6.11** shows the properties of item Name in the Enter Journal function.

Figure 6.11—Online Edit Test

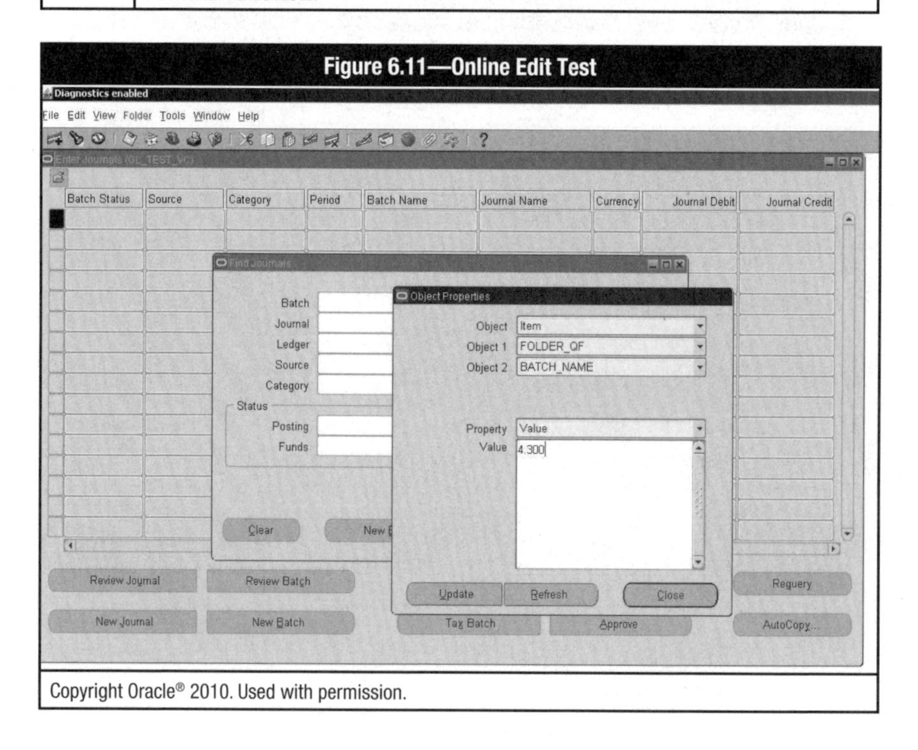

2.5.2	If suspense accounts are allowed, the COA—Suspense Accounts Listing should be run to extract a list of all suspense accounts (refer to point 1.1.5 in Master Data Maintenance: Testing Techniques), which should be reviewed by management on a regular basis. These activities should be laid down in a formal procedure. Inquire whether a staff member in the accounting department has been tasked with clearing the suspense accounts and reclassifying the expenses to the proper accounts on a regular basis.
2.5.3	When processing journal entries in the Oracle EBS GL module, the standard delivered functionality of the application will not allow a user to modify a journal entry that has been posted to the ledger. If a journal entry is posted in error, Oracle EBS recommends creating and posting an adjusting journal entry with the corrections. If a user attempts to modify a posted journal entry, the system will return an error message indicating that the fields are protected from update. Extract a list of the target high-risk functions/forms. Review the function/form listing from 2.1.1 Journal Processing: Testing Techniques to ascertain who has access privileges to the following functions: • Reverse the Enter Journals or Enter Encumbrances forms • Define journal reversal criteria • Reverse journals
2.5.4	Oracle EBS provides the ability to enter Control Totals for the journal entry batch and the individual journal entries in a batch during manual journal processing. These Control Totals allow the user to verify the total debits entered during journal entry against the defined Control Total to ensure that the data entry is complete and accurate. This is an optional feature with Oracle EBS, and Control Totals are not required to process and post a journal entry or journal batch. If the Control Totals feature is used, Oracle EBS will display a violation message if the journal does not balance (i.e., total debits do not equal the total credits) when the journal is posted. **Figure 6.12** shows the violation message.

Figure 6.12—Violation Message for an Unbalanced Journal Entry

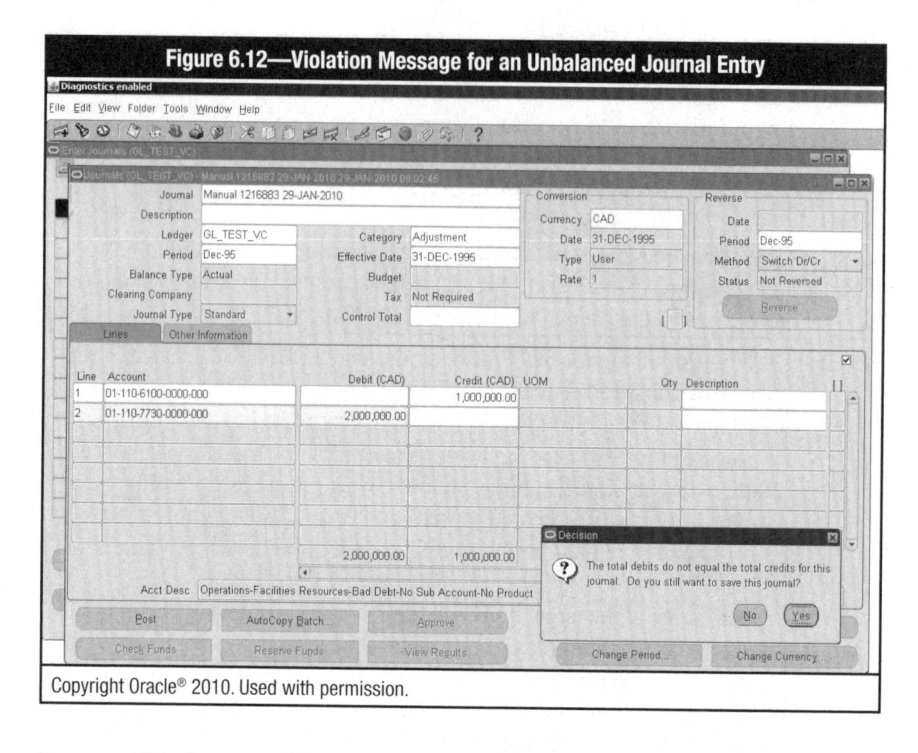

Reconciliation and Financial Reporting

Reconciliation and Financial Reporting: Risks

Significant risks associated with reconciliation and financial reporting include:

3.1	Valid GL account balances may be excluded from the financial statements. Financial statements may be inaccurate and may not reconcile to the GL.
3.2	All account reconciliations may not be performed monthly.
3.3	Financial reports may not be generated and distributed in a timely manner.
3.4	Account reconciliations may not be kept current.
3.5	Unauthorized personnel may generate financial statements.
3.6	Closing procedures may be inadequate to prevent any further postings to that period and accurately reflect the transactions that took place in a given accounting period. Unauthorized access to open/close accounting periods may exist.

Reconciliation and Financial Reporting: Key Controls

Key controls over reconciliation and financial reporting include:

3.1.1	Oracle EBS Role and Responsibility functionality should restrict to authorized personnel the ability to create, modify or generate financial statements or report criteria. Access controls over FSG report definition and generation should be restricted.
3.1.2	Oracle EBS Role and Responsibility functionality should restrict to authorized personnel the ability to add, change or delete Key Flexfield segments, Key Flexfield values, Value Sets and account combinations.
3.1.3	New Accounting Flexfield segment values should be inserted into existing ranges, and financial reports should be configured with broadly defined account ranges so that new accounts are included in appropriate financial reports.
3.1.4	All financial statements (FSG and FA) should be reconciled to the GL (assets, liabilities, equity revenue, expenses and net income) at month end, to ensure that all account numbers are included.
3.1.5	Manual controls should ensure that all changes to the COA are reflected in the financial statements.
3.2.1	Monthly reconciliation schedules should be kept and the status monitored each month to ensure that all accounts are reconciled properly. Management should monitor the reconciliation of all appropriate accounts during the closing cycle. Management should review suspense account balances on a monthly basis.
3.3.1	Management should monitor the timing and generation of financial reports during the closing cycle.
3.3.2	A monthly closing schedule, indicating who should receive individual financial reports and when they should be received, should be monitored throughout the closing.
3.4.1	See points 3.1.3 and 3.3.1.
3.5.1	Flexfield Security should be enabled and developed to ensure the accuracy of data entry.
3.5.2	Financial results (printed reports) should be kept in the accounting department and held confidential until the final results are released.
3.6.1	Closing procedures should be formalized and monitored to prevent any further postings to that period. Oracle EBS should prevent posting to closed or future periods. Access to open/close accounting periods should be restricted. Once the year-end audit is performed, accounting periods should be closed permanently.

Reconciliation and Financial Reporting: Testing Techniques

Testing techniques for reconciliation and financial reporting include:

3.1.1	Extract a list of the target high-risk functions/forms. Review the function/form listing in 2.1.1 Journal Processing: Testing Techniques to ascertain who has access privileges to the following functions: • Autocopy Report Component • Define Financial Report • Define Financial Report Set • Process Navigator definition for Define Financial Reports process • Process Navigator definition for Extract GL Balances to FA process • Run Financial reports • Define Row Set • Define Column Set If the organization is utilizing Flexfield Value Security, verify that the system Profile Option FSG: Enforce Segment Value Security is set to Yes. By setting this value appropriately, the Flexfield Value Security rules that have been developed for a given responsibility will also be enforced for any financial reports generated by that responsibility. **Figure 6.13** shows the system Profile Option FSG: Enforce Segment Value Security set to Yes.

Figure 6.13—Flexfield Value Security

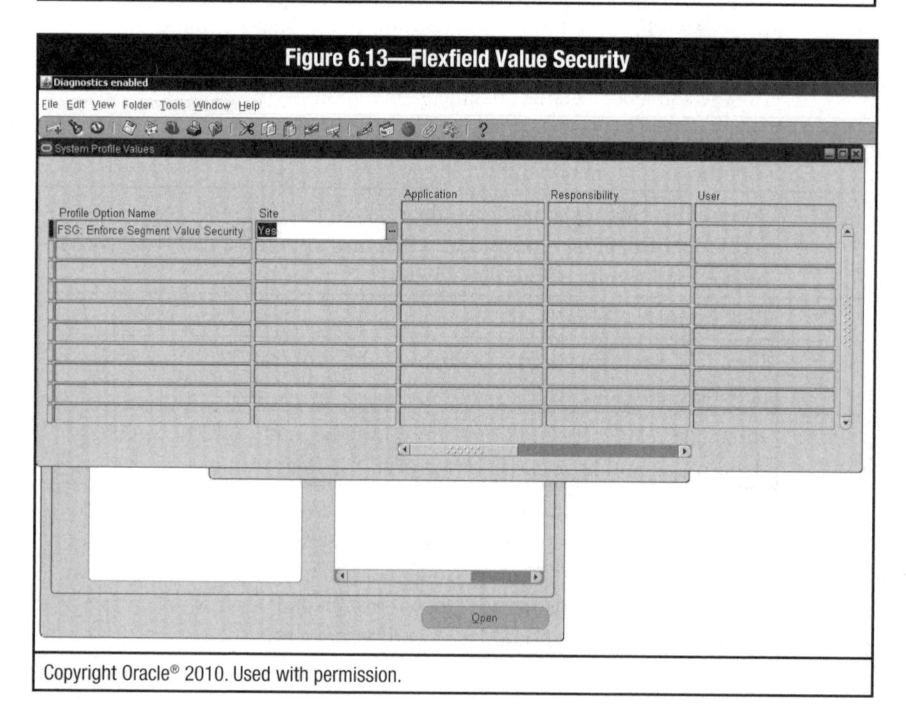

3.1.2	Extract a list of the target high-risk functions/forms. Review the function/form listing in 2.1.1 Journal Processing: Testing Techniques to ascertain who has access privilege to create, modify and delete the following functions: • Assign Descriptive Flexfield Rules • Assign Flexfield Security Rules • Assign Key Flexfield Security Rules • Cross-Validation Rules • Descriptive Flexfield Security Rules • Descriptive Flexfield Segments • Descriptive Flexfield Values • Flexfield Security Rules • Flexfield Value • Flexfield Value Sets • Key Flexfield Security Rules • Key Flexfield Segments • Key Flexfield Values
3.1.3	Ask management for the following documentation: • Policies, procedures, standards and guidelines regarding segment value and financial statement updates • System-generated reports used in performing such updates and the purpose of each report • Policies, procedures, standards and guidelines regarding the design of the COA • Policies, procedures, standards and guidelines regarding the design and configuration of financial reports Both the COA and financial reports should be designed with room to grow within the ranges of values that are used to represent financial classification. By using broad ranges, new values can be inserted behind the last value actually used in a range and still be included in the developed financial reports, since they are looking for all values within a wider range rather than actually used values. For instance, if a line item on a balance sheet is cash and it has been determined that the range of values for cash is 1000 to 1099, the balance sheet should be designed such that the line item "cash" in the report looks for all accounts 1000 to 1099, even if the only values used are 1000 and 1001.
3.1.4	Submit the COA Segment Values Listing report (Select Segment = Account). The report can be run using the Requests window in the Reports➔Request➔Standard (**figure 6.14**) in a responsibility with the relevant GL report set (e.g., GL Superuser). Use the report to determine the account ranges. Compare the FSG or FA report settings to ensure that they include the complete account range extracted in 3.1.3 and the selected account ranges, if applicable.

3.1.5	Ask management for procedural documentation that provides evidence that the reports generated in point 1.1.5 in Master Data Maintenance: Testing Techniques are cross-referenced to the report criteria used to produce the financial statements to ensure that changes to the COA are completely and accurately reflected. These reports should be reviewed on a regular basis, depending on the level of COA maintenance activity, to identify exceptions for follow-up and resolution.
3.2.1	Ask management for procedural documentation (e.g., reconciliation schedules) that provides evidence that, as a part of the monthly closing cycle, management has a process in place to reconcile activities in key accounts prior to the final close. This reconciliation is in place to make sure that the activity reflected in the GL accounts represents the activity that actually took place for the period.

Figure 6.14—Submit Request

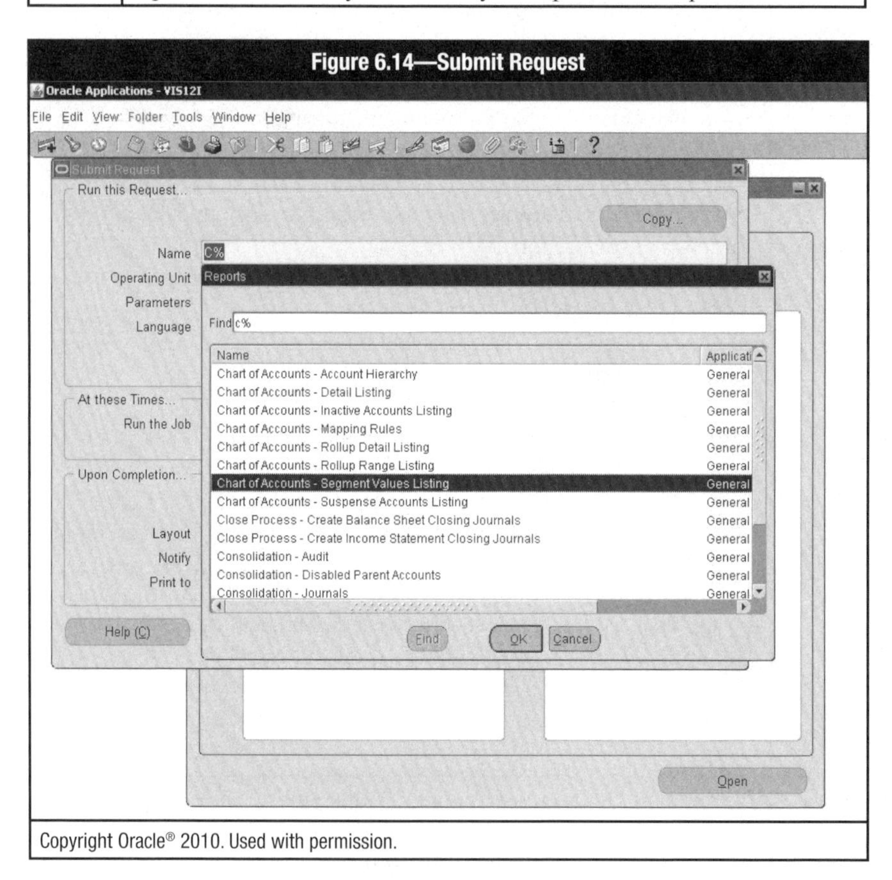

3.2.1 *(cont.)*	Accounts that reflect the activity of a subledger or another system, such as AP or accounts receivable (AR), should be reconciled to the monthly balance reflected in the subledger before closing the period. The monthly closing cycle should also include the review of monthly suspense account balances and the aging of suspense items to make sure that the reconciliation processes are effective.
3.3.1	Ask management for procedural documentation (e.g., reconciliation schedules) that provides evidence that guidance is provided by management as to when the different versions of the financial statements should be generated for review. This guidance should address both the preliminary and final financial statements for the entity. This will help ensure that the right information is available to the right people in a timely fashion so that adjustments made to the monthly activity are valid.
3.3.2	Ask management for and review procedural documentation that provides evidence of a monthly closing schedule and the monitoring of the schedule by an appropriate staff member.
3.4.1	Refer to the testing techniques in points 3.1.3 and 3.3.1.
3.5.1	Refer to the testing techniques for Flexfield Security in point 1.1.3 in Master Data Maintenance: Testing Techniques and point 3.1.1.
3.5.2	Ask management for procedural documentation that provides evidence of the retention and safekeeping of interim financial reports until they are finalized.
3.6.1	Refer to the testing techniques for closing accounting periods in points 2.4.1 and 2.4.2 in Journal Processing: Testing Techniques. In addition to testing techniques listed previously, review the function/form listing in point 3.1.2 in Reconciliation and Financial Reporting: Testing Techniques to ascertain who has access privilege to create, modify and delete Year-End Carry Forward. Review the status of the period (i.e., that it is closed for the Ledger Set) by navigating to the Open and Close Periods window (Setup➔Open/Close). This will display all the accounting periods defined within the Ledger. • Closed—A closed period can be reopened at any time. • Permanently closed—A permanently closed period cannot be reopened. Where the open/close periods option has been enabled, review the access to open/close periods (should be restricted to relevant personnel only). Ask management to review procedures for the permanent close of periods at year-end.

Summary

This chapter has outlined the risks, key controls and testing techniques surrounding the Oracle EBS R12.1 financial accounting business cycle. Among the key risks were inappropriate access to master data, inappropriate access and entry of journals, and inaccurate and incomplete financial reporting.

7. Oracle E-Business Suite—Expenditure Business Cycle

This chapter outlines the functionality of the Oracle EBS Expenditure business cycle to provide the reader with a high-level understanding of the process. From a risk and controls perspective, the Expenditure cycle has four main subprocesses (see **figure 7.1**):

- Master Data Maintenance
- Purchasing
- Invoice Processing
- Processing Disbursements

Figure 7.1—Expenditure Cycle

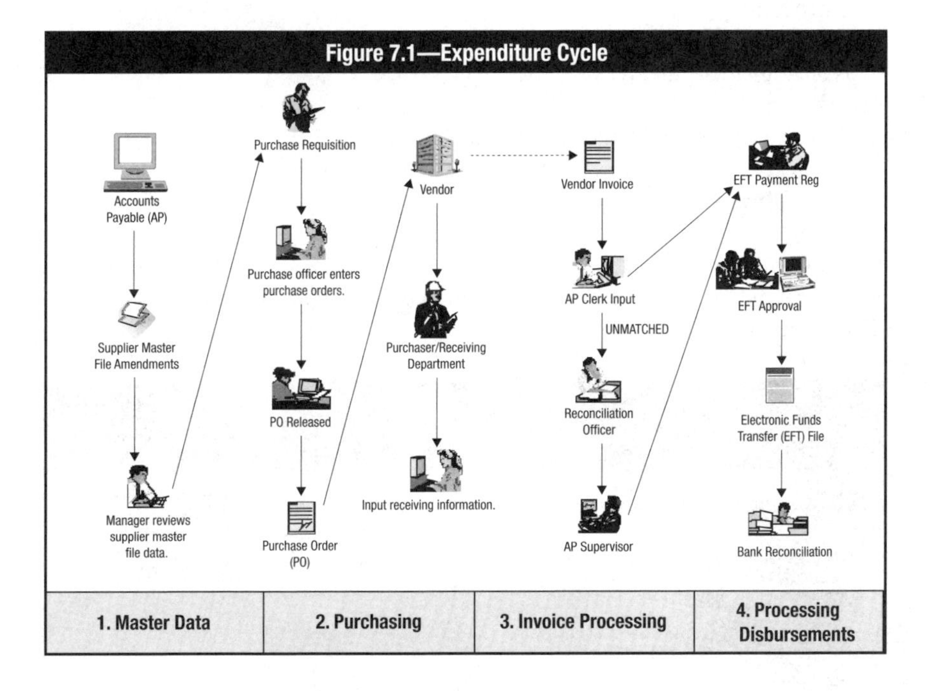

Master Data Maintenance

The supplier master file contains information about individuals and companies from whom the organization purchases goods and services, and possibly about employees who are reimbursed for expense reports. Supplier master file data typically represent standing data that are used for processing POs, payments, and the receipt of goods and services over an extended period of time.

Three Oracle Financial applications use the Suppliers window: Payables, Purchasing and Oracle Assets. If more than one of these products is used, supplier information is shared with the other product(s). In addition to the supplier name and address, Payables and Purchasing require additional information about the supplier. Oracle Assets requires no additional information.

Since the supplier master file data are used in key transactions, it is important to restrict access to the supplier master file to help ensure that only valid suppliers are maintained in the system.

A supplier master record contains information such as:

- Supplier name and number
- Taxpayer ID and tax registration number
- Supplier classification type (e.g., contractor, employee, government)
- Account setup data, including liability, prepayment and future-dated payment GL accounts
- Invoice amount limit and invoice match option
- Payment methods and terms of payment
- Bank details
- Electronic data interchange (EDI) details, if necessary
- Invoice tax
- Withholding tax

This information can be found in the tabs, which can be seen in **figure 7.2**. The Supplier window may also be used to find existing suppliers and amend their details.

Figure 7.2—Supplier Window

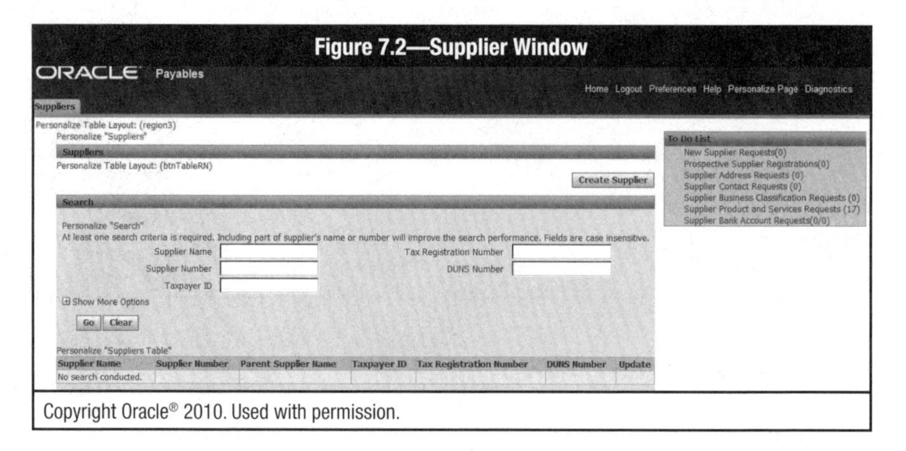

Copyright Oracle® 2010. Used with permission.

Supplier primary and alternative addresses, language, notification method, and sites used are set up in the Supplier Sites section of the Quick Update screen (**figure 7.3**). The site name will not appear on documents sent to the supplier. It is used for reference when selecting a supplier site from a list of values during transaction entry.

Figure 7.3—Supplier Sites Window

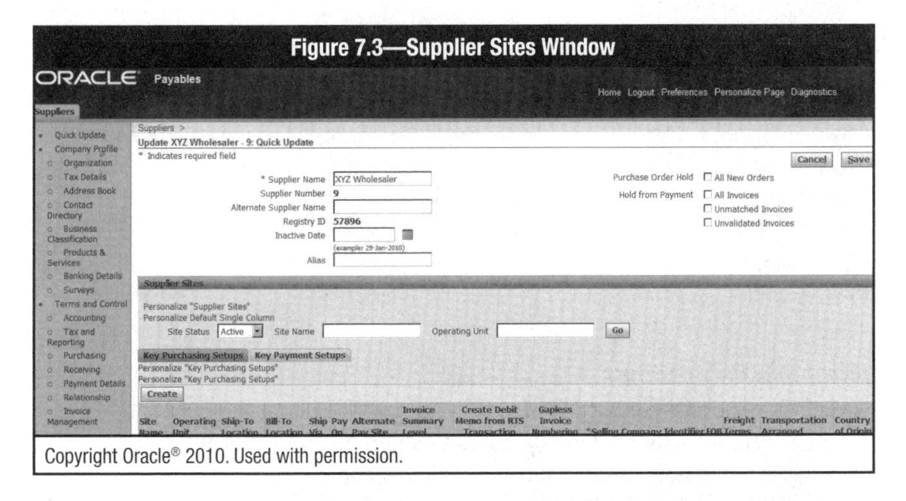

The development and use of naming conventions when creating new suppliers and restricting the ability to enter new suppliers minimize the risk of duplicate entries for the same supplier under different names. For example, the application of a consistent spelling convention of the full name of a supplier means that The Smith Corporation would not be entered as The Smith Corp.

If the Supplier Number Entry option is enabled in the Financials Options window, the Payables system automatically enters a nonupdatable supplier number. If this option is disabled, a unique supplier number must be entered when setting up the supplier.

To prevent invoice or PO entry for a supplier after a certain date, the applicable date should be entered in the Inactive On field.

One-time supplier accounts are set up for infrequently used suppliers by using the Inactive On field feature or clicking on the One Time box in the Classification tab. All one-time suppliers can be found by using the Supplier Find window. Supplier pricing information resides in the payment area (the Payment tab in the Supplier window).

Purchasing

Oracle EBS provides comprehensive functionality for purchasing to optimize relevant work processes, ranging from the generation of purchase requisitions to the printing of POs. Purchasing decides whether orders can be placed on the basis of existing quotations, or whether it is first necessary to issue additional requests for quotation. In addition, the system makes information available for:

- Supplier evaluation purposes
- Supplier selection
- Volumes tracking (with regard to a material or supplier)
- Ordering activity monitoring

Key purchasing concepts include:
- Purchase requisitions
- POs
- Autocreation
- Approved Supplier Lists
- Distribution Sets
- Oracle EBS setup parameters
- Document types
- Approvals

Purchase Requisitions

Online requisitions make it possible to centralize the purchasing department, to source purchases with the best suppliers to attain quality items and/or quantity discounts, and to help ensure that appropriate management approval is obtained before the creation of POs from requisitions.

In Oracle EBS, there are three main types of requisitions:
- **Purchase requisitions**—For goods requested from outside suppliers
- **Internal requisitions**—To transfer and request material from one inventory or expense location to another within the organization. These could also be internal requisitions if an internal sales order is created and goods are sourced from inventory.
- **Imported requisitions**—Imported or rescheduled requisitions from other systems

Requisitions created in other Oracle Applications are imported to the Purchasing module through the Requisition Interface tables. Once the requisitions have been appropriately approved, they are placed in POs.

Changes can be made to any requisition before it becomes a PO. Also, any requisition can be canceled before it has been sent to a PO or modified by the AutoCreate window. Once the requisition has been put into a PO, it cannot be canceled or modified.

The ability to create purchase requisitions can be given to the organization as a whole, or to a select group of users. It is important to conduct periodic reviews of users who have the ability to create, modify or cancel purchase requisitions to help ensure not only that valid and approved requisitions are sent through the system, but also that such permission is assigned in line with business requirements.

Purchase Orders

The Purchase Orders window in Oracle Purchasing is used to enter standard and planned POs as well as blanket and contract purchase agreements. Only users defined as buyers can use this window. Oracle Purchasing provides the following PO types:

- Standard PO
- Planned PO
- Blanket purchase agreement
- Contract purchase agreement

Generally, standard POs are created when the details of the required goods or services, estimated costs, quantities, delivery schedules, and accounting distributions are known. Standard POs are used for one-time or infrequent purchases of various items.

A planned PO is a long-term agreement where one has committed to buy items or services from a single source. Tentative delivery schedules and all details for goods or services that are to be bought, including charge account, quantities and estimated cost, are specified in the planned PO. Scheduled releases are issued against a planned PO to place the actual orders.

Blanket purchase agreements are created when the details of the goods or services that are planned to be bought from a specific supplier in a given period of time are known, but the detail of delivery schedules is not. Blanket purchase agreements are used to specify negotiated prices for items before purchasing them. A blanket release is issued against a blanket purchase agreement to place the orders, as long as the release is within the blanket agreement effective dates. Contract purchase agreements with suppliers are created to agree on specific terms and conditions without indicating the goods and services that will be purchased.

Figure 7.4 summarizes the PO types and their characteristics. The use of these PO types is site-specific.

Figure 7.4—Characteristics of Purchase Orders				
PO Types	Standard PO	Planned PO	Blanket Purchase Agreement	Contact Purchase Agreement
Terms and conditions known	Yes	Yes	Yes	Yes
Goods or services known	Yes	Yes	Yes	Maybe
Pricing known	Yes	Yes	Maybe	Maybe
Quantity known	Yes	Yes	No	No
Account distributions known	Yes	Yes	No	No
Delivery schedule known	Yes	Maybe	No	No
Can be encumbered	Yes	Yes	No	No
Can encumber releases	N/A	Yes	Yes	N/A

Purchasing allows for the cancellation of a PO or PO line after it has been approved. When a PO or PO line is canceled, payment is due only for those goods received before cancellation, and any unfilled requisitions can be reassigned to another PO.

Autocreation

Purchasing provides autocreation capabilities for documents to eliminate errors due to reentry of requisition data to POs. Buyers can create standard POs, planned POs, blanket releases and requests for quotation (RFQ) from any available purchase requisition line. Internally sourced requisition lines are supplied by internal sales orders and are not available for autocreation.

Approved Supplier Lists

The Approved Supplier List (ASL) is used to restrict the purchase of specific items to specific suppliers. Special control should be taken over ASL changes because unauthorized additions or deletions could be used to benefit a particular supplier. Examples of business reasons for using this include:

- A relationship has been established with these suppliers.
- The suppliers have proven themselves reliable.
- The organization has special deals with them.

For reasons such as these, the enterprise would want to prohibit people from buying items from different suppliers. For specified items, it is possible to restrict the allowable suppliers. Allowable suppliers are maintained in the ASL for the specified item. When a PO or outline agreement is created for items with an ASL, only suppliers on the list can be entered. Management may want ASLs to be maintained for specific (types of) items. On the other hand, for items for which free, competitive bidding should occur, no ASLs should be available in the system.

Distribution Sets

All POs, requisitions and releases require accounting distributions (i.e., accounting transactions). The distribution can be defined and linked to the item number(s) being ordered. This automates the process at the lowest level (i.e., on purchase requisitions) and flows the distribution to the invoice via matching.

Oracle Setup Parameters

Oracle EBS allows parameters to be set, such as defining buyers, Approval Groups, requisition templates, Profile Options, purchase options, document types, position hierarchy and lookup codes actions. These setup parameters are powerful settings within the Purchasing module, and unauthorized access could lead to invalid/inaccurate POs being processed.

Document Types

The Defining Document Types window can be used to define access to each document type (Oracle EBS standards include purchase agreement, PO, requisitions, quotation, RFQ and release) and the modification/control actions that users can make. **Figure 7.5** shows an example of available document types.

Figure 7.5—Document Types

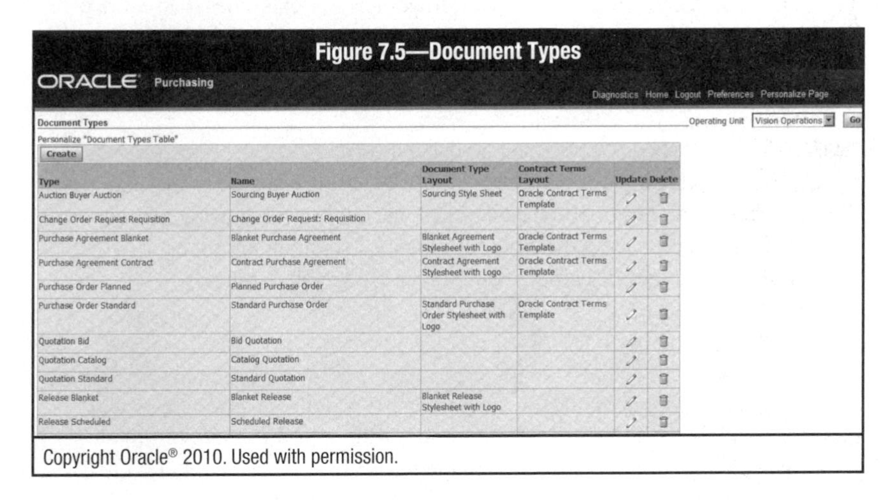

ORACLE Purchasing

Diagnostics Home Logout Preferences Personalize Page

Document Types

Operating Unit Vision Operations Go

Personalize "Document Types Table"

Create

Type	Name	Document Type Layout	Contract Terms Layout	Update	Delete
Auction Buyer Auction	Sourcing Buyer Auction	Sourcing Style Sheet	Oracle Contract Terms Template		
Change Order Request Requisition	Change Order Request: Requisition				
Purchase Agreement Blanket	Blanket Purchase Agreement	Blanket Agreement Stylesheet with Logo	Oracle Contract Terms Template		
Purchase Agreement Contract	Contract Purchase Agreement	Contract Agreement Stylesheet with Logo	Oracle Contract Terms Template		
Purchase Order Planned	Planned Purchase Order				
Purchase Order Standard	Standard Purchase Order	Standard Purchase Order Stylesheet with Logo	Oracle Contract Terms Template		
Quotation Bid	Bid Quotation				
Quotation Catalog	Catalog Quotation				
Quotation Standard	Standard Quotation				
Release Blanket	Blanket Release	Blanket Release Stylesheet with Logo			
Release Scheduled	Scheduled Release				

Document access can be restricted based on the following security-level options for each document type:

- **Public**—All system users can access the document.
- **Private**—Only the document owner and subsequent approvers can access the document.
- **Purchasing**—Only the document owner, subsequent approvers and individuals defined as buyers can access the document.
- **Hierarchy**—Only the document owner, subsequent approvers and individuals included in the security hierarchy can access the document.

Within the installation security hierarchy, individuals can access only those documents that they or their reporting employees own.

Further information relating to the security-level options for each document type is available online in the Oracle Release 12.1 Documentation Library.

In addition to the security level, Oracle EBS allows for setting the access level to control the modification/control actions that can be taken on a particular document type once access is gained. Document owners always have full access to their documents. The access-level options include:

- **View only**—Accessing employees can only view this document.
- **Modify**—Accessing employees can view, modify and freeze this document.
- **Full**—Accessing employees can view, modify, freeze, close, cancel and final-close this document.

Both the security level and access level apply to existing documents. Anyone with menu access to the document entry windows can create a document. These individuals then become the document owners.

The security level and access level work together. For example, if a security level of Public and an access level of Full is assigned to standard POs, users can access all standard POs from the Document Entry, View and Control windows.

Approvals

Oracle EBS Purchasing allows approval of requisitions, standard and planned POs, blanket and contract purchase agreements, and releases using a common process. Oracle Purchasing offers the following document approval:

- Reserve or unreserve (if using encumbrance Budgetary Control)
- Submit for approval
- Forward

Oracle Purchasing uses Workflow technology to handle the entire approval process. Workflow works in the background, using the approval controls (Approval Groups and approval rules) and hierarchies (approval and security hierarchies) that have been defined to route documents for approval.

An approval group is a set of approval rules composed of include/exclude and amount limit criteria for the following object types:

- Document total
- Account range
- Item range
- Item category range
- Location

Each position or job is ultimately associated with Approval Groups with document types to implement the approval rules.

Oracle Purchasing evaluates approval rules associated with a job or position and the document type to determine whether an individual has adequate authority. If the individual does not have the authority to approve a document, Oracle Purchasing provides default routing information based on the individual's approval setup. Depending on the level of routing flexibility specified, approvers may or may not be able to change these defaults.

Requiring management approval for POs and requisitions helps ensure that the organization is purchasing the correct items in the correct quantities, for the best prices and from authorized suppliers. Typically, different levels of management have financial purchase limits in accordance with the enterprise's policies and procedures. High-value purchases may require the approval of senior management or the board of directors, while junior management may authorize purchases for lesser amounts without having to obtain senior management approval. Setting up Oracle Purchasing approval hierarchies and Approval Groups helps ensure that only authorized users approve purchases.

Invoice Processing

Key invoice processing concepts include:
- Invoices
- Credit/debit memos
- Matching
- Invoice Validation and Approval
- Oracle EBS setup parameters
- Distribution Sets

Invoices
Function security within Oracle EBS is used to control access to the Payables functionality. Each Payables form (screen) performs one or more business functions. A function is a part of an application's functionality that is registered under a unique name for the purpose of assigning it to, or excluding it from, a Payables responsibility.

Payables come predefined with three responsibilities, each with its own set of preregistered functions assigned to it. Additional responsibilities can be added to the Payables function, as needed. There are two types of functions:
- **Form**—A form invokes an Oracle EBS Forms form.
- **Subfunction**—A subfunction is a subset of a form's functionality. In other words, a subfunction is a function executed from within a form.

Credit/Debit Memos

Credit or debit memos are entered to record a credit for goods or services purchased. They are debit balances in the GL, and are differentiated based on the source of the transaction.

A credit memo is a negative-amount invoice created by a supplier and sent to an entity to give notification of a credit. A debit memo is a negative-amount invoice created by the entity and sent to a supplier to notify the supplier of a credit that the entity is recording. **Figure 7.6** shows how credit/debit memos can be selected in the Invoices window.

Figure 7.6—Credit/Debit Memos

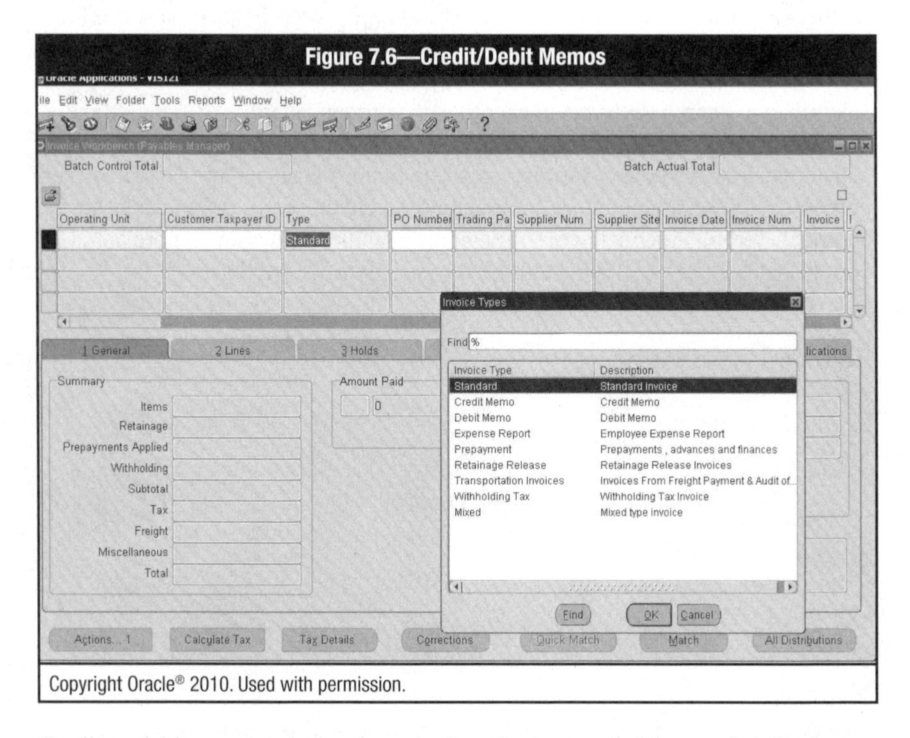

Credit or debit memos can be cleared when they are no longer needed. For example, when a supplier sends cash in lieu of a credit memo and the entity has already entered a debit memo, a positive-amount invoice can be entered to negate the impact of the credit/debit memo.

Matching of credit notes (called credit memos in Oracle EBS) and other adjustments (including debit memos) to invoices and POs helps ensure the accuracy of AP and related expense, inventory and asset accounts. When a credit/debit memo is entered in Oracle Payables, it can be matched to an existing invoice and all of the accounting information will be copied automatically. All invoice distributions are created for that credit/debit memo at the same time. The matching for credit/debit memos can take several forms in Oracle Payables:

- If the original invoice is not matched with a PO (basic invoice will not be attached to a PO or receipt), the credit/debit memo can be matched to either an invoice or specific invoice distribution (allocate the credit amount to the specific invoice distribution of the original invoice). If the credit/debit memo is matched to an invoice, Oracle Payables prorates the credit amount based on the invoice distribution amounts on the original invoice and creates the invoice distribution. If the credit/debit memo is matched to an invoice distribution, users can allocate the credit amount to a specific distribution of the original invoice, and Oracle Payables automatically creates the invoice distribution based on the original invoice distribution selected.
- If the original invoice is matched with a PO, price correction can be performed, or the credit/debit memo can be matched to invoice distributions or PO shipments. If the credit/debit memo is matched to an invoice distribution, users can allocate the credit amount to a specific distribution of the original invoice and Oracle Payables automatically creates the invoice distribution based on the original invoice distribution selected.

Price correction is done when recording a price decrease from the original invoice. The Price Correction option is selected when performing the credit/debit memo matching. If the credit/debit memo is matched to PO shipments, Oracle Payables prorates the credit amount across all PO distributions associated with the PO shipment, and automatically creates invoice distributions. Payables also updates the quantity billed for each of the originally matched PO order shipments, their corresponding distributions and the amount billed on these distributions.

In Oracle Payables, the credit/debit memos are created and matched in the same manner as the invoices since they are types of invoices.

Matching

Matching goods/services received to approved POs and/or invoices helps prevent payments for unwanted or unreceived goods/services. The Accounts Payable function confirms that the items were actually received. In addition, the Accounts Payable function might compare information listed on an invoice or other billing document to the appropriate PO to confirm that payment is made for goods that were ordered.

The control to confirm the validity of payments can take several forms in the Oracle Payables module:

- **Two-way matching**—This match verifies that PO and invoice information match within the set tolerances (quantity billed is less than or equal to quantity ordered, and invoice price is less than or equal to PO price).
- **Three-way matching**—This match verifies that the PO, invoice information and goods/services receipts match within set tolerances (quantity billed is less than or equal to quantity ordered, invoice price is less than or equal to PO price, and quantity billed is less than or equal to quantity received).

- **Four-way matching**—This match verifies that the PO, invoice information, goods/services receipts and delivery acceptance documents (quality inspection documents) match within set tolerances.

Oracle Payables automatically performs a two-way match, but if a three- or four-way match is needed, the user can select that from the Supplier, Supplier Site and Purchase Order Shipment windows.

When an invoice is entered and matched to a PO, Oracle Payables automatically creates distributions and checks to see whether the match is within the specified tolerances. Tolerances are usually created to reduce the time and cost incurred for returning items when the shipments exceed the POs. If the invoice and PO do not match within the tolerances defined for quantity and price, the Approval process places a matching hold on the invoice. Users can fix the problem/reason for the hold, release the hold and perform the Matching/Approval process again.

If the match is successful, it is saved and the PO distribution and shipment distribution are updated. This PO may also be given a status of "closed for invoice." However, matches are still allowed to this PO unless the match was indicated as a "final match." For an invoice that has been final matched to the PO and the match has cleared the approval process (which validates the matching, tax, period status, exchange rate and distribution information for invoices), Oracle Payables closes that PO permanently. Any invoice matched against a permanently closed PO is placed on a final-match hold. This hold cannot be released. Users have to reverse the distribution line to a permanently closed PO and then match the invoice to a different PO or enter the distributions manually.

Oracle EBS also provides a number of alerts that can be used to identify and resolve matching holds against invoices. These alerts include:

- **Invoices on price hold**—This queries the database for invoices currently on price hold and notifies the originating purchase agent for each held invoice.
- **Invoices on quantity received hold**—This looks for invoices on quantity received hold and notifies the originating purchasing agent for each held invoice.
- **Invoices on quantity ordered hold**—This looks for invoices on quantity ordered hold. If any exceptions occur, Oracle Alert sends a summary message to each responsible purchasing agent to resolve invoice holds caused by a difference between quantity billed and quantity ordered.
- **Receipts hold**—This looks for overdue items that have not been received. If any exceptions occur, Oracle Alert sends a detailed message to the original requestor asking for verification of receipt.

The use and configuration of these alerts are specific to a vendor site. Action should be taken on these alerts to help ensure these holds are rectified and removed from the system in a timely manner. Controls should be in place to identify long-outstanding goods receipt notes, POs and/or invoices.

Invoice Validation and Approval

The Invoice Validation process verifies the matching status and completeness of invoices and automatically applies invoice holds when exceptions are detected. These holds can then be manually released (if allowable), or the invoice can be corrected and the validation process resubmitted.

Optionally, Oracle EBS approval hierarchies may be implemented to allow for the establishment of approval processes for Oracle Payables within the system.

Oracle Payables may use the Oracle Workflow technology to handle the entire approval process. Workflow works in the background, using the approval controls (Approval Groups and approval rules) and hierarchies (approval and security hierarchies) that have been defined to route documents for approval.

An approval group is a set of approval rules comprised of include/exclude amount limit criteria for the following object types:

- Document total
- Account range
- Item range
- Item category range
- Location

Each position or job is ultimately associated with Approval Groups with document types to implement the approval rules. Oracle Payables evaluates approval rules associated with a job or position and the document type to determine whether an individual has adequate approval authority. If an individual does not have enough authority to approve a document, Oracle Payables provides default routing information based on the individual's approval setup. Depending on the level of routing flexibility specified, approvers may or may not be able to change these defaults.

Requiring management approval for Oracle Payables helps ensure that only valid invoices are paid based on negotiated prices for goods received from the appropriate suppliers. Typically, different levels of management have invoice approval limits, in accordance with the enterprise's policies and procedures.

High-value invoices require the approval of senior management or the board of directors, while management may authorize invoices for lesser amounts without having to obtain senior management approval. Setting up Oracle Payables approval hierarchies and Approval Groups helps ensure that only authorized users approve invoices.

Oracle EBS Setup Parameters

Oracle EBS allows parameters to be set, such as payment batches, payment terms, hold releases, accounting methods and payment methods. These setup parameters are powerful settings within the Oracle Payables module, and unauthorized access could lead to invalid payments.

Distribution Sets

Distribution Sets automatically enter distributions for an invoice when not matching it to a PO. For example, a Distribution Set can be created for an advertising supplier that allocates advertising expense on an invoice to four advertising departments.

Access to create, modify or delete Oracle Distribution Sets should be restricted to authorized personnel to help ensure that only accurate information is entered into the system.

In addition, Distribution Sets can be defined to simplify and speed invoice entry. A default Distribution Set can also be assigned to a supplier site so Oracle Payables will use it for every invoice and PO entered for that supplier site. Supplier sites can be designated as pay sites, purchasing sites, RFQ-only sites or procurement card sites. For example, if a supplier does business from multiple locations, the supplier information is entered only once, but supplier sites can be created for all locations.

Cross-Validation Rules

Cross-Validation Rules are an optional setup parameter. If used, the rules define the proper combinations of Accounting Flexfields to help ensure that purchases cannot be recorded improperly. For instance, certain expenses can be incurred by the plant but not by the corporate office. Using cross-field validation rules, certain expense accounts can be restricted for use by some divisions. It is especially important to set up Cross-Validation Rules when the Dynamic Insertion feature is activated because Dynamic Insertion allows flexibility to create *ad hoc* accounting combinations, and Cross-Validation Rules will help ensure that such combinations are valid.

Online Edits and Validation Tools

Oracle EBS utilizes several online edits and validation tools, including required fields (both Oracle EBS-defined and user-defined), QuickPick functionality and default field values. These tools facilitate accurate recording of payments and stop and void payments. Transactions cannot be processed if a required field is blank; for instance, the bank account from which payments are made. QuickPick functionality provides pop-up windows that contain predefined options from which the user can choose. Fields that generally require the same input can be defined to contain default values so input is not required. A password is required to view the attributes of a field. The preloaded password is "apps." If this password has not been changed, the security of the purchasing system may be compromised.

Processing Disbursements

In Oracle Payables, there are six types of payment methods (check, clearing, electronic, future-dated, manual future-dated and wire) and there are three ways to disburse a payment:

- Manual payments (for payments created outside of Oracle EBS)
- Quick payment (a single computer-generated payment)
- Payment batches (payments for multiple invoices)

Payments can be made through either the Invoice Workbench or the Payment Workbench. Users need to select the invoice(s) and generate the payment for them. Only invoices that have been validated/approved and are not on hold can be selected for payment.

Payments can be voided if the check has not been released. If the check has been released, the payment must be stopped, consistent with a stop payment through the bank, and voided.

Invoices may be paid via manual payments, quick payments or payment batches. Manual payments involve entering the data from a manually written check. Quick payments allow the user to define, within certain parameters, the vendors and/or invoices to pay. Checks may then be computer-generated or submitted for EDI. To process payment batches, the user must define the criteria for payment (i.e., payment due dates); review the Preliminary Payment register; make modifications as necessary; format the batch to an output file; print the checks or submit the electronic payments; then confirm the batch, which records the payments and updates the status of the invoices. If the entity uses EDI to make payments, Oracle EDI Gateway must be used to process the payments.

Releasing Holds

In Oracle Payables, invoices and scheduled payments can be put on hold (e.g., an invoice can be put on hold if the three-way matching is not successful or if the supplier for the invoice has been put on hold). Placing a hold on any document prevents any further action to that document. To process the invoices or scheduled payments completely, all holds must be released. Some of the holds are system-generated and others are placed manually. It is important that only authorized personnel have the ability to release those holds, especially for the payments; otherwise, invalid payments may be cleared through the system. Invoices placed on hold should be reviewed on a regular basis.

Sequential Numbering

The system administrator can enable the Sequential Numbering Profile Option. This option may be enabled at the user, responsibility, application or site level. The option may be set to one of the following:

- **Not used**—Does not enforce sequential numbering
- **Partially used**—Enforces sequential numbering for all of the sequences that have been defined and assigned to a document category. If a document type is selected (e.g., for manual payments or quick payments for which there is no active sequence), Oracle Payables will display a warning message that a sequence does not exist.
- **Always used**—Enforces sequential numbering for all document types. If a document type is selected for which there is no sequence, the user cannot proceed until the sequence is defined and assigned.

The automatic numbering sequence helps ensure that Oracle Payables automatically assigns the next available voucher number in sequence and that the user cannot enter the payment document number. In this way, validation of the document number for uniqueness and completeness is performed systematically. This helps ensure that all payment documents (i.e., vouchers) have been recorded.

If a manual numbering sequence is used, users must enter the document number in sequence. Nonetheless, a document number is still required, and Oracle Payables validates the document number for uniqueness, but not completeness.

Purging

Oracle Payables, Oracle Purchasing and Oracle Supplier Scheduling allow users to remove records to which online access is no longer needed, to free up space in the database. Users can purge POs; invoices; suppliers; and related records such as invoice payments, supplier schedules and purchase receipts. After a record is purged, it is no longer available for query and the record no longer appears on standard reports. However, the system maintains summary information of deleted records to prevent duplicate entry of invoices or POs.

The following documents that meet certain criteria can be purged:

- **Invoices**—They are posted, not generated by a recurring invoice template and without open encumbrances.
- **Payments**—They are posted, and all of the invoices to which the payment applies meet the invoice purge criteria. Unvoided payments must have a cleared date on or before the last activity date.
- **Suppliers**—They are not employees or a parent company or subsidiary of another supplier. All invoices and payments for the supplier must meet the invoice and payments purge criteria.
- **Requisitions**—They are canceled and all lines are canceled.
- **POs**—They are approved, canceled or closed, billed, or received and not referenced on an ASL.

- **Supplier's schedules**—They must have a header in which the last update must be on or before the last activity date and for the organization specified in the Submit Purge window.

A special responsibility should be created for purging information from Oracle Payables and Oracle Purchasing, and this responsibility must be assigned only to the person responsible for purging information from the database. Restricting access to purging is critical to help ensure that all Purchasing and Payables data are included in the financial statements. Therefore, procedures need to ensure that purged data are logged and backed up for future recovery.

Reporting

A large number of standard reports/requests are available in the Oracle Purchasing and Payables applications. A full listing of these can be brought up via the Purchasing Superuser and Payables Manager responsibilities. Select reports that assist in the testing techniques have been listed in chapter 8, Auditing Oracle E-Business Suite—Financial Expenditure Business Cycle.

Summary

This chapter provided an overall understanding of the operation of the Expenditure business cycle in an Oracle ERP environment. It summarized important functionality from an audit perspective—including PO approval, matching of POs, invoices and the processing of payments.

Page intentionally left blank

8. Auditing Oracle E-Business Suite—Expenditure Business Cycle

In chapter 7, the four main subprocesses of the expenditure business cycle were outlined. This chapter looks at the significant risks, key controls and techniques to test the controls for each of the following four main subprocesses:

- Master Data Maintenance
- Purchasing
- Invoice Processing
- Processing Disbursements

Refer to **figure 4.8** for the key to the numbering sequence for risks, controls and testing techniques.

Master Data Maintenance

Master Data Maintenance: Risks

Significant risks associated with supplier Master Data Maintenance include:

1.1	Entry and changes to the supplier management master data may be invalid, incomplete, inaccurate and/or untimely. For example, these errors could result in: • Ordering goods from unapproved suppliers, sending payments to incorrect addresses, valuing foreign currency payables inaccurately, or issuing unauthorized payment and discount terms • Duplicate supplier records being established, or inappropriate use of the One-Time Supplier account and subsequent duplicate or misdirected payments
1.2	Master data may not remain current and accurate. For example: • If supplier data are not up to date, payments may be made to the wrong bank account. • Obsolete supplier records may be maintained on the system, which could result in inappropriate transactions being processed.

Master Data Maintenance: Key Controls

Key controls over the maintenance of supplier master data include:

1.1.1	Relevant management, other than the initiators, should check online reports of master data changes to source documentation, on a sample basis. Recorded changes to the supplier master file should be compared to authorized source documents to ensure that they were input accurately. Requests to change the supplier master file data should be logged; the log should be reviewed to ensure that all requested changes are processed on a timely basis.
1.1.2	Access controls over the supplier master file in the Oracle Payables and Purchasing modules should be restricted appropriately. The creation and maintenance of master data should be assigned and restricted to dedicated personnel within the areas of the organization who understand how it may impact organizational processes and the importance of timely changes.
1.1.3	A naming convention should be used for supplier names (e.g., letterhead) to minimize the risk of establishing duplicated supplier master records. Alternative controls to minimize this risk may involve implementing duplicate checking controls for an existing supplier address or bank account number.
1.2.1	Standard Oracle Reports should be run and reviewed periodically by management to ensure accuracy and ongoing pertinence of the supplier master files—especially, supplier bank account, payment terms and standard pricing.

Master Data Maintenance: Testing Techniques

Testing techniques for supplier master data controls include:

1.1.1	Ask management for any relevant documentation relating to: • Policies, procedures, standards and guidance for modifying the supplier master file • Procedures for modifying supplier master file data, including forms and/or other documentation used to request/authorize such modifications • Procedures for reviewing and approving reports of supplier master file modifications • Individuals responsible for making and/or approving modifications to supplier master file details • Individuals responsible for comparing changes made to supplier master file details to authorized source documents to ensure that the changes were input accurately • Procedures for confirming the accuracy of supplier master file changes with suppliers

1.1.1 *(cont.)*	During the review of these documents, assess the existence and adequacy of the policies and procedures. Review evidence of the generation and comparison of online reports of master data changes back to source documentation (on a sample basis if there are large volumes of changes). Review evidence that recorded changes to the supplier master file are compared to authorized source documents (on a sample or sensitivity basis if there are large volumes of changes). Review evidence that requests to change the supplier master file data are logged and that the log is reviewed by authorized personnel documented in the policies, procedures, standards and guidance for modifying the supplier master file.
1.1.2	Review enterprise policy and process design specifications regarding access to maintain master data. For the principal business activity of maintaining the supplier master file, identify users who have the ability to create, modify or delete supplier master file data. Extract a list of target high-risk functions/forms, as explained in chapter 4, and review it to ascertain who has access privileges to the following functions/forms: • Suppliers • Merge Suppliers • Summary Approved Suppliers List • Supplier Statuses • Define Approved Suppliers List • Set up Approved Suppliers List • Supplier Item Catalog
1.1.3	Extract a list of supplier account names by running either the New Supplier/New Supplier Site listing or the Suppliers report explained in point 1.2.1. Review a sample for compliance with the enterprise's naming convention. The Supplier Audit report can be used to help identify potential duplicate suppliers.
1.2.1	Ask management for documentation of policies, procedures, standards and guidance related to periodic reviews of supplier master file data. Various standard reports are available in Oracle EBS to facilitate the review of supplier data to ensure ongoing pertinence and accuracy of the supplier master file, including:

1.2.1 *(cont.)*	• The New Supplier/New Supplier Site listing can be used to review new suppliers or new supplier sites. Oracle Payables provides detailed information for each supplier or supplier site, including who entered it, the pay group, terms, creation date and whether a site is a pay site. • The Suppliers report can be used to review detailed information entered for a supplier in the Suppliers and Supplier Sites windows, and indicates if a supplier is on PO hold and whether the vendor is active or inactive. • The Supplier Audit report can be used to help identify potential duplicate suppliers by listing active suppliers whose names are the same up to a specified number of characters. • After identifying duplicate suppliers, the Supplier Merge process will combine the data of duplicate suppliers and automatically print the Supplier Merge report. The report lists the new (correct) suppliers, old (duplicate) suppliers and information for invoices, updated for each set of new and old suppliers. Oracle Payables prints the total number and total amount of invoices that have been updated for each set of merged suppliers and the total number of duplicate invoices. Unpaid invoices and POs can be transferred to the new supplier; paid invoices can be transferred or not; duplicate invoices, whether paid or unpaid, cannot be transferred. This report is used to identify duplicate invoices, so the invoice number on one of the suppliers can be updated and Supplier Merge resubmitted to complete the merge. • The Supplier Paid Invoice History report can be obtained by supplier or supplier type and used to review payment history, discounts taken and frequency of partial payments. If invoices are paid in more than one currency, the report subtotals invoice amounts paid in foreign currencies and invoice amounts paid in the functional currency. • The Supplier Payment History report is used to review the payment history for a supplier or group of suppliers with the same supplier type. It also may include invoice details and voided payments. All amounts are displayed in the payment currency. Ask management whether there is a procedure for reviewing these reports. Ascertain that these reports are reviewed on a regular basis and that exceptions are identified for follow-up and resolution. Inspect physical evidence that the reports are run, approved and retained and that exceptions are resolved, as necessary. The reports can be run using the Reports window (Reports➔Run).

1.2.1 *(cont.)*	Oracle Alerts or Workflow notifications may be used to inform appropriate personnel that certain changes have been made to the master file. This alert should be used in a discriminatory manner so as not to overwhelm someone with notices, in which case the alerts will likely be so significant in number that they may be ignored. The alert notification can be restricted to additions, deletions and changes to certain key fields, such as vendor name.

Purchasing

Purchasing: Risks

Significant risks associated with purchasing include:

2.1	Entry of POs and changes may be invalid, incomplete, inaccurate and/or untimely. For example: • Inaccurate input of POs could lead to financial losses due to the purchase of incorrect goods or services. • A PO raised, but not subject to a release strategy, could result in an unauthorized purchase.
2.2	Goods (or materials, equipment) may be received for which there are no valid POs, or goods receipts may be recorded incompletely, inaccurately or in an untimely manner. For example: • If goods are received for which no valid PO exists, goods may be received for which the organization has no need. • If raw materials received are not recorded, there may be delays in production, as well as subsequent delays in supplying customers with finished goods. In addition, the incomplete recording of raw material receipts may result in a misstatement of inventory. The failure to record raw materials received also may lead to supplier disputes and/or inventory obsolescence.
2.3	Defective goods (or materials, equipment) may not be returned to suppliers in a timely manner; for example, if defective goods are not quickly returned to suppliers, credits may be lost and disputes may arise.

Purchasing: Key Controls

Key controls over purchasing include:

2.1.1	Access controls over the Purchase Requisitions, Purchase Orders, Autocreate and Setup functionalities in the Oracle Purchasing module should be restricted appropriately to authorized personnel.
2.1.2	Document security should be configured so that security levels and access levels are set by document type to authorized users. Access to document types should be restricted.
2.1.3	Oracle EBS approval hierarchies, Approval Groups, jobs and positions should be secured and configured in conformity with established policies.
2.1.4	Cross-Validation should be enabled and developed to ensure accuracy of data entry. The Dynamic Insertion feature should be used only in conjunction with Cross-Validation Rules to prevent setup of invalid account numbers.
2.1.5	Flexfield Value Security should be enabled and developed to ensure accuracy of data entry.
2.1.6	Management should monitor receiving adjustments to POs.
2.1.7	Management should review the status of POs on a timely basis.
2.1.8	The ASL functionality (if implemented) should allow specified items to be purchased only from suppliers included in the source list for the specified material.
2.1.9	POs should be numbered sequentially, and the sequence of POs processed should be verified.
2.1.10	Management should review reports detailing overrides of established PO prices, terms and conditions.
2.2.1	When goods received are matched to open POs and have receipts with no PO (unordered receipts) or that exceed the PO quantity by more than an established amount, they should be investigated. Management should review exception reports of goods not received on time for recorded purchases.
2.2.2	The ability to input, change or cancel goods received transactions should be restricted to authorized inbound logistics—raw materials personnel.
2.3.1	Rejected or defective goods (or materials, equipment) should be segregated from other goods in a quality assurance bonding area and monitored regularly to ensure the timely return to suppliers and the timely receipt of credit.

Purchasing: Testing Techniques

Testing techniques for purchasing controls include:

2.1.1	Review the following access levels needed to help ensure that only appropriate and authorized personnel have access to: • Create, modify and cancel purchase requisitions. • Autocreate POs. • Create, modify and delete POs. • Modify certain Oracle EBS setup parameters. • Perform the purging operations. Extract a list of the target high-risk functions/forms. Review the function/form listing in 1.1.2 Master Data Maintenance: Testing Techniques to ascertain who has access privileges to the following functions: • Approval Assignments • Approval Groups • Autocreate Documents • Buyers • Control Purchase Orderss • Control Requisitions • Define Position Hierarchy • Define Purchasing Lookup Types and Codes • Document Types • Lookups: Common • OM: Order Purge • Order Purge • Purchase Order Copy Document • Purchase Order Creation • Purchase Order Preferences • Purchase Order Summary: Create New Purchase Order • Purchase Order Summary: Create New Release • Purchase Order Summary: Open Document • Procure to Pay (Menu Process) • Profile System Values • Profile User Values • Purchase Order Summary • Purchase Orders • Purchase Orders: Suppliers Item Catalog • Purchasing Options • Purchasing Lookups • Purge • Requisition Preferences • Requisition Summary • Requisition Templates

2.1.1 *(cont.)*	• Requisitions • Requisitions: Supplier Item Catalog • Templates
2.1.2	Navigate to the Document Types window (Setup➔Purchasing➔ Document Types). Refer to **figure 7.4**. Select and review the security level and access level applied against each of the existing document types. Refer to **figure 8.1**, which shows an example of the security level and access level applied against a standard PO. Extract a list of the target high-risk functions/forms and review it to ascertain who has access privileges to document types.

Figure 8.1—Document Type Settings

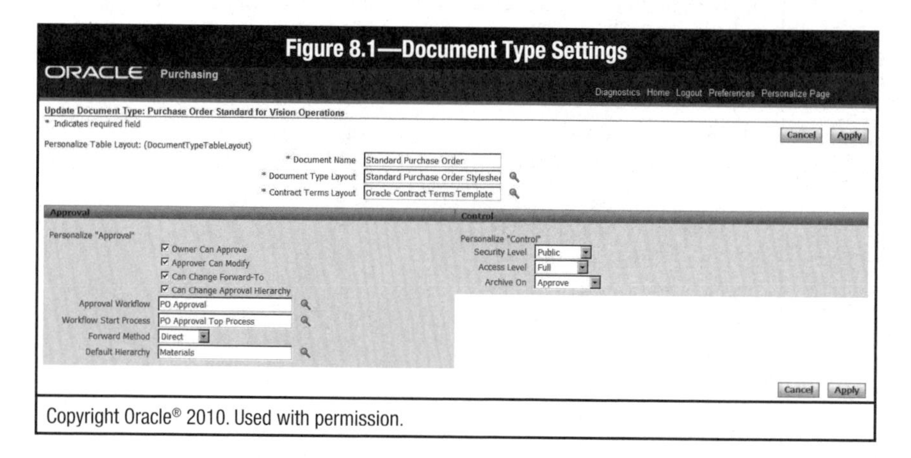

2.1.3	Obtain a sufficient understanding of the system configuration to assess the adequacy of the approval strategy, as defined and implemented by the enterprise, as well as the functioning and effectiveness of established company policies, procedures, standards and guidance. Review: • Purchasing policies, procedures, standards and guidance related to approval of POs, requisitions, capital expenditure purchases, standing orders and/or other high-value POs • The list of individual and cumulative limits for authorization of POs or requisitions • The list of individuals authorized to approve purchases, including those in excess of established limits • The setup documentation (e.g., approval group setup, approval hierarchy setup, job setup and position setup documents) Review the system settings to ensure that they conform to established organization policies, procedures, standards and guidance:

2.1.3 *(cont.)*	• Approval Groups (Setup➜Approvals➜Approval Groups)—Review the approval limits per approval group listed. **Figure 8.2** shows an example of the approval limits for a manager, as well as a listing of other Approval Groups. Note that if using approval hierarchies, the Approval Groups window must be used to establish rules for Approval Groups. Once the rules have been established, Approval Groups can be assigned to positions in the Assign Approval Group window. If approval hierarchies are not being used, the window can be used to assign Approval Groups and approval functions to jobs within the enterprise, as demonstrated in **figure 8.3**.

Figure 8.2—Approval Groups

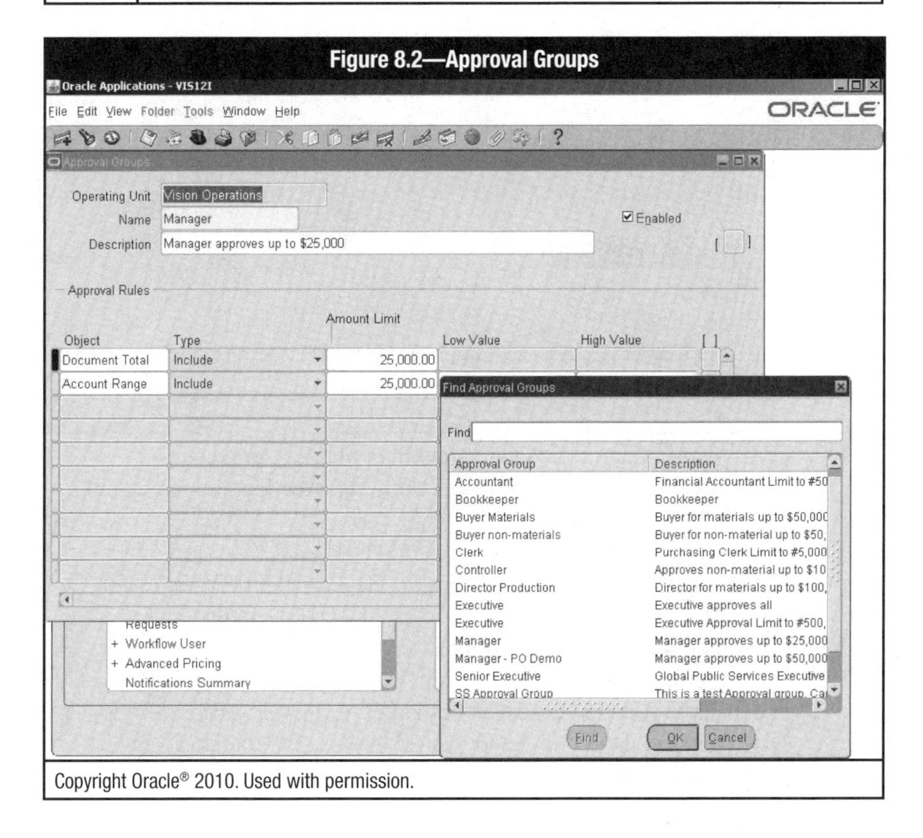

Figure 8.3—Approval Assignments

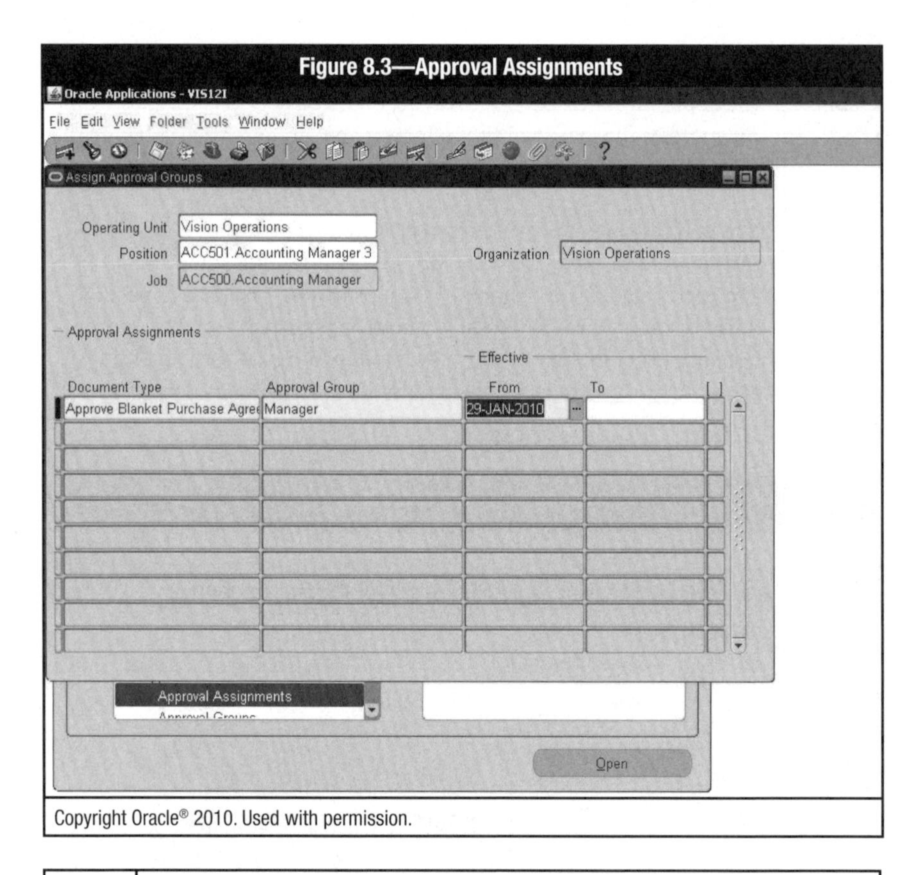

Copyright Oracle® 2010. Used with permission.

2.1.4	Ask management for documentation on policies, procedures, standards and guidance related to defining, creating and maintaining Cross-Validation Rules. This can be verified by navigating to the Key Flexfield Segments form (Setup➔Financials➔Flexfields➔Key➔Segments) and selecting the name of the application (Oracle GL) and Flexfield (Accounting Flexfield). The Allow Dynamic Inserts checkbox should be enabled if the enterprise is utilizing Dynamic Insertion. If this has been set, there will be no control in place to validate account combination creation. To address this, the Cross-Validate Segments checkbox should also be enabled. This checkbox applies Cross-Validation Rules to account combination creation. Even if the enterprise does not enable Dynamic Insertion, Cross-Validation Rules around the Accounting Flexfield should be required to identify data input errors during account combination input, rather than after the combination has been created.

2.1.5	Determine whether Flexfield Value Security is used to restrict the set of values that can be input during data entry. Check whether administrators have determined who can use Flexfield segment values. Based on responsibilities and access rules, Flexfield Value Security limits what values can be entered in Flexfield pop-up windows. For instance, a user responsible for purchasing parts for plant A can be restricted from entering and making purchases for plant B. That same user can be restricted from ordering or can be permitted to order only certain parts, types of parts or parts with a range of values. Review Flexfield error messages for relevance. When using Flexfield Value Security, error messages are defined by client personnel and should be meaningful to assist the user in understanding the reasons for being restricted from entering a value.
2.1.6	Ask management for documentation on receiving and purchasing policies, procedures, standards and guidance. There are several standard Oracle Reports available for management to use to monitor variances between goods received and goods ordered: • The Receipt Adjustments report lists PO shipments or internal requisition lines with corrections or returns to supplier. • The Receiving Exceptions report identifies receipts placed on exception hold, which can be based on quantity or monetary value. • The Unordered Receipts report lists all or selected unordered receipts, which are received items for which receivers could not be matched to POs. Unordered receipt transactions can be entered only if the system options are set up accordingly. Ask management whether there is a procedure for reviewing these reports. Ascertain that these reports are reviewed on a regular basis and that exceptions are identified for follow-up and resolution. Inspect physical evidence that the reports are run, approved and retained and that exceptions are resolved, as necessary. The reports can be run using the Reports window (Reports➔Run). **Figure 8.4** shows a list of the reports containing two of the reports explained previously.

Figure 8.4—Adjustment Reports

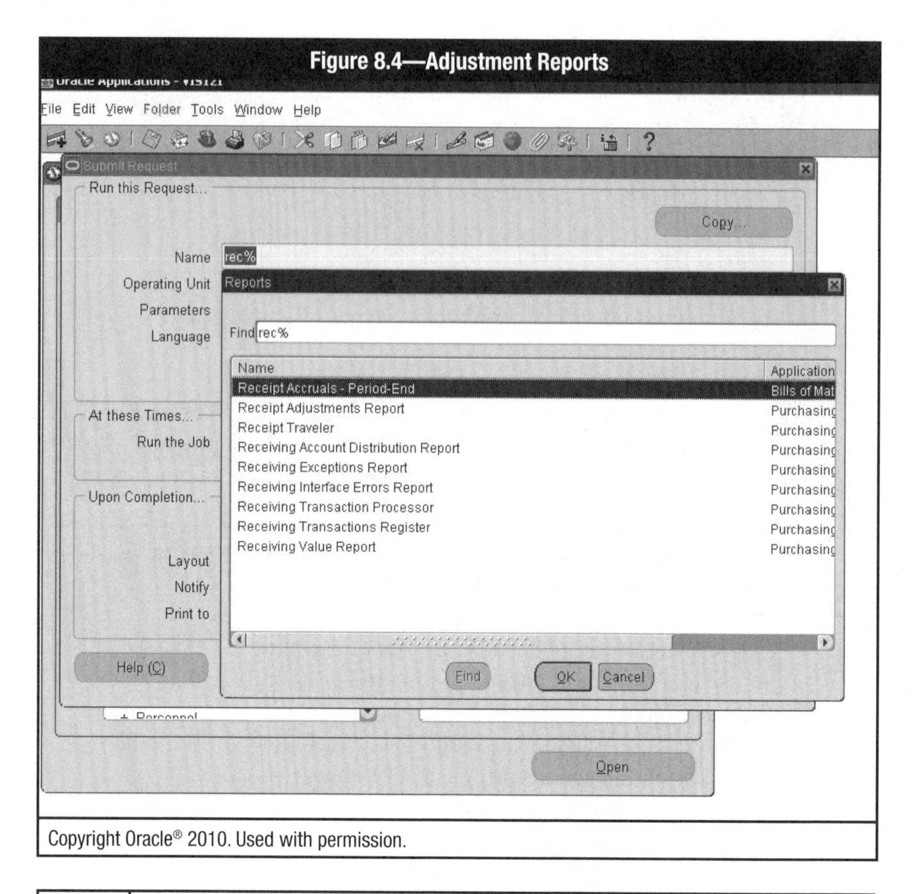

Copyright Oracle® 2010. Used with permission.

2.1.7	Ask management for documentation on policies, procedures, standards and guidance related to periodic review and monitoring of the open POs and requisitions. There are several reports available to assist management in monitoring the status of POs, including: • The Expected Receipts report, which can be used to review all or specific supplier-sourced (vs. internally sourced) expected receipts for a particular date or a range of dates • The Blanket and Planned PO Status report, which can be used to review PO transactions for items purchased using blanket purchase agreements and planned POs. For each blanket purchase agreement and planned PO created, Oracle Purchasing provides the detail of the releases created against these orders. • The Open Purchase Orders report (by buyer or cost center) lists all or specific open POs that relate to buyers or to one or more cost centers.

2.1.7 *(cont.)*	• The Overdue Vendor Shipments report provides information to assist in following up on open POs that have not been filled by the designated due date. • The Purchase Requisition Status report can be used to review the approval status of the requisitions created, including those approved, canceled, in process, incomplete, preapproved, rejected or returned. By limiting the approval status, the report can be used as a tool to monitor requisitions in the approval process or those returned and requiring further attention. Ask management whether there is a procedure for reviewing these reports. Ascertain that these reports are reviewed on a regular basis and that exceptions are identified for follow-up and resolution. Inspect physical evidence that the reports are run, approved and retained and that exceptions are resolved, as necessary. The reports can be run using the Reports window (Reports➔Run).
2.1.8	Through discussions with management, determine which suppliers and items should be reflected on the system. Compare the ASL on the system to authorized documentation (navigate to the ASL: SupplyBase➔Approved Suppliers List). **Figure 8.5** shows the ASL window.

Figure 8.5—Approved Supplier List Window

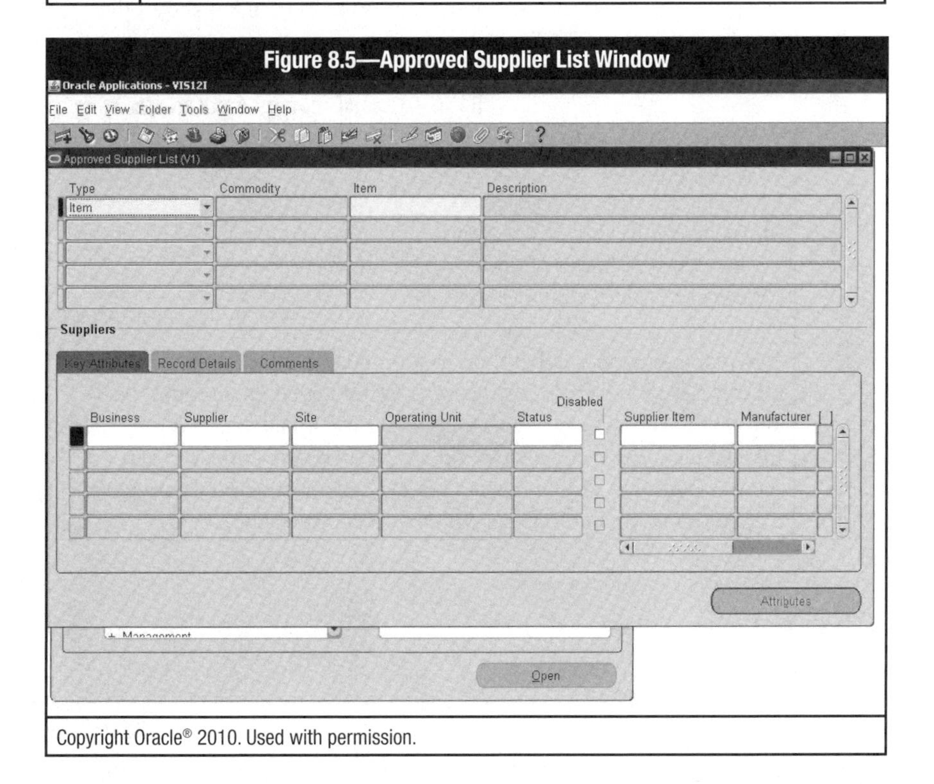

<table>
<tr>
<td>2.1.9</td>
<td>Use sequential numbering of transactions to ensure that all transactions are processed and each transaction is processed only once. Such transaction numbers enhance the audit trail and can facilitate follow-up on edit and validation errors and other exception items. Sequential numbering of POs is useful, particularly to control commitments of the enterprise and to facilitate identification of corresponding POs upon receipt of goods and services.

In Oracle EBS, systematic sequential numbering of POs, requisitions and RFQs can be set individually in the Purchasing module. This functionality is separate from the Global Sequential Numbering option in the User Profile Options of the system administrator. Numbering can be specified as numeric or alphanumeric.

If source documents are prenumbered, the document number can be input in the system, and controls should include reports missing and/or duplicate document numbers.

Sequential numbering settings should be checked as follows: Navigate (Setup➔Organizations➔Purchasing Options➔Numbering section) to the Purchasing Options window. Check the sequential numbering settings for each document type. Figure 8.6 shows the Purchasing Document Types Numbering section.

There are several reports available to assist management in monitoring the sequential numbering of transactions, including:
• Purchasing Activity Register
• Registers Activity Register
• Canceled Purchase Order Report
• Canceled Requisition Report
• Purchase Order Detail reports

Ask management whether there is a procedure for reviewing these reports. Ascertain that these reports are reviewed on a regular basis and that exceptions are identified for follow-up and resolution. Inspect physical evidence that the reports are run, approved and retained and that exceptions are resolved, as necessary. The reports can be run using the Reports window (Reports➔Run).</td>
</tr>
</table>

Figure 8.6—Sequential Numbering

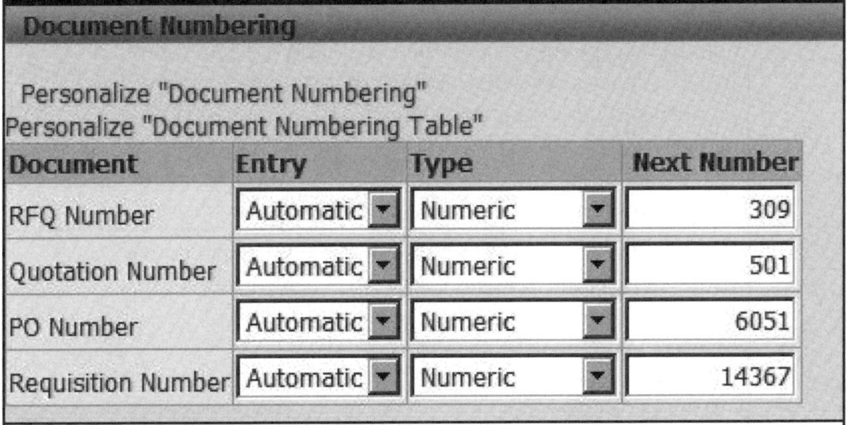

Document Numbering

Personalize "Document Numbering"
Personalize "Document Numbering Table"

Document	Entry	Type	Next Number
RFQ Number	Automatic	Numeric	309
Quotation Number	Automatic	Numeric	501
PO Number	Automatic	Numeric	6051
Requisition Number	Automatic	Numeric	14367

2.1.10	The Purchase Price Variance report shows the variance between the purchase price on a PO and the standard cost for all items received as defined in the item master, and assists management in monitoring overrides of established PO prices, terms and conditions. Ask management whether there is a procedure for reviewing this report. Ascertain that this report is reviewed on a regular basis and that exceptions are identified for follow-up and resolution. Inspect evidence that the report is run, approved and retained and that exceptions are resolved, as necessary.
2.2.1	Examine the following reports and verify that these reports are used to identify long-outstanding goods receipt notes, POs and/or invoices, or items that have been placed on hold: • Invoice Aging report (due date, payment number, days due, percent unpaid, account remaining, aging period) • Invoice on Hold report (hold, PO number, functional currency original amount, functional currency amount remaining, hold code)—Ensures that holds are reviewed by the appropriate personnel and released in a timely manner • Open Purchase Order by Cost Center Report (lists open orders, but does not list the closed, final closed and canceled orders) • Receiving Exceptions Report Ascertain from management whether there are any reasons for any long-outstanding items on the reports.

2.2.1 *(cont.)*	Ask management whether there is a procedure for reviewing these reports. Ascertain that these reports are reviewed on a regular basis and that exceptions are identified for follow-up and resolution. Inspect physical evidence that the reports are run, approved and retained and that exceptions are resolved, as necessary. Inquire about the process for clearing the holds and approving the invoices and payments once the holds have been released.
2.2.2	Extract a list of target high-risk functions/forms and review the listing to ascertain who has access privileges to the following functions: • Corrections (receiving) • Receiving options • Returns (receiving)
2.3.1	Ascertain from management the movement type used to block processing and for returning rejected goods to suppliers. Determine whether there are any long-outstanding materials pending return to suppliers/receipt of appropriate credits.

Invoice Processing

Invoice Processing: Risks

Significant risks associated with invoice processing include:

3.1	Amounts posted to AP may not represent goods or services received. For example: • If amounts posted to AP do not represent goods or services received, unauthorized payments may be made and the enterprise may incur a financial loss. • AP may be misstated, just as the relevant expense, inventory or asset accounts may be. • Payment may be made twice for the same PO.
3.2	AP amounts may not be calculated completely and accurately or may not be recorded in a timely manner. For example: • If AP amounts are not completely and accurately recorded, suppliers may not be paid in full, possibly damaging supplier relations. • If invoice verification and three-way matching are not performed effectively, large balances may build up in the goods receipt (GR)/ invoice receipt (IR) account that ultimately cannot be reconciled and must be written off.
3.3	Credit/debit notes and other adjustments are not calculated completely and accurately or are not recorded in a timely manner.

Invoice Processing: Key Controls

Key controls over invoice processing include:

3.1.1	Access controls over invoices, accounting periods, credit/debit memos, set up and Distribution Sets in the Oracle Accounts Payable module should be restricted appropriately. The ability to input supplier invoices that do not have a PO and/or goods receipt as support should be further restricted to authorized personnel.
3.1.2	The Oracle Payables module should be configured to match goods/services received to POs and/or invoices for those goods/services; if matching is unsuccessful, matching holds should be placed on the invoices. Long-outstanding goods receipt notes, POs and/or invoices should be investigated in a timely manner and accrued as appropriate; documents should be closed when final matching has been performed.
3.2.1	Online edits and validation tools should be utilized in the Data Entry process.
3.2.2	Invoice amounts should be recorded to the GL accounts based on distributions defined on POs or defined Distribution Sets.
3.2.3	Standard Oracle Reports should be reviewed periodically by management to ensure accuracy of payables processing.
3.2.4	Statements received from key/large suppliers should be regularly reconciled to the supplier accounts in the AP subledger, and differences should be investigated.
3.2.5	The Oracle EBS R12.1 software should be configured with quantity and price-tolerance limits.
3.2.6	Unposted invoices should be reviewed and moved to the current period prior to closing the Payables module and transferring data to the GL.
3.2.7	The Oracle EBS R12.1 software restricts the ability to modify the Exchange Rate table to authorized personnel. Management should approve values in the centrally maintained Exchange Rate table. The Oracle EBS R12.1 software automatically calculates foreign currency translation, based on values in the centrally maintained Exchange Rate table.
3.3.1	The ability to input, change, cancel or release credit notes should be restricted to authorized personnel.
3.3.2	Credit memos and other adjustments should be matched to corresponding invoices in Oracle Payables.
3.3.3	Management should review all recorded nonsystem debits to AP (e.g., originating from sources other than a disbursements journal).
3.3.4	Management should review all invoices with HOLD status on a regular basis.

Invoice Processing: Testing Techniques

Testing techniques for invoice processing controls include:

3.1.1	For the Processing Accounts Payable subprocess in the Expenditure business cycle, review the following access levels to help ensure that only the appropriate and authorized personnel have access to: • Create, modify and delete invoices. • Modify open and close periods in the Accounting Calendar. • Create, modify and delete credit memos/debit memos. • Modify certain Oracle EBS set up parameters relating to accounts payable (e.g., who can release holds, payment terms). • Create, modify and delete Distribution Sets. Extract a list of the target high-risk functions/forms and review it to ascertain who has access privileges to the following functions: • AP Accounting Flexfield Combinations GUI • AP Accounting Periods • Calendars to Define Periods • Close Accounts Payable Period • Distribution Sets • Invoice Accounting • Invoice Actions • Invoice Apply Prepayments • Invoice Approvals • Invoice Approve • Invoice Batch Summary • Invoice Batches • Invoice Cancel • Invoice Distributions • Invoice Distributions Prorate • Invoice Distributions Reverse • Invoice Fundscheck • Invoice Gateway • Invoice Holds • Invoice Match • Invoice Overview • Invoice Payments • Invoice Payments Schedules • Invoice Payments Workbench • Invoice Print • Invoice Release Holds • Invoices • Invoices Summary • Open and Close Periods • Payables Options

3.1.1 *(cont.)*	• Payment Terms • Period Close—Parent Company • Period Close—Subsidiary Company • Period Types • Returns
3.1.2	Ask management for the following documentation: • Payment policies, procedures, standards and guidance for matching goods received, invoices and POs • Matching procedures, including timing • Forms and other documentation used • Procedures to monitor timely completion of the matching and timely resolution of nonmatching items • Established tolerances for variances between ordered and received goods • Procedures to accrue, as appropriate, unmatched POs, receiving reports and/or invoices at period end Configure standard Oracle Payables functionality to match goods/services received to POs and/or invoices for those goods/services; if matching is unsuccessful, matching holds are placed on the invoices. **Figure 8.7** shows the Invoices window with the matching functionality showing the document types to which the invoice is matched.

Figure 8.7—Matching

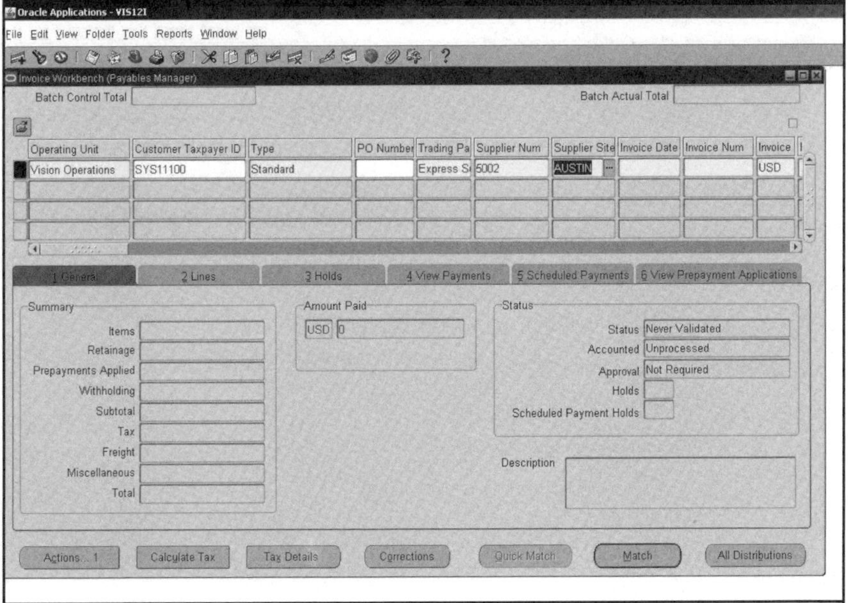

Oracle Applications - VIS12I
File Edit View Folder Tools Reports Window Help
Invoice Workbench (Payables Manager)
Batch Control Total
Batch Actual Total

Operating Unit	Customer Taxpayer ID	Type	PO Number	Trading Pa	Supplier Num	Supplier Site	Invoice Date	Invoice Num	Invoice
Vision Operations	SYS11100	Standard		Express S	5002	AUSTIN			USD

1 General | 2 Lines | 3 Holds | 4 View Payments | 5 Scheduled Payments | 6 View Prepayment Applications

Summary
Items
Retainage
Prepayments Applied
Withholding
Subtotal
Tax
Freight
Miscellaneous
Total

Amount Paid
USD 0

Status
Status Never Validated
Accounted Unprocessed
Approval Not Required
Holds
Scheduled Payment Holds

Description

Actions... 1 | Calculate Tax | Tax Details | Corrections | Quick Match | Match | All Distributions

3.2.1	Ask management for documentation on policies, procedures, standards and guidance related to the identification and development of required fields, QuickPick lists and default values. Oracle EBS utilizes several online edits and validation tools, including required fields (Oracle EBS-defined and user-defined), QuickPick functionality and default field values. These tools facilitate the accurate recording of journal entries and chart of account updates. Transactions cannot be processed if a required field is blank, e.g., the date field in the Journal Entry form. QuickPick functionality provides pop-up windows that contain predefined options from which the user can choose. Fields that generally require the same input can be defined to contain default values, so input is not required. Review the setting by navigating to the properties of a required field through the path Help➔Diagnostics➔Properties➔Item, and confirm that the null value, which defines that field as required, is set to False. A password is required to view the attributes of a field. The preloaded password is "apps." If this password has not been changed, the security of the ledger system may be compromised. Request the password to perform the test if the client has changed the password. **Figure 8.8** shows the properties of item Name in the Suppliers function.

Figure 8.8—Online Edit and Validation

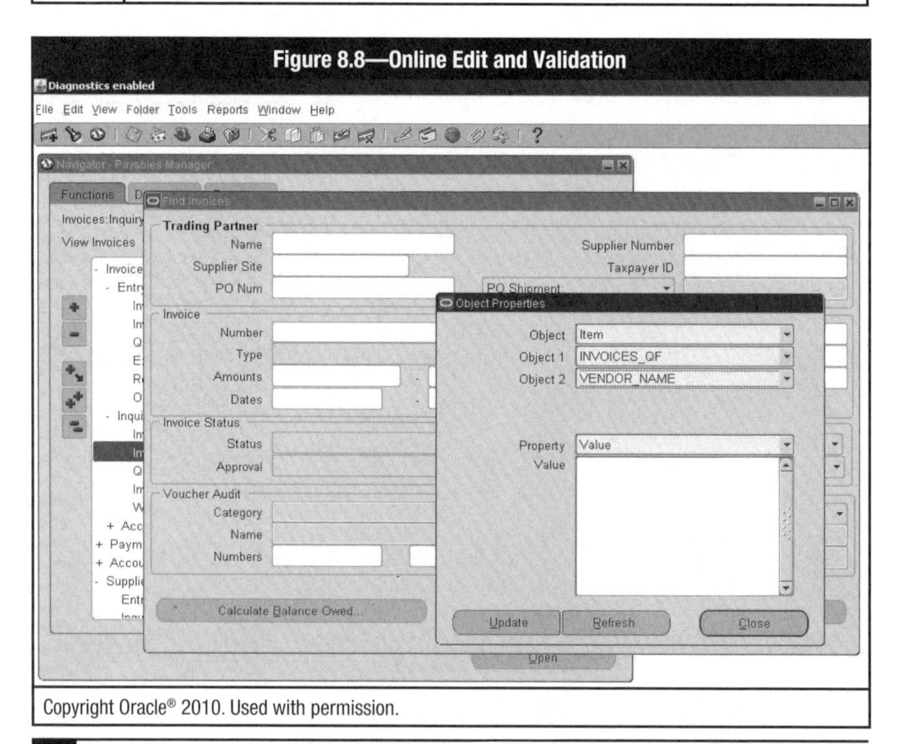

Copyright Oracle® 2010. Used with permission.

3.2.2	Use the Distribution Set Listing to review a complete list of all the Distribution Sets that have been defined by the client. This report can be run using the Reports window (Reports➔Run).
3.2.3	Ask management for the following documentation: • Policies, procedures, standards and guidance related to reviews of payables data entry • Policies, procedures, standards and guidance related to review of payment batches Various standard reports are available in Oracle EBS to facilitate the accurate processing of supplier invoices and credit memos, including: • The Expense Export Report details expense distribution information for all posted and unposted invoices that include three segments of the specified account (e.g., company, cost center and account). If the Automatic Offsets Payables option is enabled, the report will also detail the offsetting AP liability accounts for each of the invoice distributions. • The Invoice Audit Listing assists in auditing invoices for duplicates and can be sorted in various ways. For example, the report will list only invoices over a specified scope, i.e., US $1,000, and sort them by amount, supplier name and date. In addition, this report can be used to obtain a listing of invoices by invoice type, i.e., expense report invoices or prepayments. • The Invoice Audit Report can also be used to audit invoices for duplicates. This report lists invoices that appear as potential duplicates according to several criteria. One criterion that may be specified is the number of characters in the invoice number that two or more invoices have in common. The report lists invoices that meet this criterion and have the same invoice amount, invoice date and supplier. The search can be limited by checking for duplicate invoices within a specified time period. Oracle Payables sorts the report alphabetically by supplier name and lists possible duplicates together on consecutive lines. • The Invoice on Hold report provides detailed information about invoices on hold. To obtain additional detail for invoices on matching hold, obtain the Matching Hold Detail report (note that Oracle Purchasing must be installed and implemented before this report can be obtained). The Invoice on Hold report is divided into three sections: 1. The first section details the supplier, invoice, PO, amount and hold information for each invoice on hold, and provides subtotals by supplier or by hold code. The subtotal information includes the number of invoices, total original amount and total amount remaining for the invoices on hold. The report does not include canceled invoices.

3.2.3 *(cont.)*	2. The second section, invoice hold code descriptions, lists all predefined and user-defined hold codes that are in the first section of the report, descriptions of each code, and whether the hold allows posting. 3. The third section of the report, invoices with no exchange rates, lists foreign currency invoices without exchange rates. • If payables are valued in different currencies, the Open Items Revaluation report will revalue open items, e.g., invoices, prepayments, and credit/debit memos that are not fully paid. This report takes into account the changes of the value due to changes in foreign currency rates and provides three important figures: 1. Evidence of completeness, by listing the value of open items before the revaluation, which is reconcilable to the GL balances, and a complete list of all items that are included in that balance 2. The value for each open item revalued with the end-of-period rate, a field that is required to attain valid results (the total of these itemized values is needed in countries such as the US) 3. On an item-by-item basis, the higher of two values is determined—the open item before and after the revaluation, as well as the total of the higher of these values and the difference between the two (this total is used in countries such as Germany, where the higher market value of open items needs to be determined). The result of this report is used to determine the amount needed to manually adjust the GL balance to reflect the difference between the original balance and revalued balance for each liability Accounting Flexfield and for each balancing segment value. • The Invoice Approval report is used to review the total number of matching and variance holds that Oracle Payables applies and releases after approval is submitted. If Budgetary Control is used, Oracle Payables also lists any funds' control holds. Note that the same invoice may be counted in more than one category, or even twice in the same category, because each invoice can be matched to more than one PO line and more than one match rule can be violated. The same invoice can also have one or more matching holds, a tax and distribution variance hold, and a funds control hold. • The Credit Memo Matching Report lists credit/debit memos that match the specified supplier and date parameters. The report lists the following for each credit memo's distribution lines: – The distribution line amount in functional currency – The distribution line GL date – The invoice to which it is matched – Any exchange rate information

3.2.3 *(cont.)*	The report also lists the total of the distribution line amounts of each credit memo in the entered currency and functional currency. It also lists total credit memo amounts for each supplier and a total amount for the report. • The Uninvoiced Receipts Report should be run before the receipt accrual period-end process. This report shows all or specific uninvoiced receipts for period-end and online accruals. Uninvoiced receipts are goods and services that have been received, but that the supplier has not yet invoiced. This report indicates which invoices need to be accrued and for what amount, and assists in analyzing receipt accrual entries. • The Unordered Receipts Report lists all or selected unordered receipts, which are received items that could not be matched to POs. These types of transactions can be entered only if the system options are set up accordingly. Ask management whether there is a procedure for reviewing these reports. Ascertain that these reports are reviewed on a regular basis and that exceptions are identified for follow-up and resolution. Inspect physical evidence that the reports are run, approved and retained and that exceptions are resolved, as necessary. The reports can be run using the Reports window (Reports➔Run).
3.2.4	Ask management as to the procedures for the regular reconciliation of key/large supplier statements to the AP subledger. Inspect the procedures documentation and a sample of reconciliation documentation for the period under review for evidence of performance.
3.2.5	Check tolerance limits for price variances and message settings for invoice verification (online matching) as follows: • Navigate to the Purchasing Options window (Setup➔Organizations➔Purchasing Options➔Document Control section). • Check that the Price Tolerance % field is populated and the Enforce Price Tolerance box is checked. • Mark all the other boxes in accordance with organizational policy. **Figure 8.9** shows the Purchasing Options Document Control section.

Figure 8.9—Tolerance Rules

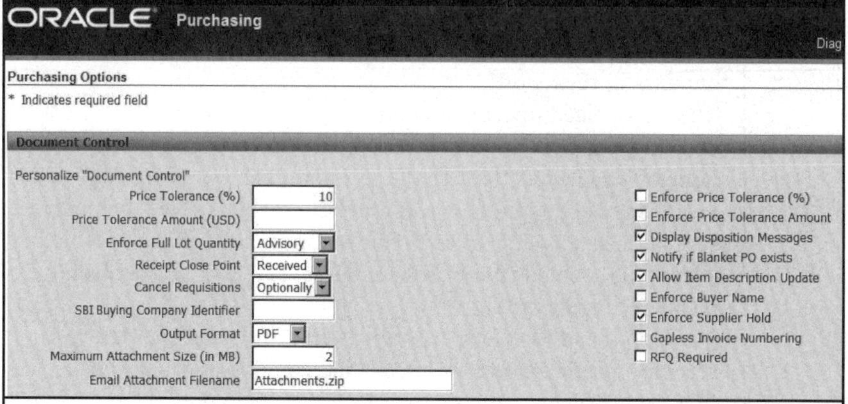

ORACLE Purchasing

Diag

Purchasing Options

* Indicates required field

Document Control

Personalize "Document Control"

Price Tolerance (%)	10
Price Tolerance Amount (USD)	
Enforce Full Lot Quantity	Advisory
Receipt Close Point	Received
Cancel Requisitions	Optionally
SBI Buying Company Identifier	
Output Format	PDF
Maximum Attachment Size (in MB)	2
Email Attachment Filename	Attachments.zip

- [ ] Enforce Price Tolerance (%)
- [ ] Enforce Price Tolerance Amount
- [x] Display Disposition Messages
- [x] Notify if Blanket PO exists
- [x] Allow Item Description Update
- [ ] Enforce Buyer Name
- [x] Enforce Supplier Hold
- [ ] Gapless Invoice Numbering
- [ ] RFQ Required

3.2.6	Entering invoices in the Oracle Payables module is a separate transaction from posting them. If the period status is updated to Closed prior to posting all invoices, Oracle Payables does not close the period and submits the Unposted Invoices report, which displays all unposted invoices for the period. Review this report to determine the invoices that do not belong to the period and that should be posted. This situation may occur when, for example, invoices have been received and processed for February, although January has not yet been closed. The Unposted Invoice Sweep program is used to transfer all unposted invoices and payments from one accounting period (January) to another (February). Once the Oracle Payables module is closed, post transactions to the GL. Prior to the execution of the Payables transfer to the GL program, review the Posting Hold report, which lists all invoices that will not be transferred, including those with posting holds, distribution variances, Budgetary Control holds and invoices not yet approved. The report is sorted by supplier name and invoice number, and displays the total number of invoices, total invoice amount and total distribution amount for each data entry person. Review policies, procedures, standards and guidance related to the closing of Oracle Payables. Make a selection of Invoice Sweep reports and Posting Hold reports throughout the period of intended reliance, and determine that management has reviewed the reports prior to closing the period. For items that have been marked as requiring action, ensure that such action has been completed in accordance with management's instructions.

3.2.7	Extract a list of the target high-risk functions/forms and review it to ascertain who has access privileges to the following functions: • Conversion Rate Types • Currencies • Daily Rates • Exchange Rate Adjustment
3.3.1	Review the access to the functions/forms for invoicing and credit/debit memos listed in testing technique 3.1.1 to confirm that access is appropriately restricted. Credit/debit notes are invoice types and should be restricted in the same manner.
3.3.2	Ask management for documentation on payment policies, procedures, standards and guidance for credit/debit notes and matching to invoices. Matching settings should be checked as follows: • Navigate to the Purchasing Options window (Setup➔Organizations➔Purchasing Options Document Defaults section). • Check the Match Approval Level setting. **Figure 8.10** shows the Purchasing Options Document Defaults section. Examine the Credit Memo Matching report and verify that it is used to check on the adjustments made to the invoices. This report lists credit/debit memos that match the supplier and date parameters as specified. The report lists the following for each credit memo's distribution lines: • The distribution line amount in functional currency • The distribution line GL date • The invoice to which it is matched • Any exchange rate information

Figure 8.10—Match Approval Level

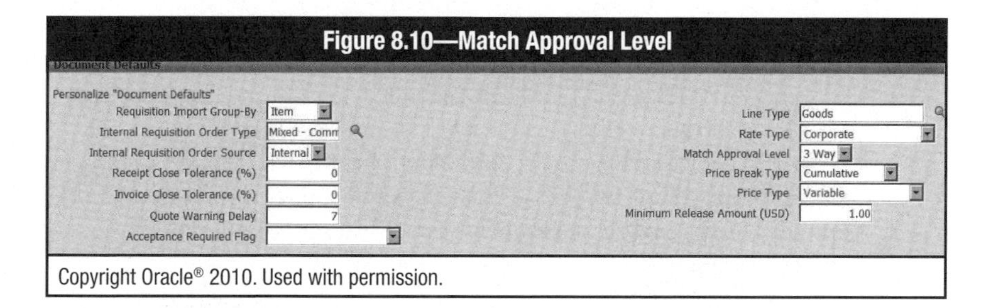

3.3.3	The Account Analysis report may be used to identify nonsystem debits to AP. This report is used to review and analyze accounting entries in Oracle Payables. The information required can be specified. Ask management whether there is a procedure for reviewing this report. Ascertain that it is reviewed on a regular basis and that exceptions are identified for follow-up and resolution. Inspect physical evidence that the report is run, approved and retained and that exceptions are resolved, as necessary.

Processing Disbursements

Processing Disbursements: Risks

Significant risks associated with processing disbursements include:

4.1	Disbursements may not be made only for goods and services received, or may not be calculated, recorded and distributed accurately or to the appropriate suppliers in a timely manner. For example: • Unauthorized payments may be made to fictitious parties if disbursements are not made to the appropriate suppliers, and such errors may not be detected. • Incorrect disbursements may result in suppliers being paid incorrect amounts, not being paid, being paid before or after due dates, or being paid for goods or services that have not been received. Accurate calculation of disbursements includes those for available discounts. Cash flow decisions could also be affected, if data on which these decisions are made are incorrect. • Disbursements that are not recorded may affect cash flow decisions and cause reconciliation difficulties. Nonrecording of disbursements may also result in duplicate payments.

Processing Disbursements: Key Controls

Key controls over processing disbursements include:

4.1.1	Access controls over payments, accounting periods and holds in the Oracle Accounts Payable module should be restricted appropriately. Oracle EBS R12.1 restricts to authorized personnel the ability to modify the payment run parameter specification or to initiate a payment run.
4.1.2	Management should review supporting documentation before approving payments.
4.1.3	Standard Oracle Reports should be reviewed periodically by management to ensure the accuracy of payables processing.

4.1.4	Management should periodically review returned paid checks for unauthorized signatures, alterations and/or endorsements.
4.1.5	Oracle EBS is configured to generate unique, sequential numbers for all payments.
4.1.6	Management should review an aged AP analysis and investigate any unusual items.
4.1.7	Adequate controls should be in place to restrict access to the drives and directories where the system generates the electronic payment flat files. The process for archiving and purging these files should be reviewed. Controls should be in place so users do not have read-and-write access to these files, except for Oracle EBS system-generated processes.

Processing Disbursements: Testing Techniques

Testing techniques for processing disbursements controls include:

4.1.1	For processing disbursements in the Expenditure business cycle, review the following access levels to ensure that only appropriate and authorized personnel have access to: • Create, modify and void/release payments and payment batches. • Modify open and closed periods in the Accounting Calendar. • Release holds on payments. Extract a list of the target high-risk functions/forms and review it to ascertain who has access privileges to the following functions: • AP Accounting Periods • Calendars, Define Periods • Close Accounts Payable Period • Open and Close Periods • Payment Accounting • Payment Actions • Payment Batch Sets • Payment Batches • Payment Batches Actions • Payment Batches Cancel • Payment Batches Confirm • Payment Batches Format • Payment Batches Modify • Payment Batches Positive Pay • Payment Batches Print • Payment Batches Summary • Payment Batches to Payment • Payment Formats

4.1.1 *(cont.)*	• Payment Invoice Rates • Payment Invoices • Payment Overview • Payment Print Check • Payment Programs • Payment Void • Payments • Payments Summary • Period Close—Parent Company • Period Close—Subsidiary Company • Period Types
4.1.2	Ask management about the procedures for the review of supporting documentation before approving or canceling payments. Inspect the procedures documentation and a sample of canceled supporting documentation for the period under review for evidence of performance.
4.1.3	Various standard reports are available in Oracle EBS to facilitate the accurate processing of payments, including: • The Cash Requirement Report is used to forecast cash needs for invoice payments for any number of future payment batches. The report may be obtained for a single currency or for all currencies, and provides an alert if any invoices in the currency are missing exchange rates. The report lists unpaid or partially paid invoices that match specified parameters and does not include canceled or fully paid invoices. • The Discounts Available report identifies payments where advantageous discounts may be taken and should be run prior to processing a payment batch. • The Discounts Taken and Lost report identifies payments for which discounts could have been taken but were not. Based on review of this report, if it is determined that discounts are generally lost, system and supplier defaults can be changed and payment batch selection criteria modified to ensure that all valid discounts are utilized. • The Payment Register is printed when a user initiates, formats or modifies payments for the invoices in the payment batch and lists the number of documents set up, nonpayment documents (documents that exceed the maximum payment amount or are below the minimum payment amount, zero payments allowed, zero invoices allowed, or payment documents that were deselected while making modifications to the payment batch), overflow documents and negotiable documents. It also prints the total outlay required for the payment batch.

4.1.3 *(cont.)*	• The Batch Control by Batch Name Report shows detailed information about the payments printed in one or more payment batches and may be submitted to the check signer to show comprehensive information about each payment document and eliminate the need to review paper invoices. The report displays payment, supplier, invoice and expense distribution information for all payment documents in a payment batch. Some clients may use a third party or custom Positive Pay program to notify the bank of negotiable and nonnegotiable checks. Review and approve the Payment Register prior to the release of payments, to ensure that the following are identified, investigated and resolved in a timely manner before a payment is released: • Unusual payment amounts to valid suppliers • Payments made to invalid or unusual suppliers • Expected or emergency payments not on the listing • Payments made prior to the required date Ask management whether there is a procedure for reviewing these reports. Ascertain that these reports are reviewed on a regular basis and that exceptions are identified for follow-up and resolution. Inspect physical evidence that the reports are run, approved and retained and that exceptions are resolved, as necessary. The reports can be run using the Reports window (Reports➔Run).
4.1.4	Ask management about the procedures for the periodic review of returned paid checks for unauthorized signatures, alterations and/or endorsements. Inspect the procedures documentation and a sample of returned paid checks for the period under review for evidence of performance.
4.1.5	Ask management for policies, procedures, standards and guidance related to defining, creating and modifying sequential numbering rules. There are various reports available to help manage the sequential integrity of payment documents, including: • The Payables Posted Payments Register lists each payment created for a payment batch, including setup and overflow payment documents, in ascending order by payment number.

4.1.5 *(cont.)*	• The Audit Report by Document Number lists payments with assigned sequential document numbers and helps identify missing document sequence numbers. If the Sequential Numbering Profile Option is enabled, a unique, sequential number is assigned (systematically or by the user) to each payment. This report can also be used to review assigned and available document numbers for the specified sequence name, as well as sequential numbers that have been deleted. Ask management whether there is a procedure for reviewing these reports. Ascertain that these reports are reviewed on a regular basis and that exceptions are identified for follow-up and resolution. Inspect physical evidence that the reports are run, approved and retained and that exceptions are resolved, as necessary.
4.1.6	Use the standard Invoice Aging report to review an aged AP analysis. List all unpaid invoices. This report provides information about invoice payments due within four time periods that can be specified. Ask management whether this report is reviewed on a regular basis and that exceptions are identified for follow-up and resolution. Inspect physical evidence that the report is run, approved and retained and that exceptions are resolved, as necessary.
4.1.7	List the users who have access to the drives and directories where the pay files are located. Review the process for archiving and purging these files. Users should not have write access to these files, and read access should be restricted.

Summary

This chapter outlined the risks, key controls and testing techniques surrounding the Oracle EBS R12.1 expenditure business cycle. Among the key risks were setup of master data and expenditure-related functions; access to supplier master data; and inappropriate access to approve requisitions, POs, blocked invoices or payment runs.

9. Oracle E-Business Suite Security

The security of an ERP system is a major consideration in any audit. Without appropriate security there is a potential risk for the financial data residing in the ERP system to be compromised.

While this guide focuses on the risks and controls in relation to Oracle EBS Security, weaknesses in the other interrelated components, such as database, operating system (O/S) and network level security, could compromise the security implemented at the application level.

The components of Oracle EBS security that need to be considered are depicted in **figure 9.1**. The following sections of this chapter will examine each component in more detail, together with the audit functions built into Oracle EBS:

- Components of Oracle EBS and underlying database and infrastructure security
- Oracle EBS application security
  - User management
  - User security structure and configuration
- User and data auditing

Figure 9.1—ERP Audit Security Considerations

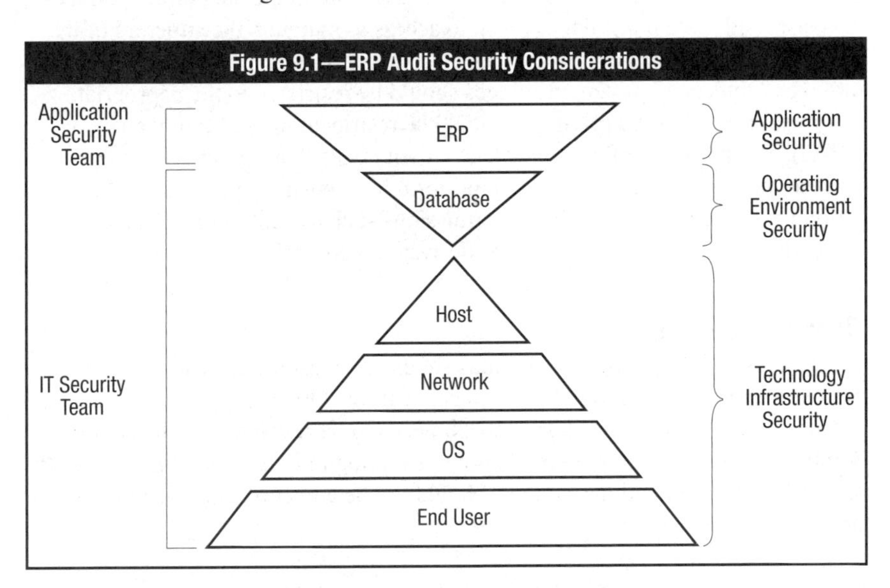

Components of Oracle EBS and Underlying Database and Infrastructure Security

This section describes the components of Oracle EBS and underlying database and infrastructure security.

User IDs and Roles

Business user access to the underlying data should be controlled and restricted through Oracle EBS Security. The primary concern will always be on the production client of Oracle EBS, but if there are sensitive data stored on Development or in the test environment, then access to these clients should be restricted.

IDs exist for gaining access to an Oracle EBS system application. It should be noted that individual business user IDs are not required at both the Oracle EBS and database layer for that user to be able to access EBS; in fact, normal business users should not have any direct database access since that provides opportunities for direct manipulation of application data, bypassing application-enabled validations and processing logic. However, it is not unheard of for business users to have read-only database access as well as a normal Oracle EBS application system ID assigned. It is highly recommended to limit SQL access to the database only to the database administrator (DBA). All end users should use approved reports for interfaces.

Object and System Privileges

Object and system privileges determine what a user can do in the Oracle Database. Object privileges relate to access to maintain the data directly in the database, and system privileges relate to access to maintain the structure of the database tables as well as users, database configuration and performance, batch jobs, etc. Object and system privileges should be restricted to the appropriate IDs/roles (i.e., sensitive privileges should be restricted to application and DBAs). The principle of least privilege recommends that accounts have the least amount of privilege required to perform their business processes. This encompasses user rights, resource permissions such as central processing unit (CPU) limits, memory, network and file system permissions.

Operating System

User access and server file permissions should be appropriate. The operating system command line allows users to access all the Oracle Applications start-up profiles and configuration files; hence, it needs to be restricted. In the Oracle Applications environment, users do not need to log in to the operating system to access Oracle Applications, so there should not be any application user accounts on the operating system.

Access to the following executables should be restricted:
- dbshut
- dbstart
- lsnrctl
- orapwd
- svrmgrl
- svrmgrm

- tnslsnr
- tnsping
- trcroute
- tstshm
- sqlplus
- sqlload
- srvctl

The most common operating system seen with Oracle EBS is UNIX. UNIX system administrators should have access; however, ROOT authorization should be restricted to authorized personnel, and the ROOT password must be kept confidential. Default administrators such as SYSADMIN should have passwords changed immediately after installation.

File and directory permissions should be appropriately restricted for the Oracle EBS control files, log files and Oracle Database files on the operating system server. For example, the Oracle Database files are protected so only authorized accounts can read and write to these files.

There should be a clear segregation of duties between the DBA and the system administrator at the operating system level.

Backups and Restoration Testing

Backups should be performed on a scheduled/regular basis. Audit programs need to include review of a backup schedule, storage at an offsite location, and backup release and retention documentation.

Physical/Environmental Security

Servers should be located in a physically secure environment. Environmental controls such as temperature/humidity monitors, cooling and fire suppressants should also be in place. Policies and procedures should be defined and enforced for each of the security components.

Network Security

Network security can be controlled via one or a combination of controls depending on the environment. A list of common network security controls are:

- **Secure Sockets Layer (SSL)**—For Hypertext Transmission Protocol (HTTP) connections, not the thin clients, SSL secures communication between the client and server. The SSL protocol provides data encryption, server authentication, message integrity and optional client authentication for a Transmission Control Protocol (TCP)/Internet Protocol (IP) connection. SSL is built into all major browsers and web servers; therefore, installing a digital certificate turns on their SSL capabilities. SSL needs to be configured and enabled as a part of the implementation.

- **Encryption of password**—Encryption is the process of coding or scrambling the data before transmission over the network.
- **Session expiration**—Session management features include:
  - Each session is assigned a unique identifier, which is stored in a table.
  - Session identifiers are encrypted and returned to the client via cookies.
  - Session expiration is based on the number of hours or number of hits.

Oracle EBS Application Security

User Management

Application Users

A user can sign on to Oracle EBS only after having been set up as an application user. The application user is assigned a unique username and an initial password. The Oracle EBS login connects the user to his/her responsibilities, which controls access to the menus and reports within a specified Oracle Database.

A user can be assigned with a single role or multiple roles. Depending on what role is assigned and how the role has been configured, the user with a single role will enter the application assigned in the role immediately upon signing on, while a user with two or more roles or with a role assigned to more than one responsibility will bring up a window listing the available responsibilities.

User accounts cannot be deleted once they are created. To remove a user's access, the account needs to be deactivated. The following subsections explain how user accounts are deactivated.

As shown in **figure 9.2**, there are three layers of user administration in Oracle EBS. These are Self Service and Approvals, Provisioning Services, and Delegated Administration.

Figure 9.2—User Administration

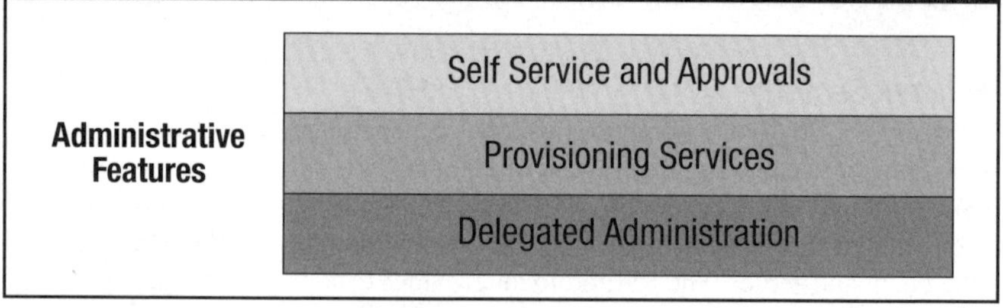

Administrative Features	Self Service and Approvals
	Provisioning Services
	Delegated Administration

Self Service and Approvals

As mentioned previously, Self Service Requests is the method for users to obtain new user accounts and request additional access. Enterprises can also use the Oracle Approvals Management engine to create customized approval routing for such requests. For example, an enterprise may enable users to request a particularly sensitive role and, prior to the user being granted access to the role, the enterprise can require that two senior staff members approve the request.

Provisioning Services

Provisioning Services are modeled as registration processes that enable end users to perform some of their own registration tasks, such as requesting new accounts or additional access to the system. They also provide administrators with a more efficient method of creating new user accounts, as well as assigning roles. There are essentially two provisioning processes available—Traditional and Self Service:

- **Traditional**—Account Creation by Administrators: Administrators benefit from registration processes having been designed to streamline the process of creating and maintaining user access. Each account creation registration process can be made available to selected administrators.
- **Self Service:**
  - Self Service Account Requests—Provides a method for individuals to request a new user account
  - Requests for Additional Access—Users can also request additional access through the Oracle User Management Access Request Tool (ART), available in the Global Preferences menu.

Delegated Administration

Delegated Administration is a privilege model that builds on the Role-based Access Control (RBAC) system to provide enterprises with the ability to assign the required access rights for managing roles and user accounts. Instead of relying on a central administrator to manage all of its users, an enterprise can create local administrators and grant them sufficient privileges to manage a specific subset of the enterprise's users and roles.

Administration Privileges determine the users, roles and organization information that delegated (local) administrators can manage. Although each privilege is granted separately, the three work together to provide the complete set of abilities for the delegated administrator:

- **User Administration Privileges**—Required to determine the users and people who the local administrator is able to manage. Local administrators can be granted different privileges for different subsets of users.
- **Role Administration Privileges**—Define the roles that local administrators can directly assign to and remove from the set of users they manage.
- **Organization Administration Privileges**—Define the external organizations that a local administrator can view in Oracle User Management.

User Creation Window

Figure 9.3 shows the options that an administrator has to define roles and update users via the User Management Page. **Figure 9.4** shows the search option to find a person on the system to either add a user account or update an account if one is already assigned.

Figure 9.3—User Maintenance

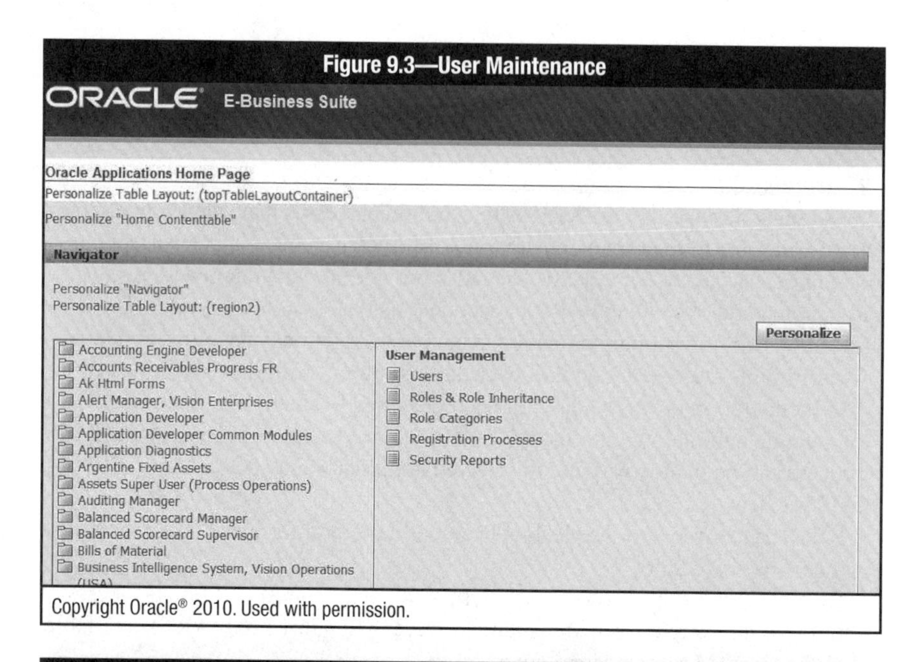

Figure 9.4—Search Option

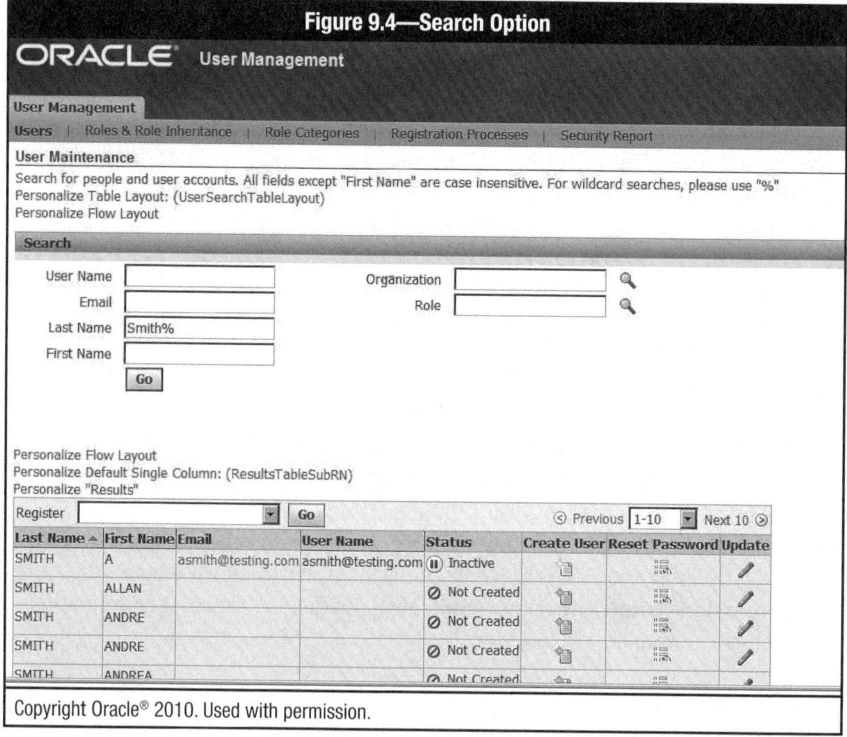

Note: To create a new user, the user must first be a person/employee in the system.

The key areas within the User window are described as follows:

- **User Name field**—Contains the name of the application user. Naming conventions should be used, such as the user's first initial followed by surname. Naming rules include the requirements that the username:
  - Must be one word
  - Can contain only alphanumeric characters (A through Z, 0 through 9). All other characters are invalid.
  - Should be limited to the set of characters that the operating system supports for file names

  The Description field should be populated with the user's real name and job title/role, location, or entity description to make the review and maintenance of user accounts effective and efficient. The Person, Customer and Supplier fields should be populated with the employee (person), customer or supplier contact's last and first names (separated by a comma).
- **Password area**—An initial password should be set up when defining a user. Users will be forced to change the password the first time they sign on. Password expiration can be set using one of three options: days, accesses or none. Password options in R12 also include complexity because the passwords can be set as hard-to-guess.
- **Effective dates**—This area allows a time frame to be set for the user accounts to be active. The default value for the start date is the current date. If no end date is entered, the user account is valid indefinitely. Users are deactivated by setting the end date to the current date. Users can be reactivated by changing the end date to a date after the current date or clearing the end date.

Figure 9.5 shows the new user screen.

Figure 9.5—New User

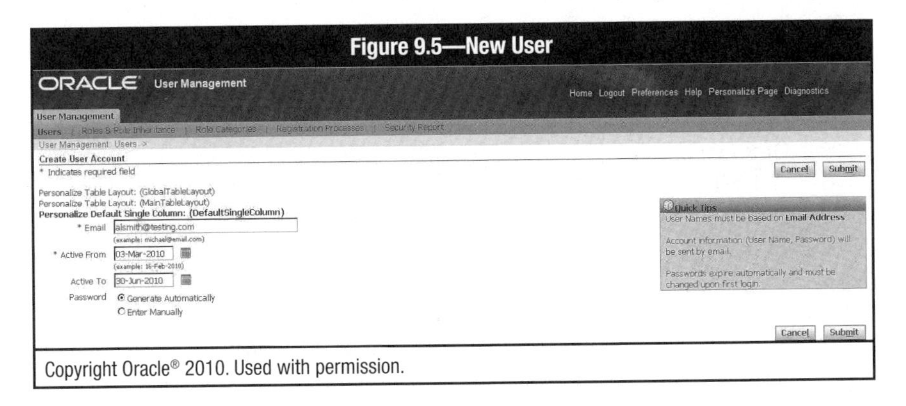

Figure 9.6 shows the user update screen where roles can be added to the user.

Figure 9.6—Add User Roles

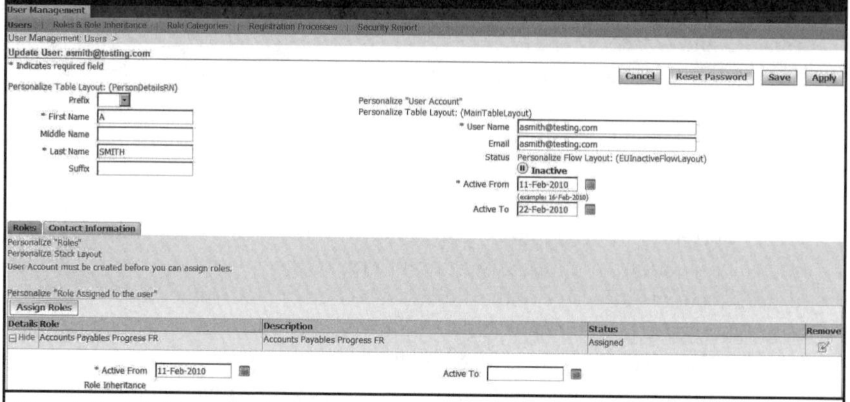

Copyright Oracle® 2010. Used with permission.

- **Roles**—The set of responsibilities that relate to particular access rights in Oracle EBS, usually aligning to the user's job role. They are defined in the next section of this chapter. The roles are added by searching for the predefined roles established in Oracle EBS and then assigning one or more roles to the user. A justification for why the user is being assigned the role must also be entered.

  Roles assigned to a user account may be valid for only a specified time period. The From/To field allows the start and end dates for a role assigned to a user to be specified. Roles assigned to user accounts can be removed from the user, but only the system administrator can modify the role.
- **Securing Attributes tab**—The following abbreviations are used in **figure 9.7** and in figures in upcoming sections:
  - R = Required
  - O = Optional
  - D = Default
  - Q = QuickPick (i.e., whether the setting can be selected from a QuickPick selection list)

The key fields shown in **figure 9.7** relate to the definition of configurations for a particular application user in Oracle EBS. User details are usually entered by HR with the user account being set up by system administration or through provision or Self Service options.

Figure 9.7—Application User Definition				
Field Name	**Description**	**Status**	**Recommended Setting**	**Comments**
User Name	Name of application user	R	Use the same username as the LAN ID (e.g., first and second initials followed by surname with no spaces, e.g., jxsmith).	The username must be one word; it can use only alphanumeric characters and must be unique.
Password	Initial password for user	R	Use a word that is familiar to the user, but not obvious (e.g., username), and is at least five characters long (maximum is 100 characters). It can be alphanumeric. The form does not display the password on the screen. If the password complexity function has been enabled, the initial password needs to meet those requirements. Refer to chapter 4 for Oracle EBS default password rules.	If the user already exists and the two passwords required do not match, a warning message is displayed and the user is asked to reenter the password. SYSADMIN cannot access a user's password, but can overwrite it with a new one (e.g., if the user forgets). Users will be prompted to change it the first time they log in.
Description	Description of the application user	O	Recommend entry of: • Full name of user • User Job/position title • Purpose of account • Whether it is a temporary or permanent account	Provides useful business function information to the security administrator at a later date (can be defined and obtained from the Administration Access Request form)
Password Expiration	Maximum number of days or sign-ons allowed between password changes	O	Choose the Days button, and set 30 days between password changes.	A pop-up window will prompt the user to alter the password after the number of days specified has elapsed. The default setting of None provides poor security because the system will never require the user to change the password.

Figure 9.7—Application User Definition *(cont.)*				
Field Name	**Description**	**Status**	**Recommended Setting**	**Comments**
Person	Employee name	O	Choose LastName, FirstName.	Information can be entered in this field only if employees have already been defined. This would be the case if either the HR, PO or ICM, or optionally the AP (expense reports), GL (Journal Approvals) and AR (adjustment approvals), modules are being used. This field must be completed for many of the approvals and notifications to work.
Customer	Customer name	O		If the user is already created in the customer master, the customer name can be selected from the list of values.
Supplier	Supplier name	O		If the user is already created in the supplier master, the supplier name can be selected from the list of values.
E-mail	The user's e-mail address	O	An e-mail address is required for workflow notifications and alerts. It will default from employee information, but it can also be entered/altered manually in this form.	
Fax	The user's fax number	O	The fax number is not needed at present.	This may be relevant for future or third-party applications not currently used.
Effective Start Date	Date on which the user account becomes active	R, D	Enter the date on which the user will require access (no earlier).	The default value is the current date.

Figure 9.7—Application User Definition *(cont.)*				
Field Name	**Description**	**Status**	**Recommended Setting**	**Comments**
Effective End Date	Date on which the user can no longer sign on to Oracle EBS	O	Provide an end date to disable a user account. No entry in this field assumes that the account is valid indefinitely.	Disabling a user account by placing an end date in this field is preferable to user account deletion (since it helps to provide audit trail information). To reactivate a user, change/clear the End Date field.

User Security Structure and Configuration

Security within Oracle EBS is defined at four key access levels:

- Roles
- Responsibilities
- Menus or Functions, Data Groups and Request Groups
- Flexfields

As part of application security, the users' responsibilities should be assigned appropriately. As mentioned earlier, a user can have a number of responsibilities assigned to his/her username. The responsibilities determine which application functions the user can access. The user must enter his/her username and password and select the appropriate responsibility to gain particular access to the system. This responsibility setup is a function of the security administration and should be dealt with carefully to ensure that users are receiving appropriate access to the system. Since the release of Oracle EBS R12, EBS security has three levels: RBAC, Data Security and Function Security (Menu Level), as shown in **figure 9.8**.

Figure 9.8—User Access Structure

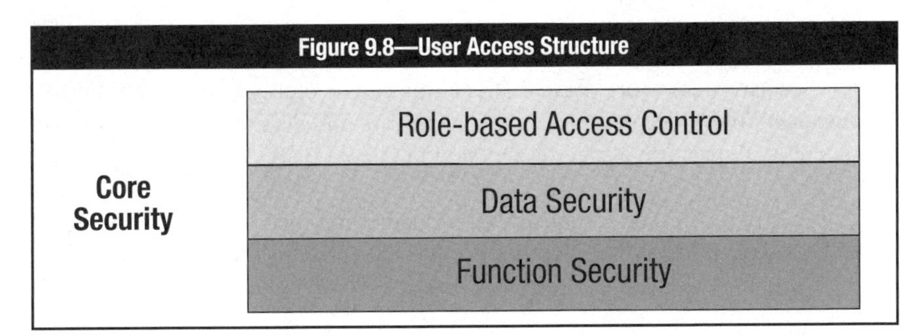

To ensure that security can be adapted easily to changes in the organizational structure and business needs, menu and function security should be designed in a modular and generic fashion. In addition, RBAC individual user privileges are less common, with users no longer needing to be directly assigned the lower-level permissions and responsibilities since these can be implicitly inherited based on the roles assigned to the user. This modularity enables flexible and controlled building of access capabilities, tailored to specific user requirements. This generic nature increases reusability of the security elements, thereby improving administrative cost effectiveness.

To ensure that the previously mentioned security components, and more, have been administered adequately, it is important to consider:

- How users are set up, modified and assigned
- The authorization process in place for change control (in particular, to allow specific modifications to users' roles and the underlying responsibilities)

The security administrator has a vital role in maintaining the security setup in Oracle EBS. By administering the setup of users within the system, the administrator helps ensure that the confidentiality and integrity of the system are not compromised. This role should be performed only by the administrator or a select handful of competently trained IT staff. The security administrator's role includes creating user profiles, grouping users into roles, and modifying and assigning roles and permissions lists.

The security administrator should be responsible for security administration in both the nonproduction and production instances of Oracle EBS. In addition, any development of initial security authorizations is the responsibility of the security administrator.

The data guardians define/approve the overall approach to security, providing role descriptions and associated access requirements. They also approve the addition of application user accounts and responsibilities on an ongoing basis.

Role-based Security

As mentioned previously, the introduction of roles via RBAC is on top of the current security framework. Essentially, a role can be configured to consolidate the responsibilities, permissions, function security and data security policies that users require to perform a specific function in Oracle EBS.

Roles can group common responsibilities together into one role, e.g., "AR clerk" could be assigned to any AR clerk user in the system. This provides easier and more central management of user access assignment in Oracle EBS.

Setting Up Roles

Apart from user/profile creation and/or modifications, the role of the system administrator also includes designing roles, as well as the underlying responsibilities that are assigned to the role. Oracle EBS R12.1 through RBAC allows users to be assigned access to many system functions via a single Role. A role can align to a job function and be assigned to any users who perform the function, simplifying the user administration process.

The use of roles has allowed enterprises to move away from the previous approach of assigning system functions and responsibilities individually to users. Responsibilities now specifically refer to the Functional security level within Oracle EBS, whereby a responsibility basically allows a user access to an EBS function via either allowing or denying a user to view certain menus and menu paths within EBS.

Overall Role Settings

The same role or roles can be assigned to a number of application users, and one application user can have more than one role. Examples of the different relationships include:

- **One user, one role**—An AP clerk may be assigned to a role that limits access rights to entering invoices. Another role may include the ability to make check payments or invoice adjustments and be assigned to the AP supervisor.
- **One user, multiple roles**—The AP clerk may also be given a GL inquiry role that allows inquiry about GL account balances.
- **One role, multiple users**—Several AP clerks may require invoice entry functionality (defined within the same role).

The system administrator is normally required to set up a role, although there are also predefined standard roles available in Oracle EBS.

Roles are set up in Oracle EBS within categories or groups (such as Security Administration) to make it easier to manage the many roles in the system. Roles can be assigned privileges/system functions and responsibilities.

Figure 9.9 shows the Update page for a System Administrator Role.

Figure 9.9—Update Role

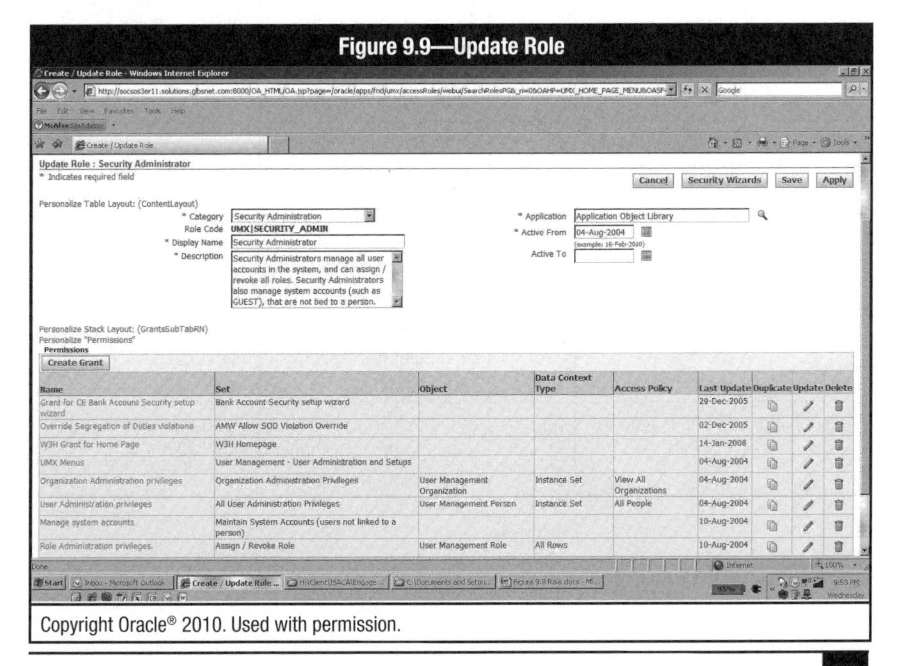

Name	Set	Object	Data Context Type	Access Policy	Last Update	Duplicate	Update	Delete
Grant for CE Bank Account Security setup wizard	Bank Account Security setup wizard				29-Dec-2005			
Override Segregation of Duties violations	AMW Allow SOD Violation Override				02-Dec-2005			
W3H Grant for Home Page	W3H Homepage				14-Jan-2008			
UMX Menus	User Management - User Administration and Setups				04-Aug-2004			
Organization Administration privileges	Organization Administration Privileges	User Management Organization	Instance Set	View All Organizations	04-Aug-2004			
User Administration privileges	All User Administration Privileges	User Management Person	Instance Set	All People	04-Aug-2004			
Manage system accounts	Maintain System Accounts (users not linked to a person)				10-Aug-2004			
Role Administration privileges.	Assign / Revoke Role	User Management Role	All Rows		10-Aug-2004			

Figure 9.10 shows an attached Grant to the role and the details it contains.

Figure 9.10— Role Privileges Grant

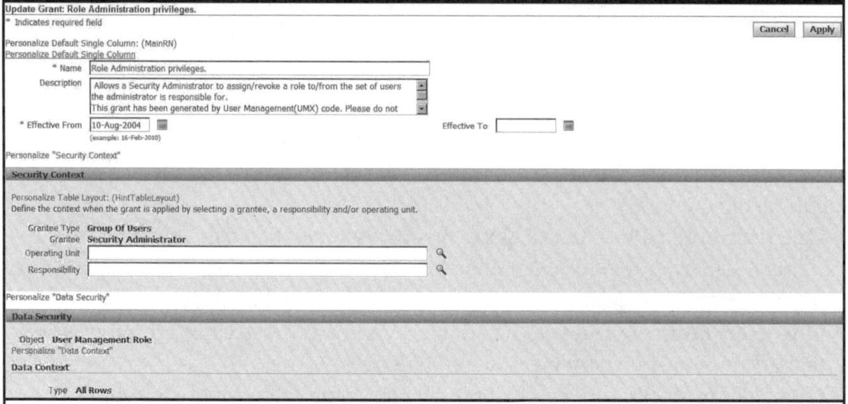

Update Grant: Role Administration privileges.
* Indicates required field
Cancel Apply
Personalize Default Single Column: (MainRN)
Personalize Default Single Column
* Name: Role Administration privileges.
Description: Allows a Security Administrator to assign/revoke a role to/from the set of users the administrator is responsible for. This grant has been generated by User Management(UMX) code. Please do not
* Effective From: 10-Aug-2004 (example: 16-Feb-2010)
Effective To:
Personalize "Security Context"
Security Context
Personalize Table Layout: (HintTableLayout)
Define the context when the grant is applied by selecting a grantee, a responsibility and/or operating unit.
Grantee Type: Group Of Users
Grantee: Security Administrator
Operating Unit
Responsibility
Personalize "Data Security"
Data Security
Object: User Management Role
Personalize "Data Context"
Data Context
Type: All Rows

The key fields shown in **figure 9.11** relate to the definition of a role when creating or editing one in Oracle EBS.

Figure 9.11—Defining Roles

Field Name	Description	Status	Recommended Setting	Comments
Category	Name of the category that the role will come under	R, Q	Dependent on the role type and the users it will be assigned to, e.g., System Administrator should be assigned a role under the category system administration	The categories established in the system to assign to the role will display in a drop-down list.
Role Code	The code that the role is given	R	Normally a short name or system-oriented name for the role	
Role Name	Name of the role	R	An easy-to-understand name that is assigned to the role	
Description	Description of the role	R	A long description explaining the reason behind the role	
Active From	Effective date for the role	R	Set Active From to the date that the role should be active in the system.	The default value is the current date.

Figure 9.11—Defining Roles *(cont.)*				
Field Name	**Description**	**Status**	**Recommended Setting**	**Comments**
Active To	Expiry date for the role	0	Set the expiry date if a role should be set up for a limited time.	The default value is blank.
Grant	Assigned grant	0	Set the various responsibilities and other system privileges to the role via grants.	

Responsibility Configuration

The following sections outline the field names, description and recommended settings in defining and configuring responsibilities.

Responsibilities are at the functional security level groupings of access privileges. A responsibility restricts the user's access to functions and data appropriate to the user's job role in the enterprise. As mentioned previously, the key change for responsibilities in Oracle EBS since the R12 release is that instead of assigning responsibilities directly to users, they are now assigned to roles.

Oracle EBS is installed with predefined responsibilities. These predefined responsibilities are often referred to as superuser responsibilities and typically grant the user broad access to application functionality. The predefined responsibilities are not consistent with segregation of duties, and the use of predefined responsibilities should be avoided.

The names of the predefined responsibilities are contained in the Oracle EBS product reference guide. Standard Oracle Responsibilities, which are preset in the Oracle EBS, can be assigned, or custom responsibilities can be created and tailored to the enterprise's needs.

Note: Access to superuser responsibilities (especially the system administrator responsibility) should be restricted since these responsibilities may confer the ability to change the application configuration if they include administration access or result in inadequate segregation of duties.

Recommended best practice is to define and use only custom responsibilities and roles. Default superuser responsibilities are only required for original setup and exceptional situations, not for daily operations.

Each responsibility provides access to:
- A specific application or applications (e.g., Assets or Payables)
- Ledgers/Ledger Sets of an enterprise (e.g., UK sales or Canadian operations)

- The Restricted Windows list in which a user can navigate (e.g., a responsibility may allow a purchasing user to enter Supplier Registration, but not Supplier Services)
- Reports in a specific application. Groups of reports or individual reports should be assigned to one or more responsibilities.

Defining a Responsibility

The key fields shown in **figure 9.12** relate to the definition of a responsibility when creating or editing one in Oracle EBS.

Figure 9.12—Defining Responsibilities

Field Name	Description	Status	Recommended Setting	Comments
Responsibility Name	Name of the Responsibility	R	Long-form, descriptive name	A pop-up window will appear at sign-on if the user has multiple responsibilities.
Application	Name of the application associated with the responsibility	R	Appropriate Oracle Applications module name, e.g., General Ledger, Oracle Payables	This name will appear in the left side of the title bar for every form that the user accesses. This does not restrict the responsibility from having menus, functions or reports for other applications.
Responsibility Key	Internal name of the responsibility	R	Short-form name for internal Oracle EBS use	Appropriate naming conventions must be followed.
Description	Description of the responsibility	O	Recommended description for future recall and for quality assurance (QA), audit and other documentation purposes	The business function owner's position could be entered for future reference/changes.
Effective Start Date	Date on which the responsibility becomes active	R, D	Enter date on which the user will require access (no earlier).	The default value is the current date.
Effective End Date	Date on which the responsibility is no longer active	O	Provide an end date to disable a responsibility. No entry in this field assumes that the responsibility is valid indefinitely.	Disable a responsibility by inserting an end date. Note: if a responsibility is end-dated in this form, it closes access to that responsibility for all users.

Figure 9.12—Defining Responsibilities *(cont.)*				
Field Name	**Description**	**Status**	**Recommended Setting**	**Comments**
Available From	Application system with which the responsibility is associated	R	Typically, either Oracle Self Service Web Applications or Oracle EBS	
Data Group	Data group with which the responsibility is associated	R	The default data group is Standard, which is used in most cases.	
Menu Name	Name of the main menu assigned to the responsibility	R	Menu already defined within Oracle Applications (e.g., ALR_OAM_WAV_GUI)	The main menu is the top-level menu that the user sees after selecting a responsibility.
Request Group Name	Name of the request (report) group assigned to the responsibility	0	Request group already defined within Oracle EBS	The request group is the set of reports to which the user has access within this responsibility.
Exclusion Type	Selects the type of menu or function exclusion to be applied to the responsibility	0	Select either Menu or Function.	This allows the administrator to restrict access to a specific submenu or function. Functions often correspond to forms or elements within forms, such as buttons and tabs. Submenus are collections of forms and often roll up into levels in a tree structure. Exclusion of a parent submenu excludes all child submenus and functions down the hierarchy.
Exclusion Name	Name of the menu or function; access is restricted for the responsibility	0	File name of a menu or function already defined within Oracle EBS	Select the appropriate menu or function from a pop-up list.
Description	Description of the menu or function selected	0	Automatically provided when the menu or function is selected	This is a long-form, descriptive name of the menu or function selected.
Name	Description of name in Securing Attributes	0		The application displays once the appropriate name is selected from the list of values.

Figure 9.13 shows the Responsibility Setup screen in Oracle EBS.

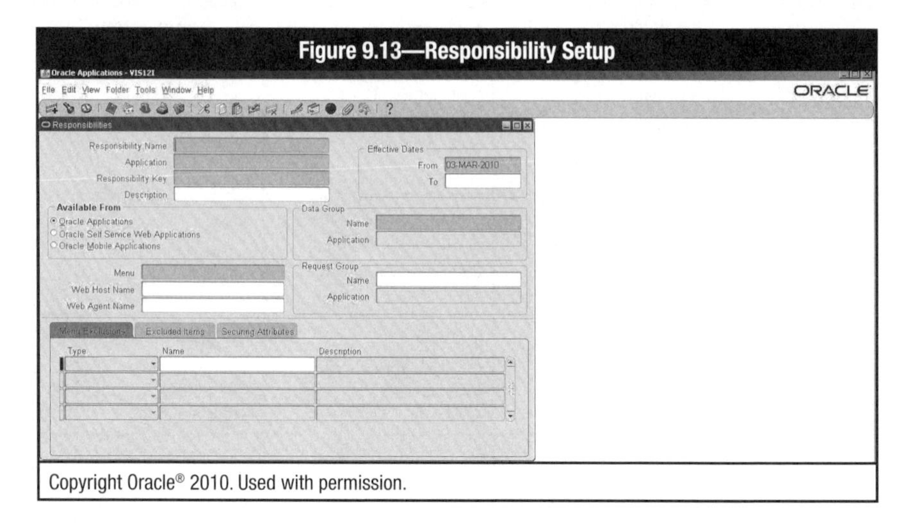

Figure 9.13—Responsibility Setup

Defining a Data Group

The pairing of this internal Oracle ID and the application itself defines a data group as shown in **figure 9.14**.

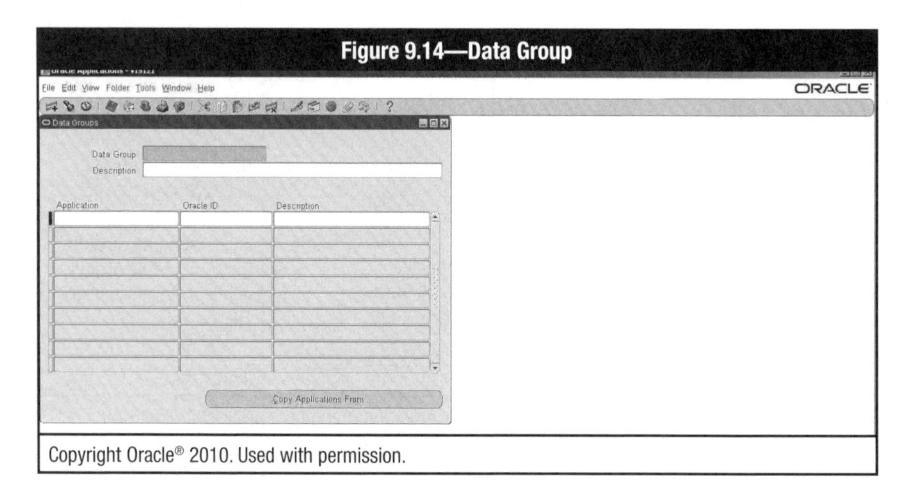

Figure 9.14—Data Group

The fields in **figure 9.15** relate to the definition of data groups that can be assigned to a responsibility.

Note: The default data group is Standard, which is also the most common.

Figure 9.15—Defining Data Groups				
Field Name	**Description**	**Status**	**Recommended Setting**	**Comments**
Data Group Name	Name of the data group	R	Long-form, descriptive name	This appears in the list of values while setting up a responsibility and selecting the data group.
Description	Description of the data group	O	Recommended description for future recall and for QA, audit and other documentation purposes	The purpose of this data group can be entered for future reference/ changes.
Application	Selects the application to which Oracle EBS forms connect when signed on with this data group	R, Q	Appropriate Oracle Applications module name	The Data Group defines the pairing of the Oracle ID and application.
Oracle ID	The display-only field indicating the Oracle ID chosen (character version only)	N/A	Display-only field	The Oracle ID determines the database tables and table privileges accessible by this responsibility.
Description	Description of the application and Oracle ID pair	O	Recommended description for future recall and for QA, audit and other documentation purposes	The purpose of this application and Oracle ID pair could be entered for future reference/ changes.

Defining a Request Group

Note: The Request Security Groups feature is for backward compatibility only in Oracle EBS R12 and later.

When a request group is assigned to a responsibility, it becomes a request security group. From a standard submission form, such as the Submit Requests form, the choice of concurrent programs and request sets to run are those in the user's responsibility's request security group.

If the Submit Requests form is not included on the menu for a responsibility, a Request Security Group does not need to be assigned to the responsibility.

The Request Groups form (Security➔Responsibility➔Request) allows the definition and management of Oracle Request Groups, as shown in **figure 9.16**.

Figure 9.16—Responsibility Groups

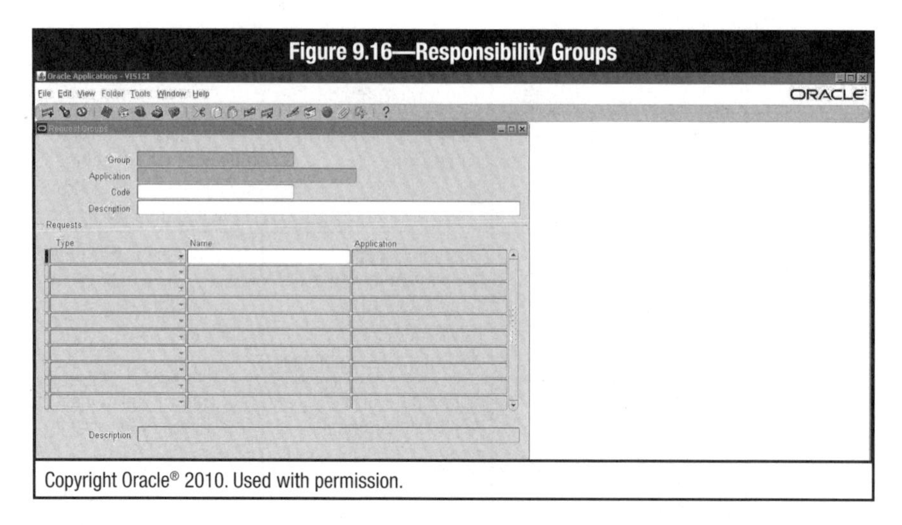

The fields in **figure 9.17** relate to the definition of a Request Group in Oracle EBS. Request Groups should be defined, where possible, prior to the creation of the responsibilities that will use them.

Figure 9.17—Defining Request Groups

Field Name	Description	Status	Recommended Setting	Comments
Group	Name of the request group	R	Use the request group's name to assign the request group to a responsibility on the responsibilities form.	An application name and request group uniquely identify those to be included in a specific request group.
Application	Name of the application associated with the request group	R, Q	Select the name of the application to associate with the request group.	This does not prevent requests from being assigned to the request group or sets from other applications being requested by the request group.
Code	The request identification parameter used by other applications	R	Set the appropriate code name (see Oracle EBS administration and development manuals). The code should follow corporate naming conventions (e.g., CST_PO_ REPORTS).	This is a parameter that can be passed to other requests or programs, allowing the nesting of reports and inquiries.
Description	The request group functionality description	O	Use the description of the report functionality provided by the request group.	

Figure 9.17—Defining Request Groups *(cont.)*				
Field Name	**Description**	**Status**	**Recommended Setting**	**Comments**
Request Type	Type of report grouping applied	R, Q	Select Program to add one item or set to Add Report Sets, or choose Stage Function Request or Application to include all requests in an application.	A single report in a program, a set of reports or all reports in the application program can be selected.
Request Name	Name of report, report set or application	R, Q		
Request Application	Name of application related to the request selected	D	Automatically displayed when the request name is entered	The application to which the request/program belongs
Description	Other description information	O	Include optional comments or details.	

Attribute Security

This feature is used by the Oracle Self Service Web Applications and any HTML view to allow rows (records) of data to be visible to specified users or responsibilities, based on the specific data contained in the row. The authority to view records should be based on the individual's job role. One or more values may be allocated to any of the securing attributes assigned to the user.

In the User and Responsibilities windows in the Securing Attributes tab, the following settings need to be populated:

- **Attribute**—An attribute should be selected that will determine which records the user/responsibility can access. Any attribute assigned to the user/responsibility can be selected.
- **Value**—A value should be selected that will determine which records the user/responsibility can access.

Flexfield Security

Two types of Flexfields exist:

- **Key Flexfield**—A field that can be customized to enter multisegment values, e.g., account numbers. It is a mandatory field that must be populated during data entry.
- **Descriptive Flexfield**—A field that can be customized to enter additional information that is not already provided in the field. It is an optional field that should be populated during data entry, if it has been set up as a required field.

Flexfields can be used to meet the following organizational needs by allowing the:

- Use of intelligent fields. An intelligent field is composed of one or more segments, in which each segment has both a value and a meaning.
- Application to validate the values or combination of values entered into an intelligent field
- Changing of the structure of an intelligent field depending on the data entered
- Capture of additional information, if necessary
- Customization of data fields without programming
- Queries of intelligent fields for specific information

A Flexfield is a flexible data field that an enterprise can customize to its business needs without programming. Security rules can be implemented to restrict data entry or reporting on specific accounting segments (e.g., cost centers, account ranges). Each security rule is composed of one or more rule elements, restricting or allowing user access to specific ranges of Flexfield values. A Key Flexfield definition is required by applications as part of the setup. A Descriptive Flexfield is set up to capture additional information related to the transaction.

Once a rule has been defined in the Flexfield Security window, it must be assigned to a responsibility before the rule can be enforced. These rules are active for every user in that responsibility. Rules can be assigned to different responsibilities and can be shared across responsibilities. Flexfield Value Security can be altered at any time, but the changes will not be in effect until users log out and back into the responsibility. Appropriate message text is entered when the security rule is defined to alert users attempting an action not allowed by their responsibilities.

It is recommended that a security representative be selected from the user community to define and maintain this security for the application. This person will have to identify the appropriate Flexfield values to be secured and the departments or individuals requiring access to the various segments. A unique responsibility (e.g., journal entry clerk—transportation, journal entry clerk—country services) will have to be created by the security administrator for each user group for which Flexfield Value Security has been implemented.

The use of Flexfield Value Security may negatively affect system performance depending on the number of security rules implemented. It will also require some maintenance at the user management level.

User and Data Auditing

Effective security techniques provide security administrators with the ability to monitor the activities of users and assist in resolving security weaknesses. Oracle EBS has the ability to log security-related events, such as unauthorized access attempts and user sign-on, and activities (if the audit function is activated). Reports should be generated and reviewed routinely by security management and system owners. Unauthorized users may access Oracle EBS and process transactions without detection, if violations are not researched in a timely manner.

There are two types of auditing in Oracle EBS:
1. Auditing user activity, including:
   - Sign-On: Audit Level Profile Option setting
   - Sign-On: Notification
   - Sign-On: Audit Reports
   - Monitoring User Activity
2. Auditing database row changes, including:
   - Audit trail: Activate Profile Option setting
   - About This Record... (from the Help menu)

Auditing User Activity

The Sign-On Audit feature provides the ability to:
- Record the usernames, dates and times of user access to Oracle EBS:
  - Select auditing by allowing the selection of users and the type of information to track.
  - Notify users of any unsuccessful attempt to sign on with their application username since their last sign-on, if any concurrent requests failed since their last session or when the default printer identified in their user profile is unregistered or not specified.
- Provide reports that contain detailed historic information about user activity based on search criteria.

Other auditing user functionalities include the ability to monitor users in real time and online, or to observe what responsibilities, forms, concurrent processes and terminals they are using.

Note: It is important that audited users include administrators such as DBAs and O/S users because most internal incidents occur from such users.

Sign-On: Audit Level Profile Option Setting

When determining whether to use the Sign-On Audit functionality, the additional system overhead required to monitor and audit users should be considered. Adding to the number of users selected to audit and increasing the level of auditing result in additional system overhead.

The Sign-On: Audit Level User Profile Option Setting determines which users the Sign-On Audit function tracks and the level at which they are audited. The System Profile Values form is used to enable Sign-On Audit and select Sign-On: Audit Level parameters. The type of sign-on audit to be set is determined using the Sign-On: Audit Level Profile Options Setting. The System Profile Values form allows the implementation of auditing user activities. A site profile search for Sign% will return sign-on profiles.

The Sign-On: Audit level can be set to the following:
- None is the default value, meaning that no users are audited.
- Auditing at the user level tracks the:
  - Users who sign on to the system
  - Times at which users log in and out
- Auditing at the responsibility level performs the user-level audit functions and tracks:
  - The responsibilities that users select
  - How much time users spend using each responsibility
- Auditing at the form level performs the responsibility-level audit functions and tracks:
  - The forms users select
  - How long users spend using each form

Note: Only forms-based access is audited, not web-based access, such as those with the newly developed modules and Self Service modules. This is a major limitation of the Sign-On Audit functionality. Self Service access, for example, shows up as using the guest account, which makes all audit data unusable.

Figure 9.18 shows the Sign-On: Audit Level screen and the audit-level options. Users cannot see or change this Profile Option. After audit levels are set or changed, the new audit levels for a user will only take effect the next time the user signs on to Oracle Applications. This option can also be set to track the Sign-On activity for a particular application.

Figure 9.18—Sign-On Auditing Options

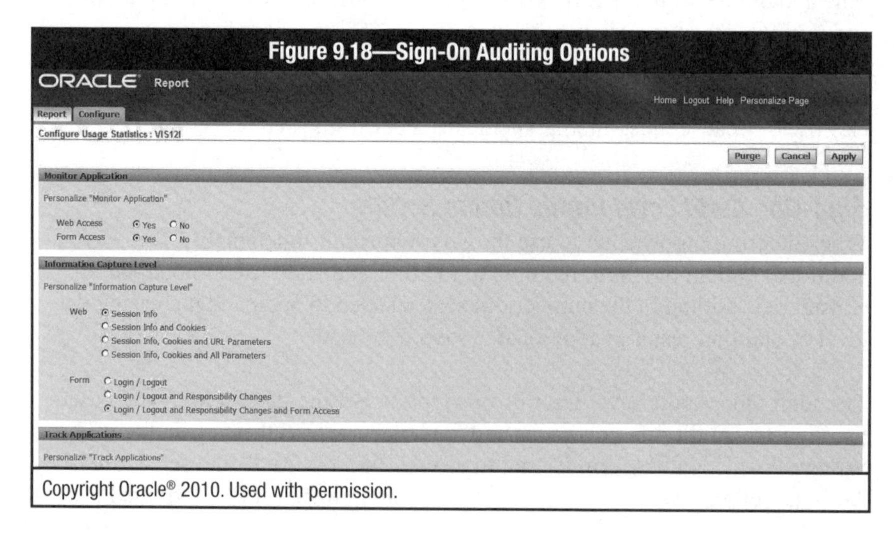

Copyright Oracle® 2010. Used with permission.

Sign-On: Notification

The Sign-On: Notification function can track user logins and provide users with a warning message when:

- Someone makes an unsuccessful attempt to sign on with the user's application username since the user's last successful sign-on attempt.
- Concurrent requests fail since the user's last session.
- The default printer identified in the user profile is unregistered or not specified.

This feature, in effect, notifies a user if an unauthorized user has attempted to guess the user's password since the user last used the system, concurrent requests have failed or the user is unable to print. This warning message appears after the user signs on. This feature can be activated by using the Personal Profile Values form by setting the Sign-On: Notification User Profile Option to Yes.

Please note that this functionality is useful only if the enterprise has developed and communicated to users a process for reporting, and dealing with, potential impersonations and access violations. Without this process, there is no reporting of this information to management and the control feature is rendered ineffective.

Sign-On: Audit Reports

This feature provides the ability to review the security of Oracle Applications using Sign-On: Audit Reports. Search criteria can be specified to customize each report. These reports provide detailed historic information about user activity based on search criteria. Sign-On: Audit Reports is selected and run using the Submit Requests (Request→Run) form. A request search for Sign% will return sign-on reports.

To generate Sign-On: Audit Reports on users within the system, audit settings must be enabled at the relevant profile level (via the Sign-On: Audit Level Profile option setting). The following is a list of the available Sign-On: Audit Reports and a brief explanation as to what they contain. Use the Submit Requests form to print these standard Oracle EBS audit reports:

- **Sign-On Audit Concurrent Requests**—Identifies which users are requesting concurrent requests and through which responsibilities and forms
- **Sign-On Audit Forms**—Identifies which Forms were navigated to and at what times by audited users
- **Sign-On Audit Responsibilities**—Identifies the responsibilities users being audited have logged in under and when
- **Sign-On Audit Unsuccessful Logins**—Identifies any unsuccessful attempts to log in to Oracle Applications (where the user ID is correct but the password is incorrect). These reports can be generated for any user, regardless of whether the user is being audited.
- **Sign-On Audit Users**—Identifies the users, time and Oracle EBS processes accessed by the users

Monitoring User Activity

The Monitor Users form assists in the monitoring and auditing of user activities for which the Sign-On: Audit function is enabled. Use the Monitor Users form to view online what users are doing. The Monitor Users form can be used at any time. The application monitor displays which users are signed on, what responsibilities they have, what forms they are using, how long they have been working on forms and what Oracle EBS processes they are using.

Only users that are being audited can be monitored using Sign-On: Audit. The application monitor also reflects the level of auditing defined for the users.

Auditing Database Row Changes

Oracle EBS provides data-level audit/monitoring via the following three features:

1. Record History—About this record
2. Oracle Alerts
3. AuditTrail for tables and columns

Record History—About This Record

Oracle EBS tracks the following information for each transaction entered in the system:

- Name of the user who created/updated the row
- Date the user created/updated the row
- Name of the table containing the row
- Name of the user who last updated the row

Note: Only the user who originally created the row and the user who last updated it are displayed in this window. If tracking of all changes is required, AuditTrail needs to be enabled. See the AuditTrail section for additional information on tables and columns.

Oracle Alerts

Oracle Alerts can be used to detect exceptions (or errors) in database processing. An alert is a mechanism that checks the database for a specific exception condition that is predefined by the security administrator. An alert tells the application what database exception to identify, as well as what output to produce for that exception. For example, an alert can be defined to flag POs exceeding US $10,000, and have that alert output the name of the individual who requested the PO, as well as the name of the individual's manager. Alerts are often configured to automatically send out e-mail messages when the exception is detected.

Two types of alerts can be defined: an event alert or a periodic alert. An event alert immediately notifies the user of activity in the database as it occurs. For example, if a PO exceeding US $10,000 is entered, Oracle Alerts can notify the purchasing manager immediately.

A periodic alert, on the other hand, checks the database for information according to a defined schedule. For example, a periodic alert can be defined for Oracle Purchasing to send a message to the purchasing manager listing the number of approved requisition lines that each purchasing agent placed on POs. This alert can be defined to run weekly.

AuditTrail for Tables and Columns

AuditTrail reporting allows the security administrator to report on the history of changes to specific data, identifying which users altered the data, and when/if customized reports or queries have been developed. An AuditTrail data report can determine how any data row or element obtained its current values, and which users made the changes, at what time and with what forms. Database triggers and stored procedures can also be audited. Auditing is accomplished via defined audit groups. Most types of information can be tracked, such as characters, numbers and date fields. The AuditTrail functionality provides detailed information about changes to the record and the shadow table. AuditTrails provide a mechanism, based on Oracle Database triggers, to store change information in a shadow table of the audited table. The enterprise must carefully plan what tables are to be audited because the overhead of audit trails can become significantly heavy. Furthermore, this information needs to be purged manually, as the standard Oracle EBS purge routines do not look at the shadow audit tables.

By setting up AuditTrail, a history of all changes made to a table can be selected and stored using audit groups. Audit groups functionally are group tables to be audited. Steps for setting up AuditTrail include:

- **Define audit groups**—Define groups of tables or selected columns to audit. Tables that belong to the same business process can be grouped together.
- **Define audit installations**—The registered Oracle EBS IDs at the client site should be selected to audit. This allows the auditing of multiple application installations. When adding a table to an audit group, auditing will be enabled automatically for all installations of the table for which audit is enabled.
- **Run the AuditTrail Update report to enable auditing**—The program must be rerun for any changes made to definitions. This can be done using the standard submission (Submit Reports) form.

The AuditTrail Update Tables report program creates Database triggers on the tables in the audit groups for the installation as well as shadow tables (one for each audited table) to contain the audit information. Audit triggers and shadow tables are modified to take into account any changes to audit definitions. The program also builds special views to retrieve the audit data for reporting.

Summary

This chapter provided an overall understanding of the configuration of security in Oracle EBS, including user administration. It detailed the important addition of the RBAC security layer and how this has changed the security landscape setup in Oracle EBS.

10. Auditing Oracle E-Business Suite Security

Security Administration Testing

In chapter 9, application security was discussed. This chapter looks at the significant risks, key controls and techniques to test security administration within Oracle EBS.

Refer to **figure 4.8** for the key to the numbering sequence for risks, controls and testing techniques.

Security Administration: Risks

Significant risks include:

1.1	Oracle EBS security/control parameters may not be defined adequately. Refer to chapter 9 for further details on security parameters.
1.2	A formal or informal security administration function may not exist. Oracle EBS security administration policies and procedures may not have been developed or implemented. User access change management procedures may be nonexistent or inadequate.
1.3	Access to programs, data and other information resources may not be restricted or monitored.
1.4	Users may perform incompatible functions and/or user access rights/ responsibilities may not match the users' organizational duties.
1.5	System and start-up profiles may be defined inadequately. The ability to change the profiles may be unrestricted, and changes to profiles may not be monitored.
1.6	Access to system output may be unrestricted.
1.7	User IDs, responsibilities and preferences documentation may be inadequate.
1.8	Remote access by external vendors may not be controlled adequately.
1.9	Direct access to the production environment by application developers may not be controlled, and as a result, modifications that are inconsistent with management intentions may be developed/ implemented in production.
1.10	Unauthorized or inaccurate changes to the database schemas or setups may affect the integrity of the data in the system.

Security Administration: Key Controls

Key controls over security administration include:

1.1.1	Oracle EBS security/control parameters should be defined adequately. Refer to chapter 9 for further details on security parameters. The testing of the parameter setting in Oracle EBS could be extended into the password parameter setting to ensure that passwords and user access are adequately controlled by the system.
1.1.2	An organizational structure for security administration should be established and staffed. Oracle EBS responsibilities related to information security administration should be defined and assigned. Information security responsibilities related to Oracle EBS security administration policies and procedures should be developed and implemented. User access change management procedures should be formal and implemented. New users must be appropriately authorized. Terminated employee access must be deactivated in a timely manner.
1.1.3	Logical security tools and techniques should be implemented and configured to enable restriction of access to programs, data and other information resources. Management should review and approve the implementation and configuration of Oracle EBS security techniques. Information security techniques (such as Oracle Alerts, user auditing or other monitoring reports) should be activated and monitored to record and report security events as defined in information security policies. Reports generated should be reviewed regularly, and necessary action should be taken.

1.1.3 *(cont.)*	Default Oracle EBS passwords should be modified. Oracle Applications comes with an initial user ID and password used to set up all other users on the system. The default user ID and password for these accounts are widely known and should be changed. These pose a significant threat to security because the user IDs and passwords are extremely obvious and are among the first to be tried by internal or external hackers to gain access to systems. In particular, some of the default user ID and password combinations provide powerful access, such as system administrator or supervisor access. To counter the risk of these being used by malicious systems hackers or other extremely curious parties, these profiles should be eliminated or disabled, or the passwords should be changed from the Oracle EBS default. The identity of users (local and remote) should be authenticated to the system through passwords or other authentication mechanisms. The policies relating to the use of passwords incorporate periodic password changes, confidentiality requirements and password format (e.g., password length, alphanumeric content). Users should be required to have a unique user identifier to distinguish one user from another and establish accountability. In addition, standards should be defined to govern the construction and use of passwords to ensure that the password is not guessed easily. Standards often include prohibiting the creation and use of passwords that may be associated easily with the respective user, establishing minimum and maximum length, requiring a combination of alphanumeric characters, mandating periodic password changes, and preventing the reuse of passwords. The lack of password criteria or standards increases the risk that a password may be guessed or obtained by an unauthorized individual and used to gain access to the system. Sensitive data should be encrypted. Use of privileged responsibilities (i.e., superuser, manager GUI, security administrator and system administrator) should be limited to appropriate personnel, monitored and reviewed. In the case of the production client, it is always recommended that, where possible, custom responsibilities and roles should be created and assigned appropriately, with Oracle EBS standard high-privilege roles and responsibilities not being assigned in the production client. If these standard high-privilege responsibilities are assigned, then special attention should also be applied to the assignment of these responsibilities within the User Roles setup within Oracle EBS. Any high-privilege responsibility assigned to users, regardless of being custom or standard, should be monitored and reviewed regularly by management.

1.1.3 *(cont.)*	In Oracle EBS, privileged access is granted through the use of assigning users to roles that contain the access to privileged responsibilities and are considered to be on the same level as the security administrator or system administrator. Examples of privileged responsibilities in Oracle EBS are Payables Superuser and Assets Manager. Each application module should have a privileged responsibility. Each manager/superuser role has the highest level of access within each application module, allowing access to such functions as Setup and Configuration. As a result, use of privileged responsibilities should be restricted to only the functional managers or data owners. The use of these privileged responsibilities should be on an as-needed basis and should be monitored and reviewed by management. Inappropriate use of the privileged responsibilities can result in system access beyond the minimum level necessary. This could result in entry of unauthorized or invalid transactions, loss of data, changes to setup configuration, and loss of functionality of Oracle Applications. Access to sensitive data should be restricted to authorized users only, and access privileges should be reviewed regularly to assess whether such access is appropriate. The data would typically include items that are confidential in nature, such as product designs and personnel records that include personal details and payroll information. Oracle Security has a facility to record authorized access to specific data items. This is performed through the Sign-On: Audit Level Profile option setting. With Sign-On: Audit, usernames and the dates and times that users access Oracle Applications can be recorded. Sign-On: Audit can also track the responsibilities and forms used. Unauthorized attempts to log in to Oracle Applications should be logged and included in a security violation report; the logs and reports should be reviewed regularly, and any necessary action taken.
1.1.4	SoD within Oracle Applications should exist, and duties should be assigned to authorized users, as appropriate. The responsibilities should match the organizational functions performed by the system users.
1.1.5	System and start-up profiles should be defined adequately. The ability to change the profiles may be restricted, and changes should be monitored.
1.1.6	System and start-up profiles should be defined appropriately, access to them should be restricted, and changes should be monitored. Oracle EBS profiles should be appropriately defined and restricted, and actively administered.

1.1.7	System output should be managed so that all expected information is produced and received by the intended recipient.
1.1.8	Documentation on user IDs, roles, responsibilities, preferences and their use should be developed.
1.1.9	Remote access by external vendors for software maintenance should be restricted to the production environment, activated only on request and logged.
1.1.10	Access to the test and production environments should be restricted appropriately.

Security Administration: Testing Techniques

Testing techniques for security administration controls include:

1.1.1	Check that the password parameters are in line with the enterprise's policies on password strength. These password System Profile Options are: • Sign-On: Password Length • Sign-On: Password Failure Limit • Sign-On: Password Hard to Guess • Sign-On: Password No Reuse Check that all user accounts have the password expiry set to the enterprise policy (e.g., 90 days). Verify that the time-out/background disconnect interval in the System Profile Option ICX (Inter-Cartridge Exchange) Session Timeout is set to the enterprise policy (e.g., 1,800 seconds/ 30 minutes). FAILED_LOGIN_ATTEMPTS is the number of failed login attempts before the account is locked (3). PASSWORD_LIFE_TIME is the number of days that the same password can be used for authentication (30). When an Oracle Database is also utilized, some common parameters to check on the database side are: • PASSWORD_REUSE_TIME (180) • PASSWORD_REUSE_MAX (unlimited) • PASSWORD_LOCK_TIME (7) • PASSWORD_GRACE_TIME (14) • PASSWORD_VERIFY_FUNCTION (recommended)

1.1.2	Verify the effectiveness of a data administration group's activities. Verify whether a security administration group or administrator exists and controls access to the Oracle EBS system. Ask IT management or the security administrator about the: • Policies, procedures, standards and guidance for security administration • Degree to which such policies, procedures, standards and guidance are implemented • Management commitment and communication regarding the security administration function • Understanding of security administrator roles and responsibilities Obtain and review evidence that corroborates the responses to the inquiries in the previous bullet list by examining client documentation, such as: • Information security policies, procedures, standards and guidance • Organizational structure of the information systems group • Security administrator job descriptions • Security administration department charter and mission statement • The Active Responsibilities report, to determine whether the client is using a security administrator responsibility and who is assigned the responsibility • The Function Security Function report, to determine the configuration of the security administrator responsibility Evaluate the User Access Change Management process (i.e., what reporting and documentation are in place to provide assurance that users are assigned the appropriate responsibilities and profiles). Determine what assessment is made of user roles and responsibilities and the assignment of access levels. Determine what approval and update procedures are in place to administer changes to user access. Review evidence that changes to access authority are tested by the security administrators. Through observation, witness a sample of access requests being tested by system security administrators. The auditor should review a listing of users and verify whether there is a process to automatically disable the account and any access for employees and users leaving the enterprise.
1.1.3	For implementation of logical tools and techniques, ask IT management or the security administrator about the: • Method of defining and implementing Oracle EBS security • Role of security within the information systems plans

1.1.3 *(cont.)*	• Techniques used to prevent unauthorized access • Communication of security strategies to system users • Process used to identify and evaluate security-related risks • Process of evaluation when implementing a security tool or technique • Initial definition of security configuration parameters • Monitoring and reporting controls for the customization of Oracle EBS and access rules, and how these controls ensure that security policies and standards are followed • Reports and other information used, such as the security and internal control framework • System security configuration reports, including how they are used • Procedures performed when unauthorized access is encountered Obtain and review evidence that corroborates the responses to the inquiries. Examine client documentation of the review and approval of information security tools and techniques used by management to ensure that they were performed in accordance with established policies and procedures. Use online inquiry and/or audit tools to test the system, for example: • Information systems plans • Security policies, procedures, standards and guidance • Security and internal control framework • Minutes of planning/steering committee meetings • Review and approval analysis documentation • Oracle EBS reports: Function Security Function reports, Function Security Menu report, Function Security Navigator report For management reviews, ask management whether it reviews and approves the implementation and configuration of Oracle EBS security techniques. Identify issues for follow-up and resolution. Inspect physical evidence that the reviews occur; the configuration is approved and retained; and issues are resolved, as necessary. For information security techniques, ask IT management or the security administrator about the: • Policies and procedures related to the activation of information security tools to record and report security events (such as security violation reports) as defined in information security policies • Steps involved, including how security tools are activated and how security events are reported, reviewed and resolved • Reports and other information used, such as available security reports and how they are used • Procedures performed when security violations are encountered

1.1.3 *(cont.)*	Obtain and review evidence that corroborates the responses to the previous inquiries by examining client documentation, such as system configuration settings and security reports, and that the control activity was performed in accordance with established policies and procedures, for example: • Oracle User Profile Option Values report for settings on: – Sign-On: Notification (see chapter 9) – Sign-On: Audit Level (see chapter 9) – AuditTrail: Activate (see chapter 9) • Oracle Sign-On: Audit reports (see chapter 9): – Sign-On: Audit Concurrent Requests – Sign-On: Audit Forms – Sign-On: Audit Responsibilities – Sign-On: Audit Unsuccessful Logins – Sign-On: Audit Users For default passwords, test attempts to log in to Oracle EBS with default user IDs and passwords. For default application passwords, test the following default user IDs and passwords by attempting to log in to Oracle EBS: • SYSADMIN/SYSADMIN • GUEST/ORACLE • WIZARD/WELCOME • AUTOINSTALL/DATAMERGE • IEXADMIN/COLLECTIONS • IRC_EMP_GUEST/WELCOME • IRC_EXT_GUEST/WELCOME • MOBILEADM/WELCOME • OP_CUST_CARE_ADMIN/OP_CUST_CARE_ADMIN • OP_SYSADMIN/OP_SYSADMIN Oracle EBS is seeded with a number of accounts that should be deactivated and have their passwords changed. Deactivate and change the passwords to a random string for these accounts: • AME_INVALID_APPROVER • ANONYMOUS • APPSMGR • ASGADM • ASGUEST • AUTOINSTALL • CONCURRENT MANAGER • FEEDER SYSTEM • GUEST • IBE_ADMIN

1.1.3 *(cont.)*	• IBE_GUEST • IBEGUEST • IEXADMIN • INDUSTRY DATA • INITIAL SETUP • IRC_EMP_GUEST • IRC_EXT_GUEST • MOBADM • MOBDEV • MOBILEADM • OP_CUST_CARE_ADMIN • OP_SYSADMIN • ORACLE12.0.0 • PORTAL30 • PORTAL30_SSO • STANDALONE BATCH PROCESS • SYSADMIN • WIZARD • XML_USER For default database passwords, Oracle EBS has more than 150 associated database user accounts. The default password is usually the name of the database user account. The best method for testing all of the account passwords is by using an SQL script or other automated tool. Examples of the default accounts and passwords are: • SYS/CHANGE_ON_INSTALL • SYSTEM/MANAGER • APPS/APPS • APPLSYS/APPS • GL/GL • AR/AR • AP/AP • FA/FA • PO/PO • CTXSYS/CTXSYS There are a significant number of default accounts in addition to those listed. The correct approach would be to obtain a list of Oracle EBS user accounts (select username from DBA users) and use that as the basis for validating accounts.

1.1.3 *(cont.)*	The SYSADMIN system account cannot be deleted because no user accounts can be deleted from the system once they have been created. The SYSADMIN is also unique in that it cannot be deactivated since it must be used for a few system administration tasks and several internal processes rely on the account always being active. As recommended by Oracle Corp., the best practice is to change the SYSADMIN account settings by changing the password and allow it to be used only in extreme circumstances. System administrators should sign on with their unique user IDs and be assigned to the system administrator responsibility. For password controls, by default, a password should be at least five characters and can extend up to 100 characters. The Sign-On Password Length parameter is saved in the user profiles Option Values; however, only the system administrator can change this value. User Profile Options can be set at four different levels: Site, Application, Responsibility and User. The values set at each level provide run-time values for each user's Profile Options. An option's run-time value becomes the highest-level setting for that option. For instance, when a Profile Option is set at more than one level, Site has the lowest priority, superseded by Application, then Responsibility, with the User level having the highest priority. Basically, a value entered at the Site level may be overridden by values entered at any other level. A value entered at the User level has the highest priority and overrides values entered at any other level. Using the Submit Requests form, view user Profile Options by printing the User Profile Option Values report. Review the Profile Options for the Sign-On: Password Length Profile Option. Sign-On: Password Length sets the minimum length of an Applications Sign-On: Password. If no value is entered, the minimum length defaults to five characters. Only responsibilities with system administrator rights can change this Profile Option. Inquire whether the Profile Options were initially set by the system administrator at the Site level, which becomes the default setting if options have not been set at the Application or Responsibility level. Alternatively, the Profile Options can be displayed using the System Profile Values window. A search can be performed using the Find System Profile Values window. First, view the Site-level Profile Option by selecting Site and viewing the default. Ensure that the Profile Option is set at the Application, Responsibility and User level.

1.1.3 *(cont.)*	For the password area (User window), the system administrator should set up the initial password when defining a user. The user should change the password the first time he/she signs on. Setting the number of accesses to "1" will require changing the password. Username and password rules include: • Passwords must be at least five characters long and may extend up to 100 characters. • The username must contain alphanumeric characters (A through Z, 0 through 9). All other characters are invalid. • Passwords are not displayed when entered. The system administrator can set an initial password or change a password if a user forgets it, but cannot access the user's password. Password expiration can be set using one of three options: Days, Accesses or None. The initial password should be set to expire after one access, thereafter to 30 days. A window will pop up prompting the application user to change the password after it has expired. For encryption of sensitive information, in Oracle R11*i*, sensitive data (i.e., passwords) are encrypted automatically. Prior to the application sign-on process, part of the application setup requires the security administrator to create users of the system. When the security administrator creates new users, the program specifies the creation of two random strings secured by the same username/password 40-bit key. The two random strings are: • The "apps" password encrypted using a combination of the user ID and user's password as the key • The user's password encrypted using a combination of the user ID and "apps" password as the key Whenever the security administrator updates an application user in terms of username or password, the security (encryption) routine is called automatically to perform the eight encryption steps again. For the transmission of other sensitive data, ask IT management or the security administrator about the: • Policies, procedures, standards and guidance regarding encryption of sensitive data over internal and external networks • Steps involved. Consider the specifications of required cryptographic techniques and strengths. • Design of the cryptographic infrastructure • Selection of appropriate encryption software and modules • Use of private- and public-key cryptographic solutions

1.1.3 *(cont.)*	• Selection of suitable certification authority; key management, including generation, distribution, storage, entry, use and archiving; and protection of encryption software and hardware • Reports and other information used, including how they are used Obtain and review evidence that corroborates the responses to the previous inquiries by examining client documentation, such as: • Policies, procedures, standards and guidance relating to encryption of sensitive data • Graphical representation of cryptographic structure • Statements of certification by independent reviewers of cryptographic software • Lists of data items and their corresponding classification • Statements of certification by independent reviewers of certificate authorities For privileged responsibilities, determine through inquiry with the security administration that powerful user IDs and passwords are strictly confidential and that there are specific procedures to ensure confidentiality. Ask IT management or the security administrator about the: • Policies and procedures related to the limitation of usage • Organizational structure of the security administration function • Monitoring and reviewing of privileged responsibilities • Steps involved, including the procedures for assignment of privileged access • Reports, such as security administration reports for privileged responsibilities, and other information used, including how they are used Obtain and review evidence that corroborates the responses to the previous inquiries by examining client documentation, such as: • Information security policies, procedures, standards and guidance • Security administration reports (e.g., Active Responsibilities report) for privileged responsibilities • System logs that identify use of privileged system access For sensitive data access, ask IT management or the security administrator about the: • Policies and procedures related to the fact that authorized access to sensitive data is audited • Audit logs, which should be regularly reviewed to assess whether the access and use of such data were appropriate

1.1.3 *(cont.)*	• Steps involved, such as configuration and audit capabilities, generation of regular logs and distribution to authorized staff, responsibility to perform a review and resolve concerns, selecting criteria used to review access logs, independent data to which management compares the logs (e.g., job descriptions), and level of detail in the access log • Reports and other information used, such as copies of logs of access to sensitive data, a list of sensitive data items and their corresponding classification, job descriptions and responsibilities, and documentation of action taken to resolve inappropriate access or use of sensitive data • Procedures performed when inappropriate access is encountered Obtain and review evidence that corroborates the responses to the previous inquiries by examining client documentation, such as: • Policies, procedures, standards and guidance relating to access to sensitive data • Copies of logs of access to sensitive data • Job descriptions and responsibilities • Lists of sensitive data items and their corresponding classification • Documentation of action taken to resolve inappropriate access or use of sensitive data For unauthorized login attempts, ask IT management or the security administrator about the: • Policies and procedures related to the inspection of logs and reports to identify and prevent unauthorized access to Oracle Applications • Steps involved (consider frequency of inspection of logs) • Action to be taken immediately to prevent unauthorized access and, subsequently, to identify the perpetrator of the unauthorized access attempt and assess the adequacy of security configuration • Security violation reports and other information used, including how they are used • Procedures performed when exceptions, misstatements or unusual items are encountered • Extent of unauthorized access Obtain and review evidence that corroborates the responses to the previous inquiries by examining client documentation, such as: • Policies, procedures, standards and guidance relating to the inspection of logs to identify and prevent unauthorized access to information resources • Security system configuration • Lists of security events that are captured

1.1.3 *(cont.)*	• The Sign-On: Audit Unsuccessful Logins report. Use this report to view who unsuccessfully attempted to sign on to Oracle Applications as another user. An unsuccessful login occurs when a user enters a correct username, but an incorrect password. Select an appropriate number of unauthorized access attempts from the audit population, and examine documented evidence that unauthorized access attempts are properly identified and reviewed and that necessary action is taken.
1.1.4	Duties are segregated through the use of roles and the assigned responsibilities within these roles. Standard menus are created for each type of user through standard Oracle EBS responsibility creation. Roles should be based on approval by the application owner for each corresponding application. When determining a user's access privileges, the application owner should ensure that SoD is maintained and job requirements are fulfilled. Determine how user IDs and roles are defined. Ask application owners and the security administrator if the SoD is considered when defining or changing the access of users. Review evidence that the SoD has been addressed when assigning user access or creating roles so that users are unable to initiate, record, approve and post a transaction (e.g., roles/responsibilities matrix). The security administration staff members should control access to the Oracle EBS system. They should not be authorized to enter transaction data or update master file data. For large implementations (more than 800 users), determine whether the security administrator tasks are segregated. Verify that roles do not include SoD conflicts by selecting a sample of key business and administration users and reviewing their responsibilities. Compare the access granted by the role to their job function/tasks. Determine whether system administrators and application developers are assigned business function menus or forms.

1.1.5	Ascertain the extent to which Oracle EBS profile settings have been utilized at the user, responsibility, application and site levels. The site profile should include the following settings: • Sign-On Notification: Yes • Sign-On Password Length: 6 • Sign-On Password Failure Limit: 3 • Sign-On Password Hard to Guess • Sign-On Password No Reuse The password hard to guess verification routine performs the following checks: • The password has a minimum length of four characters. • The password is not the same as the user ID. • The password has at least one alpha and one numeric character, and one punctuation mark. • The password does not match simple words such as welcome, account, database and user. • The password differs from the previous password by at least three letters.
1.1.6	Inquire as to the manner in which system output is controlled so that the intended recipient receives the information produced. Each user may be assigned a respective print location. These print locations can be assigned based on the user's department and relative geographic location within the building. Review the assignment of printer locations, if used. Report printing is controlled by the Oracle EBS user. Only the user who prints a job can view the report online, and that user is the only one who can reprint the report. Ensure that user access to drives and directories containing output files is restricted.
1.1.7	Verify with security administration, and inspect evidence, that user ID, responsibility and profile documentation exists and that a process for maintenance of the documentation process is in place.
1.1.8	Check with security administration to determine how vendor remote access is managed. Review the remote access setting to ensure that access is controlled and logged.

1.1.9	Ask the development team and security administrator about the process involved for application change management. In addition, review: • Policies and procedures for controlling access to test and production environments • Access control listings for the test and production environments • Lists of users • Violation reports
1.1.10	Understand management's policies and procedures regarding the review of data dictionary changes. Assess the adequacy of such policies, procedures, standards and guidance, taking into account the frequency with which the review is performed, the level of detail in the reports, other independent data to which management compares the reports, and the likelihood that the people performing the review will be able to identify exception items and the nature of exception items. Review a listing of users with access to view and modify the data dictionary, and discuss with the DBA whether these users are appropriate.

Summary

This chapter outlined the risks, key controls and testing techniques surrounding Oracle EBS R12.1 security administration. Among the key risks were inadequately defined security/control parameters and inappropriately granted access.

11. Continuous Control Monitoring in an Oracle E-Business Suite Environment

This chapter introduces the concept of continuous control monitoring and how governance risk and compliance (GRC), and Oracle's GRC suite in particular, is changing the way businesses manage both their compliance responsibilities and the way they are audited. GRC tools allow an integrated and efficient control governance framework to be implemented within an application system, which can be monitored and managed continually. The ability to access ready-made reports and information on the system and application-enforced controls with ease makes the compliance burden more manageable and the work of an auditor more efficient and effective.

Continuous Monitoring and the Evolution of GRC Tools

The concept of continuous control monitoring is becoming increasingly attractive as an approach to auditing information systems. This is partly attributable to the recent introduction of sophisticated tool sets that can automate the generation and management of real-time information and be used to assess system and business risks accurately on a continuous basis.

Over the last 10 years, the automated auditing and monitoring tools utilized by businesses in an Oracle EBS environment have evolved dramatically. The first tools used in the market were heavily focused on auditing system security configuration and user access management. The tools were built in, complex and required security access rights to run reports, discouraging use by business users and limiting the effectiveness of the tools. This approach evolved into utilizing third-party tools, often created by professional services firms or external applications developed by Oracle Corp. These externals tools were easier to use and downloaded point-in-time data from an Oracle EBS system to generate automated reports that could be analyzed by relevant personnel.

The products provided by Oracle Corp. for automated monitoring of an EBS system have also evolved over time. The first tool provided was Oracle Internal Controls Manager (ICM), developed due to the increased pressure on internal control environments as a result of the US Sarbanes-Oxley Act of 2002. ICM initially acted as a repository for process narratives and controls. Audit management capabilities, including basic control monitoring functionality, were included in version B. An automated Segregation of Duties (SoD) conflict identification tool and Application Controls Monitoring tool were also added to the suite in version B.

The advent of GRC tools on the market has gone a long way to truly automating the monitoring of controls in a preventive and detective way. GRC as a concept and the products provided by Oracle Corp. will be outlined further in the following sections.

What Is GRC?

GRC is a concept that brings together three disparate activities under one banner, in part because they share much of their input data, but more importantly because they are performed more effectively and efficiently together:

- **Governance**—The oversight, direction, and high-level monitoring and control of an enterprise to achieve defined objectives
- **Risk**—Managing the combination of the likelihood and impact of an event to influence the outcome in the context of an enterprise
- **Compliance**—The processes to achieve legislative and regulatory requirements and meet enterprise policies and procedures

Many enterprises believe efficiencies should be driven out of their Sarbanes-Oxley and other compliance efforts, but they sometimes see their GRC responsibilities as a burden. Approximately 85 percent of internal controls at an average enterprise are manual. With a heavy reliance on manual conrols with multiple compliance and audit requirements, there is an impact on the auditor's time/budget, thus diverting staff away from focusing on the core issues.

One aim of an effective GRC suite is to allow a business to manage and monitor internal controls in a harmonized way, whereby a control can be effectively tested once to satisfy many stakeholders. This removes the need for multiple assurance activities and lessens the burden placed on frontline staff.

An integrated and comprehensive GRC suite moves away from decentralized governance and compliance frameworks to a centralized approach, enabling internal control auditing, management and monitoring to be automated and effectively managed on a continuous basis and allowing for the establishment of a strong governance culture in the enterprise, as shown in **figure 11.1**.

Oracle GRC Solutions

The Oracle GRC product suite has evolved since it was first released in early 2007. Initially, the suite was released with three main components:

- **GRC Manager**—Used to monitor business process risk and control performance
- **Application Access Controls**—Offered a library of segregation of duties controls as well as the ability to prevent and detect control violations
- **Application Configuration Controls**—Monitored more than 500 internal controls to the source application, providing continuous monitoring for changes in configuration controls and providing the ability to set up auditing parameters

Figure 11.1—GRC Evolution

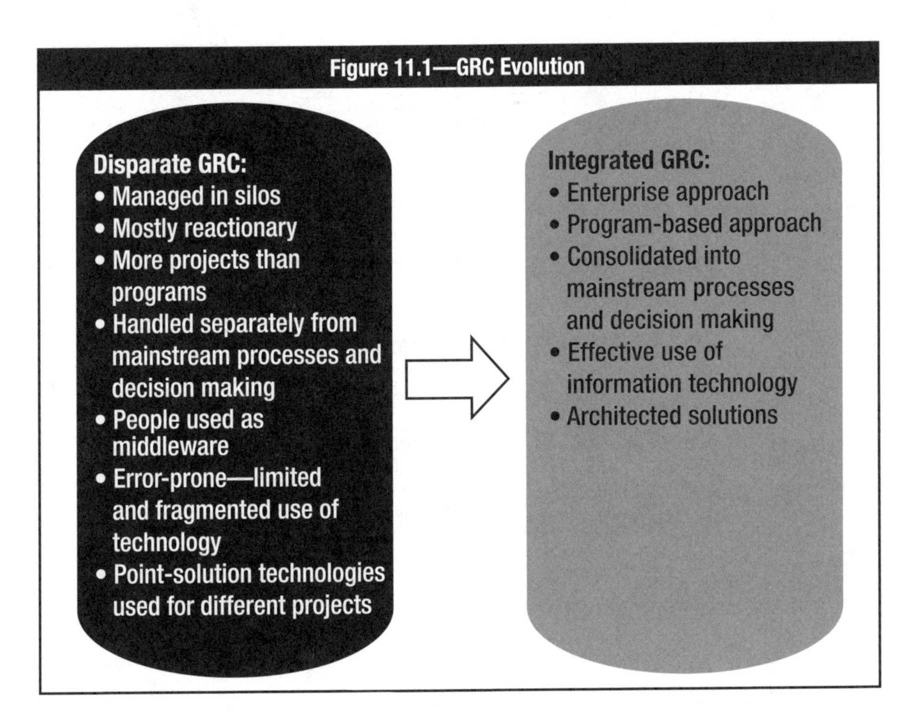

Disparate GRC:
• Managed in silos
• Mostly reactionary
• More projects than programs
• Handled separately from mainstream processes and decision making
• People used as middleware
• Error-prone—limited and fragmented use of technology
• Point-solution technologies used for different projects
Integrated GRC:
• Enterprise approach
• Program-based approach
• Consolidated into mainstream processes and decision making
• Effective use of information technology
• Architected solutions

The suite had a focus on compliance and content management after the acquisition of content management software provider Stellent. At the time, the release was also only compatible with Oracle EBS.

The next major release of the suite came in late 2007 after the acquisition of LogicalApps, a provider of GRC solutions. The acquisition allowed Oracle Corp. to combine the LogicalApps suite with their current GRC solution to add greater functionality and integration with a suite capable of running in any major ERP application environment.

The most recent version of the Oracle GRC suite has been built on newer architecture. GRC Intelligence and GRC Manager, as shown in **figure 11.2,** have been built on Oracle's Fusion architecture to allow the applications to easily integrate into the forthcoming Fusion application space. The control modules of the GRC suite have been built on the Oracle Fusion architecture as well, allowing it operate on any platform. The GRC suite is an area of focus for Oracle Corp. and will continue to be enhanced, upgraded and become more integrated with other applications in the future.

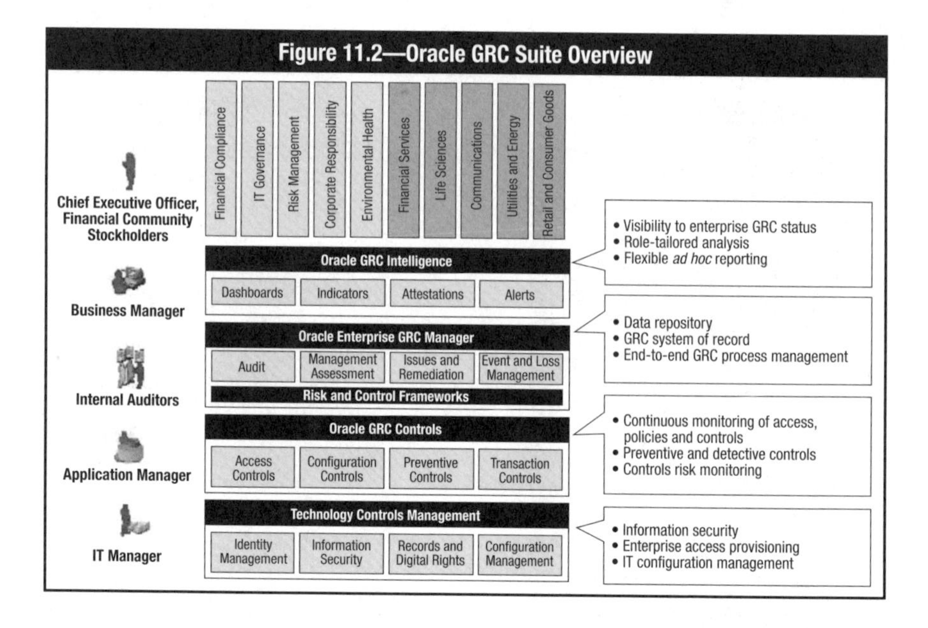

The current Oracle GRC suite includes three key layers:

- **GRC Intelligence**—Provides dashboards and strategic reporting capabilities
- **Enterprise GRC Manager**—Provides a central repository to establish, organize and maintain risk policies, internal controls and audit plans; tools for collecting employee assessments and reports, and workflows to manage issue review and audit tasks
- **GRC Controls Products**—Includes:
  - Application Access Controls Governor—Ability for real-time enforcement, monitoring and detailed analysis of crucial user access policies and controls
  - Configuration Controls Governor—Ability to track any changes to key application configuration data, and identify differences between various environment configurations
  - Enterprise Transaction Controls Governor—Ability to continuously monitor transactions against established policies to detect suspicious transactions or redundant business practices
  - Preventive Controls Governor—Ability to configure preventive controls in the EBS user interface on items such as transaction entry forms

Each of the key parts of the Oracle GRC suite identified previously are detailed in the following sections of this chapter.

GRC Intelligence

The GRC Intelligence module is built upon Oracle Corp.'s latest fusion-architecture-based business intelligence technology. The module provides dashboards and strategic reporting capabilities to business process owners, compliance officers and risk managers so that they can better understand the effectiveness of GRC programs across the enterprise, as shown in **figure 11.3**.

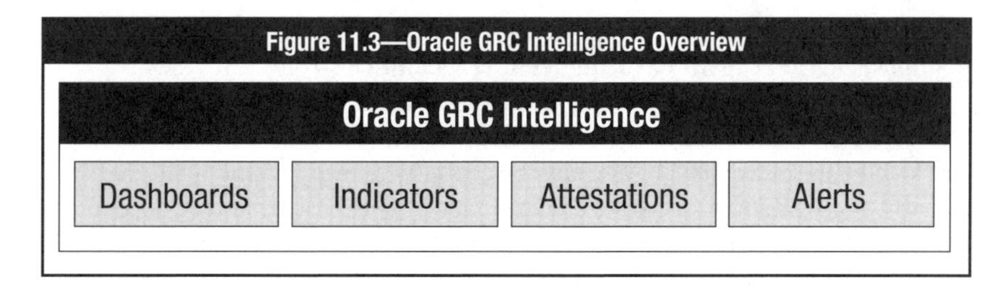

Figure 11.3—Oracle GRC Intelligence Overview

Description of the main features:

- **Real-time insight into status of risk and controls**—Via heat maps and other analytical reporting from a single integrated source of information, the intelligence module can provide a consistent view and can be used to measure and monitor business risks on a continuous basis.
- **Ability to provide tailored intelligence**—Via predefined or customized reports, role-based dashboards allow for the ability to drill down across different data views.
- **Improve risk response and decision making**—Alerts can be defined against key performance indicators (KPIs), allowing users to identify control issues and their root causes. GRC Intelligence integrates with performance management solutions to assist financial consolidation and reporting.

Enterprise GRC Manager

With the increasing burden of compliance and ever-present risk of fraud, businesses are pressured to provide evidence of regulatory compliance, effective risk management, and the existence of a governance and internal control framework. Oracle Enterprise GRC Manager provides enterprises with a central repository to establish, organize and maintain risk policies, internal controls and audit plans, as shown in **figure 11.4**.

Figure 11.4—Oracle Enterprise GRC Manager Overview

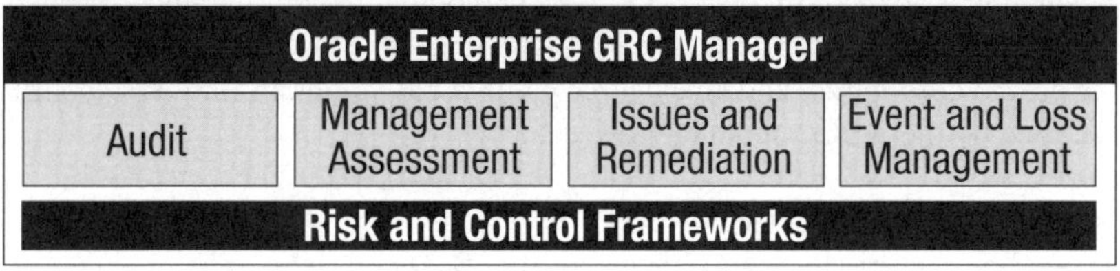

Oracle Enterprise GRC Manager

Audit | Management Assessment | Issues and Remediation | Event and Loss Management

Risk and Control Frameworks

Oracle Enterprise GRC Manager also provides dynamic workflows for compliance and risk management processes, and pre-built integration to automated application controls.

The main features of GRM Manager are the:

- **Single central repository for policies, risks and controls**—Provides enterprises with the ability to implement customized risk matrices, utilize date-effective audit trails, and use version control. The entire organization's risk and control framework can now be centrally managed rather than the yearly aggregation of disparate documentation. This provides much needed efficiencies and a holistic perspective to the organization's overall risk status.
- **Automation of compliance and risk management workflows**—Streamlines the compliance life cycle, potentially reducing the complexity and resources required for external and internal parties to complete routine audits; enables tracking of activities and focus on key issues for timely management.
- **Oversight of automated and manual control activities**—Provides operational reporting capability with standard reports leveraging the COSO framework for audit, control and risk monitoring activities

GRC Controls Products

Oracle's GRC Controls products provide enterprises with four independent automated tools that allow for integrated central management and monitoring of controls across user access, SoD, transactional anomalies, security and key configuration changes. The products also offer preventive controls at a user level, such as limits on transaction authorizations and mandatory fields when entering transactions. These automated tools can help to enhance overall security and decrease the likelihood of unauthorized and unwanted access, configuration change, or transaction processing.

Application Access Control Governor

Application Access Control Governor (AACG) provides the ability for real-time enforcement, monitoring and detailed analysis of crucial user access controls, including those surrounding SoD, as shown in **figure 11.5**. The figure shows the various steps a typical user will go through when identifying, analyzing and remediating a SoD conflict using the AACG module. In addition, there are two highlighted areas showing where the policy library of SoD conflict rules can be viewed and configured, and an example conflict path diagram for a SoD conflict identified by the AACG module.

The main features of AACG are the:

- **Real-time monitoring and enforcement of SoD controls**—The module can perform real-time monitoring and enforcement of SoD and access controls.
- **Detailed User Access, including SoD reporting**—The module can act in a detective capacity by providing detailed user access conflict reporting, including graphical representation via heat maps. The analytics in the reports allow managers to trace conflicts to their root causes and to decide on effective resolution strategies.

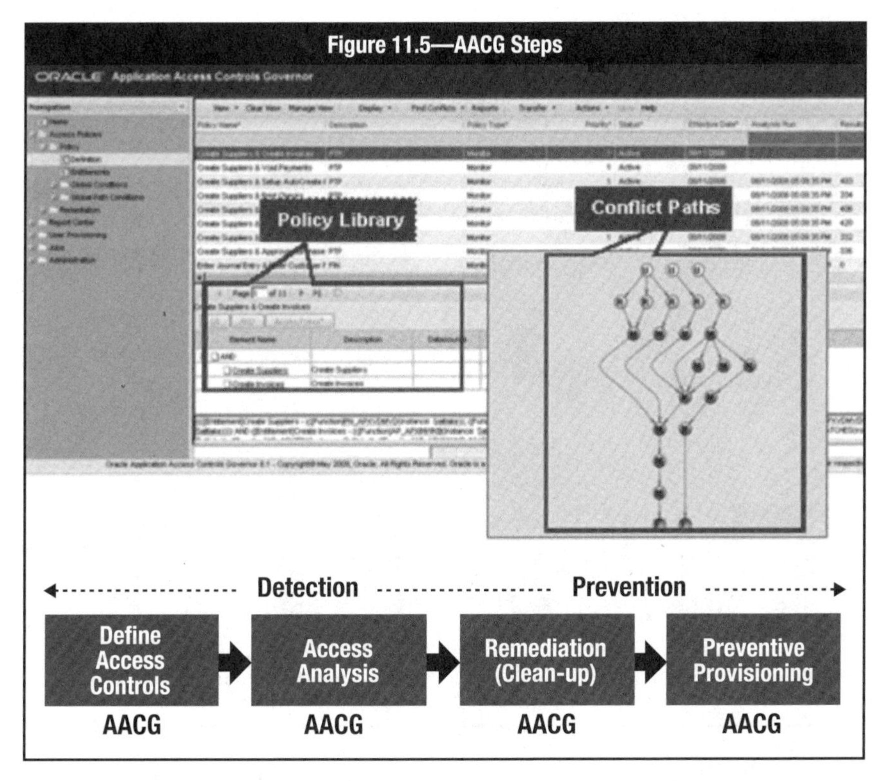

Figure 11.5—AACG Steps

- **Available Graphical Remediation Tools**—A key objective of SoD analysis is to remediate existing conflicts to an acceptable level. AACG offers visual tools to better understand the conflict issues and to perform "what if" simulated remediation activies.
- **Comprehensive Library of SoD controls**—Better practice SoD conflict libraries including those provided by Oracle EBS can be set up within AACG, saving significant setup time for the enterprise.
- **Extensibility and Seeded Connectors**—Oracle EBS provides prepackaged connectors to both EBS and PeopleSoft environments to reduce the amount of time to begin analysis and remediation activities. Custom connectors can also be built for organizations with critical applications on other platforms.
- **Multi-Platform and Cross-Platform Support**—These help manage user access across multiple platforms, instances and applications, as shown in **figure 11.6**.

Configuration Controls Governor

Ensuring data and application integrity is a fundamental business requirement to implement an effective IT governance strategy. Oracle Configuration Controls Governor (CCG) is an automated application with the ability to track any changes to key application setup and configuration data. The solution tracks all changes, providing a detailed audit history. By utilizing CCG, application integrity can be greatly improved through the ability to audit and continuously monitor setup configuration changes. As a result, a business can typically reduce regulatory cost, audit effort, and the risks associated with inappropriate or unauthorized application changes.

Figure 11.6—AACG Cross- and Multi-Platform

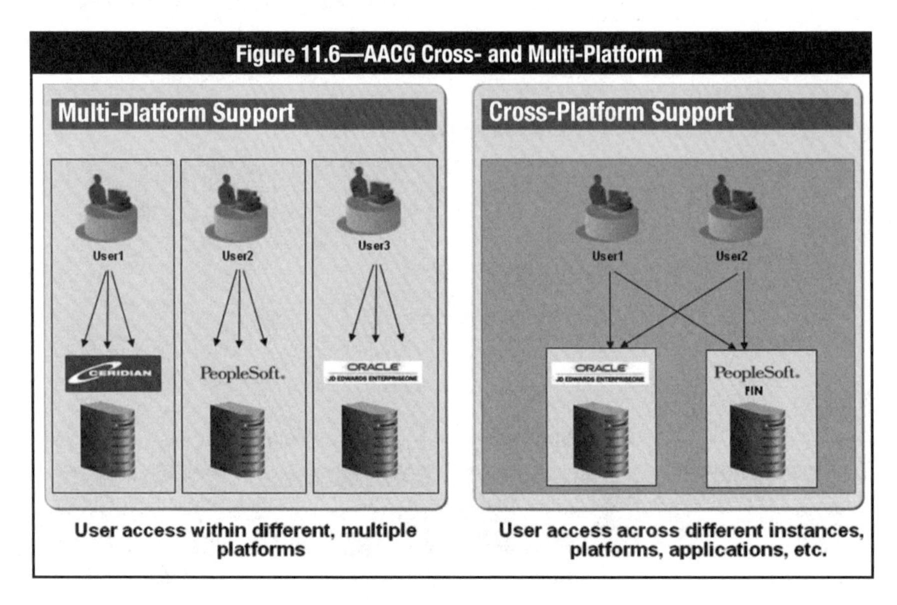

Legacy audit solutions can cause a significant impact in performance. Oracle CCG is set up to minimize the impact on daily business operations and performance of the system, as shown in **figure 11.7**. The figure shows an example of what kind of details (e.g., who, what, where) can be tracked on a data record as well as the key steps a user goes through to establish and implement a configuration control within CCG.

The main features of CCG are:

- **Monitor key setups for any change**—Monitor changes to set up configuration in a business application via a complete audit trail, including the "who, what, where and when" of the change made. In addition, security and system administration staff can receive automated notifications when a critical setup is changed in the system.
- **Utilize point-in-time configuration snapshots**—The system provides the ability for the user to define and run specific point-in-time audit snapshots.
- **Consistent setup and operating standards across multiple environments**—CCG has the ability to compare configuration settings across different environments.

Figure 11.7—CCG Steps

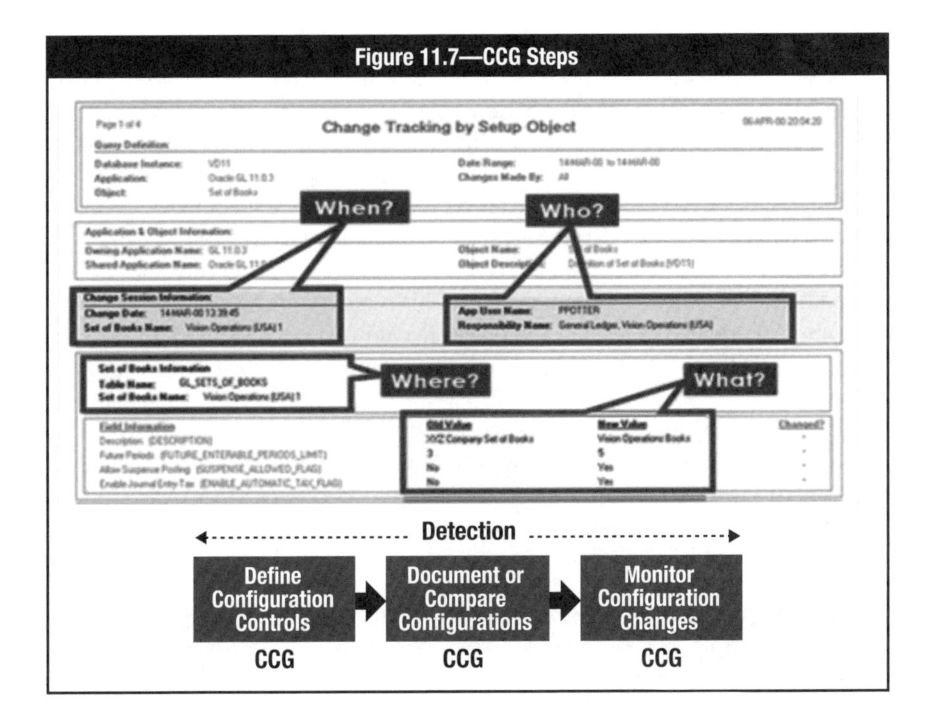

Enterprise Transaction Controls Governor

The latest version of the controls suite includes Oracle Enterprise Transaction Controls Governor (ETCG) product, which continuously monitors transactions against established policies to detect suspicious transactions or redundant business practices that could affect business performance. While all other control types deal with potential risk based on assessed likelihood and related impact, transactions reflect the actual risk that exists based on events that have already occurred. ETCG can assist management in identifying anomalies in everyday business transactions (such as those related to order-to-cash), helping to prevent cash leakage and detect potentially fraudulent transactions.

Events that—ETCG is typically able to detect are:
- Potential violation of internal controls
- Heightened levels of risk
- Validation of the effectiveness of up-stream controls
- Significant reportable transaction events

The main features of ETCG are:
- **Continuous monitoring of transactions across processes including Procure-to-Pay, Order-to-Cash, Hire-to-Retire and Record-to-Report**—This allows a business to detect actual operations of controls across these common processes and provide transactional information and notifications when a suspicious transaction occurs within and across these processes.

- **Statistical logic to uncover inappropriate or suspicious transactions and control violations systematically**—The product allows a user to mitigate unknown exposure to error, misuse and fraud from out-of-policy business transactions. The ability to identify transaction anomalies by applying pattern analysis against the data can assist in uncovering issues that would have gone undetected.
- **Intuitive authoring of access, master data and transaction controls**—A user can create and analyze controls using an interactive interface. The construct of a control is based on common business terminology and defining the conditional logic in an intuitive manner. This removes the need to involve technically oriented personnel and places the ability to define the controls in the hands of those that understand the control objectives—the business users.

Figure 11.8 shows an example of a predelivered transaction control that can be configured within the ETCG module as well as an example screen of some suspect transactions that can be pulled out and reported to users by the module. The figure also identifies they key steps a user goes through when configuring and utilizing the ETCG module.

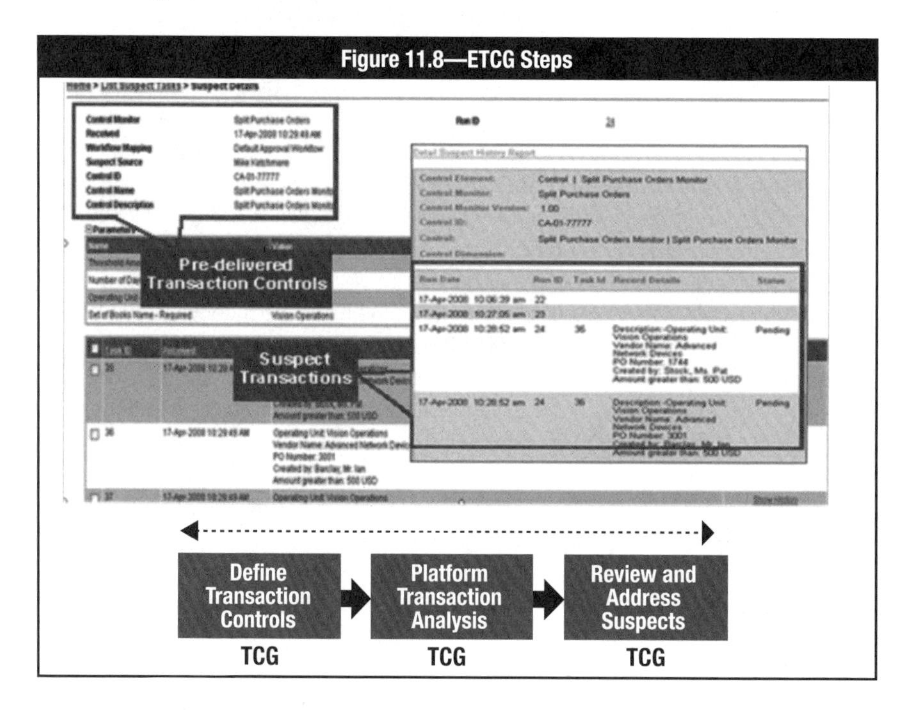

Figure 11.8—ETCG Steps

Preventive Controls Governor

Oracle Preventive Controls Governor (PCG) provides the ability to configure preventive controls in the system on items such as transaction entry forms. This provides control over the ability of certain users to enter or change certain key data on a transaction. The data fields that applications users can change or see can be limited or controlled, the types of data that can be input in various fields can be defined, and the values of certain data fields for specific user types can be limited. In addition, approval workflows and notifications can be implemented to incorporate the necessary approvals and visibility to the transactions that are being created or changed. With PCG automatically enforcing controls around data entry and manipulation, data quality is increased and the risk of incorrect or inappropriate transactions being entered is reduced. PCG can act as a key mitigating control for when additional, at-risk access must be granted.

PCG can also be integrated into AACG for EBS via the user provisioning functionality, enforcing that access policies as roles or responsibilities are assigned to users.

The main features of PCG are the:

- **Enforcement of data quality policies**—Such as mandatory fields, customizable lists of values and default values. In addition, the tool can enforce data integrity with field, block and form change control.
- **Enforcement of tolerance limits**—Such as maximum values for transactions to meet a business process or regulatory control
- **Audit trail of transaction changes**—Customizable audit logs that make keeping track of high-risk areas the focus
- **Restriction of user views**—Views of only the fields needed for the user to complete a legitimate transaction indicated by the user's setup, which can protect sensitive information
- **Audit trail of access to specific information**—Customizable audit logs of who accessed which risk-area information, facilitating compliance with the relevant privacy and data protection legislation
- **Customized approval workflows and notifications**—In addition to affecting the user's interaction with the forms of the application, workflows and notifications can be created to obtain the necessary approvals as well as to inform management of activities that are occurring.

Additional Oracle Enterprise Tools

In addition to the core GRC Suite and its underlying components, Oracle EBS provides a number of auditing and security tools that can be applied across an IT environment to assist auditing and continuous monitoring objectives further.

Oracle EBS has a robust access and security tools suite (grouped under GRC technology tools) across all levels of the IT infrastructure, including databases, applications and middleware. These enterprise tools can be used in alignment with the core GRC suite to provide enterprisewide compliance and governance.

There are four main control components available to organizations, which sit at the Oracle Database level:

- Oracle Audit Vault
- Oracle Database Vault
- Oracle Enterprise Manager
- Oracle Identity Manager

The key benefits these tools can provide include:

- Enterprise user access control and segregation of duties with business-driven rules to prevent breaches to critical corporate and customer information
- Restricted access to sensitive information in the database, based upon data classification and user clearance
- Sensitive information secured within the database with column-level database encryption
- Reduced fraud and IT risk with continuous monitoring and automated enforcement of configuration policies

Key Auditing Considerations

When auditors use tools to assist in testing Oracle EBS, they will need to confirm first that the tool is appropriately configured and that the processes for managing the tool are appropriately defined to ensure the completeness, accuracy and confidentiality of data.

Some audit procedures should be performed at least annually, whereas others may be configured as part of the initial installation of the tool.

The table in **figure 11.9** describes some overall audit considerations for the GRC Suite. These audit considerations may assist and guide auditors in reviewing the reporting and monitoring provided by the suite and its underlying tools.

Figure 11.9—Audit Considerations

Oracle GRC Suite	Audit considerations
All modules and components Note: The considerations outlined in this section should be assessed for all components each year.	**Security:** • Review logical access controls for each Oracle GRC tool to confirm that user access, in particular administrator privileges, is restricted to appropriate personnel. • Ensure that formal procedures exist for granting and removing access to the tool. • Verify that logical access controls (e.g., passwords) are in place to enhance user security of the tool. • Ensure that appropriate security audit logging is configured for changes to the tool's configuration. Logs should be reviewed on a periodic basis, and follow-up action taken as necessary. • Verify that adequate segregation of duties exist between the administrators who configure the tool and the end users. End-user privileges should be assessed to ensure that conflicting functionality is not assigned (e.g., administrators should not have the opportunity to take advantage of the tool, which would breach internal controls). • Review who has access to maintain the tables relating to the Oracle GRC solution and ensure appropriateness. These users may be different from those nominated as administrators of the tools.

Figure 11.9—Audit Considerations *(cont.)*	
Oracle GRC Suite	**Audit considerations**
All modules and components *(cont.)*	**Change Management:** • Ensure that a formal change management process (including approval for change initiation and testing procedures) is implemented. • Where configuration change history is available from the tool sets, extract a list of changes for the financial period and use this to assess the configuration changes made to the tool.
	Operations: • Confirm that the processes for administering and using the tool are documented, up-to-date and have been communicated to affected parties. • Verify that all functionalities of the GRC components required to be used by the enterprise have been appropriately tested and are operating as expected. • Confirm that the outputs of the GRC solution meet the needs of key stakeholders and that key stakeholders understand the information being presented to them. • Check that users of the GRC solution have received appropriate training. • Ensure that all the information subject to privacy and data protection legislation is accessed only by authorized personnel.

Summary

This chapter introduced the concept of continuous monitoring in an ERP environment. An overall understanding of the Oracle GRC suite and its components for implementation in an Oracle EBS environment was provided as well as a summary of the key components of the Oracle GRC suite.

Page intentionally left blank

12. Trends and Discussions Around Oracle ERP

This chapter looks ahead to major new directions emerging within the security, audit and control arena, with specific reference to tools and products from Oracle Corp. to support these new directions. This chapter also focuses on providing information related to ERP audit discussions that continue to evolve, namely:

- The implications of the changing compliance landscape for ERP control over financial reporting
- An extension to the traditional ERP control framework to encompass the integrated ERP environment

Oracle Corp. Product and Technology Changes

Over the past several years, Oracle Corp. has undertaken an aggressive corporate acquisition strategy, acquiring leading enterprise software application vendors such as PeopleSoft (which included JD Edwards) and Siebel. In addition, Oracle Corp. significantly improved its continuous control monitoring and GRC tool set by acquiring one of the leading GRC tool providers, LogicalApps. The early 2010 acquisition of Sun brings more depth and capability to the middleware application suite from Oracle Corp., including identity and enterprise management solutions.

Project Fusion

Having acquired a variety of software application vendors and their application suites, Oracle Corp. embarked on the ambitious software engineering exercise of combining each of the disparate applications into a consolidated application suite. The name given to this next-generation enterprise application suite is Oracle Fusion Applications. Fusion Applications plans to combine the best features of their current enterprise applications, including Oracle EBS, the Siebel CRM Suite, PeopleSoft Enterprise Applications, and the JD Edwards World and EnterpriseOne Application Suites along with other smaller software developments.

This suite will utilize the Oracle Fusion middleware technologies, currently at version 11*g*. To allay current customers' fears, Oracle Corp. also indicates that the new Fusion Suite will not directly replace any current products. Customers running supported Fusion versions will have the flexibility to upgrade the Fusion Applications they wish to use in conjunction with their current enterprise solutions. This allows for a manageable upgrade path to Fusion Applications. Current projections expect the Fusion application suite release in 2010.

Figure 12.1 shows the integration of current product suites into the Fusion middleware, which will allow for an upgrade path where the customer chooses when to upgrade to the Fusion applications. **Figure 12.2** shows the replacement of current suites with the Fusion applications.

Figure 12.1—Fusion Middleware Current Product Integration

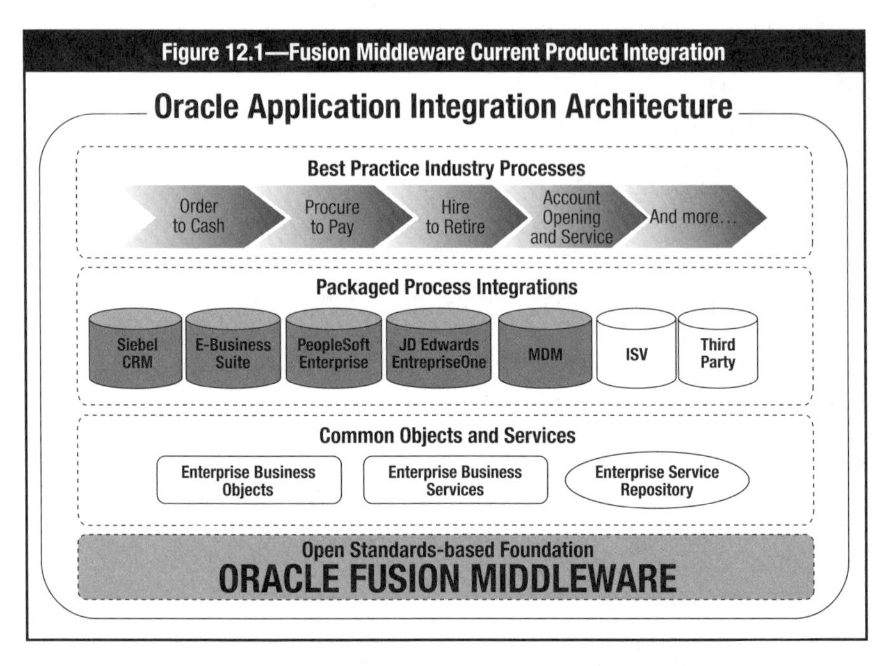

Figure 12.2—Replacement of Current Suites With the Fusion Applications

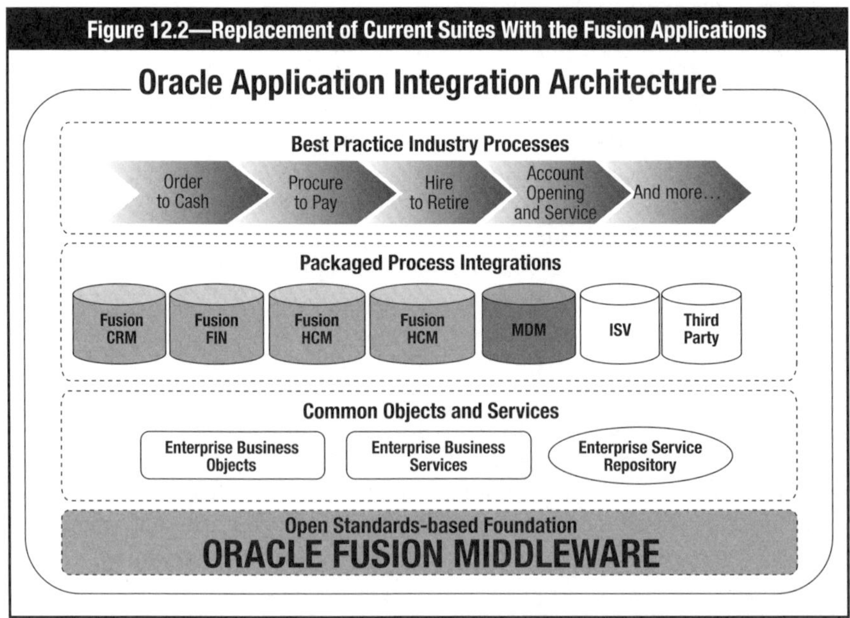

Supported Upgrade Paths to Fusion Applications

To encourage clients to upgrade existing installed base applications to current versions, Oracle Corp. has also committed to direct migrations paths to Fusion Applications. The key application versions supported are:

- Oracle EBS R11*i*10 and R12
- PeopleSoft Enterprise 8.8, 8.9, 9.0
- JD Edwards EnterpiseOne 8.11 and 8.12
- JD Edwards World A7.3, A8.1 and A9.1

Post-Fusion Application Support

Despite the impending release of the Fusion enterprise suite, Oracle Corp. intends to continue to support each of its major enterprise solutions including EBS, Siebel, PeopleSoft and JD Edwards. It has committed to continually supporting and improving each of these applications into the near future. Specifically, through a lifetime support policy, Oracle Corp. has established detailed timelines for each major application or technology in its portfolio, including the planned support, enhancement and fix releases for these applications/technologies, as well as which versions will upgrade to an application built on Fusion middleware and when.

The support in place for current EBS solutions is:

- **Oracle EBS R11*i*10**—Direct upgrade to application built on Fusion available. Extended Support until November 2013.
- **Oracle EBS R12**—Direct upgrade to application built on Fusion available. Extended Support until January 2015.
- **Oracle EBS R12.1**—Built on Fusion middleware. Extended Support until May 2017.

Monitoring the Adoption of Fusion Middleware in Preparation for Fusion Applications

This publication addresses security, audit and control features for Oracle EBS. As the migration to Fusion Applications accelerates, the implications for auditors will become clearer. It is likely that ERP audit frameworks and audit techniques will have to be reconsidered. Particular emphasis should be given to the cross-platform business process management technology and consolidated identity management infrastructure being implemented within the Fusion Applications architecture.

As Oracle Corp.'s portfolio of application suites has evolved, enhancements have enabled business process integration beyond enterprise boundaries. The inherent design of Fusion Applications, combined with new Fusion functionalities, enables extended and collaborative business process integration with suppliers, business partners, customers and internal organizations. This integration is made possible through the Oracle Fusion middleware platform architecture and its associated functions. Their adoption presents a variety of implications for security and audit processes:

- **Business process orchestration (BPO)**—Tools to define, automate and execute enterprise business processes to interact with multiple technology systems, workflows, manual operations and human interactions in a standards-based manner. BPO encapsulates business processes and increases automation of systems and process integration. Accordingly, auditors must be aware of the risks of inappropriately designed or implemented processes, and data integrity risks associated with complex processes.
- **Business activity monitoring (BAM)**—Tools to collect data from multiple operational and analytical sources of information from across the enterprise and present data to decision makers in a graphical format to assist real-time decision making. BAM provides auditors with a mechanism for capturing the status of operational processes and risk profiles across the enterprise.
- **Identity management (IM)**—Tools to define and administer user accounts and their authorization and access policies centrally, and provisioning and deprovisioning access across enterprise processes and systems. The implementation of IM suites, such as the one provided by Oracle Fusion middleware, enables auditors to attest to the compliance of security policies across a larger breadth of enterprise activities more readily, particularly those that relate to system and information access.

The Changing Compliance Landscape

Increased pressures from government regulations, financial markets and shareholders, combined with corporate collapses and acts of management fraud in recent years, have resulted in a continuing focus on corporate governance and risk management. The events of the global financial crisis in 2009 have led to numerous updates to regulations and the possibility of further changes.

The overall result has been a rising tide of corporate regulations around the world over the last decade, with significant implications for financial and nonfinancial reporting, such as:

- **US Sarbanes-Oxley Act of 2002**—One of the most far-reaching pieces of legislation to impact financial reporting. Sarbanes-Oxley compliance efforts have focused on sections 302 and 404 of the Act and have brought about a major increase in internal control frameworks and corporate governance. Other regulations such as JSOX, the Japanese equivalent to Sarbanes-Oxley, and the French Financial Security Law, *Loi sur la Securite Financiere* (LSF), have had similar impacts on enterprises.
- **Accelerated and automated reporting requirements**—At the same time that the Sarbanes-Oxley Act was introduced by the US Securities and Exchange Commission (SEC), a requirement was introduced that progressively shortened the grace period between the end of the financial year and the due date for the

lodging of financial statements with the SEC. This accelerated the reporting schedule and subsequently introduced changes such as the automation of significant aspects of the financial reporting process. ERP systems, in particular, have had an important role in this regard.

- **New Basel Capital Accord (Basel II)**—Requires banks to maintain minimum levels of capital with the objective of better aligning regulatory capital measures with the inherent risk profile of a bank, considering credit risk, market risk, operational risk and other risks. If the bank can demonstrate a risk management and control regime that reduces its risk profile, the accord allows for a reduction in the level of capital held.
- **International Financial Reporting Standards (IFRS)**—Many of the standards existing in IFRS were known as the International Audit Standard (IAS) prior to 2001. Although not a recent regulation, the focus has now shifted toward the global implementation of IFRS standards. IFRS promotes the adoption of a single set of global accounting standards requiring high-quality, transparent and comparable information in an enterprise's financial statements. The European Commission (EC) adopted the consolidated text of the IFRS in the European Union (EU) in 2008 (for detailed information about IAS/IFRS, refer to *www.iasplus.com*). Based on continuing convergence, the SEC announced that US enterprises will be allowed to report under these new standards as early as 2010 and that compliance for all US enterprises may be required by 2014, assuming certain project milestones are met. This means that enterprises may be required to make changes to the way in which account balances are presented in their financial statements (e.g., elimination of internally generated goodwill) and provision of additional disclosures in the notes to their annual report.
- **Sustainability and Corporate Responsibility**—Enterprises worldwide are under increasing pressure by stakeholders to conduct themselves in a more socially responsible manner. This includes environmental regulations, such as International Organization for Standardization (ISO) 14064 and the Kyoto protocol, enforced in a number of countries through local regulations such as the National Greenhouse and Energy Reporting Act 2007 in Australia. The energy sector is particularly impacted and will need to monitor and report accurate information about greenhouse gases and other emissions, as well as ensuring that the information reported is audited. This will continue to become an increasingly important regulatory requirement as there is a global push for a decrease in carbon emissions.
- **Data Privacy**—There is a global spotlight on privacy, with more countries adopting data privacy regulations. Data privacy is becoming a priority for a number of enterprises to ensure that information—in particular, sensitive personal information (e.g., credit card numbers, health records)—is handled with concern, to ensure information confidentiality, integrity and protection. With the new social media landscape and the increasing value of data to businesses, data privacy has not only become a regulatory requirement, but a key risk that enterprises must consider.

To support corporate governance reforms, many enterprises around the world have adopted an internal control framework such as COSO, COCO[6] or Turnbull.[7] The COSO framework (discussed in chapter 3) is commonly used by enterprises to define an internal control framework and methodology and can be adapted to manage the financial reporting risks associated with an ERP environment. The framework covers five interrelated components:

1. Control environment (tone at the top)
2. Risk assessment (risk identification and analysis)
3. Control activities (process-level controls in support of significant accounts in the financial statements, e.g., review of the control system and segregation of duties)
4. Information and communication (identification, capture and reporting of financial and operating information that is useful to control the enterprise's activities)
5. Monitoring (assessment of the robustness and relevance of the internal control activities)

An internal control system is judged to be effective if the five components are present and functioning effectively for operations, financial reporting and compliance.

As enterprises have scrambled their way toward compliance with a plethora of regulations, the role of IT in meeting the requirements of these regulations has increased. **Figure 12.3** illustrates how IT controls are embedded within each element of an enterprise's business. The role of ERP systems is particularly important at the IT application control level, as data from numerous business processes are integrated and directly input into financial statements. As such, the ERP system is at the heart of the control requirements of corporate governance and risk management regulations that affect the financial reporting process.

When designing internal controls using the COSO framework, enterprises should consider enabling certain functionalities in their ERP systems or implementing new technologies to achieve the following objectives:

- Effectiveness and efficiency of operations
- Reliability of financial reporting
- Compliance with applicable laws and regulations

The key areas that should be considered by enterprises to ensure compliance with financial reporting regulations include:

- **Data capturing capability**—The ability of the ERP system to capture sufficient information to meet the requirements of corporate governance regulations, such as IFRS. A risk management system designed to identify,

[6] Canadian Institute of Chartered Accountants (CICA), *Criteria of Control* (COCO), Canada, 1995
[7] Turnbull, Nigel; Institute of Chartered Accountants in England and Wales (ICAEW); *Turnbull Report*, UK, 2005

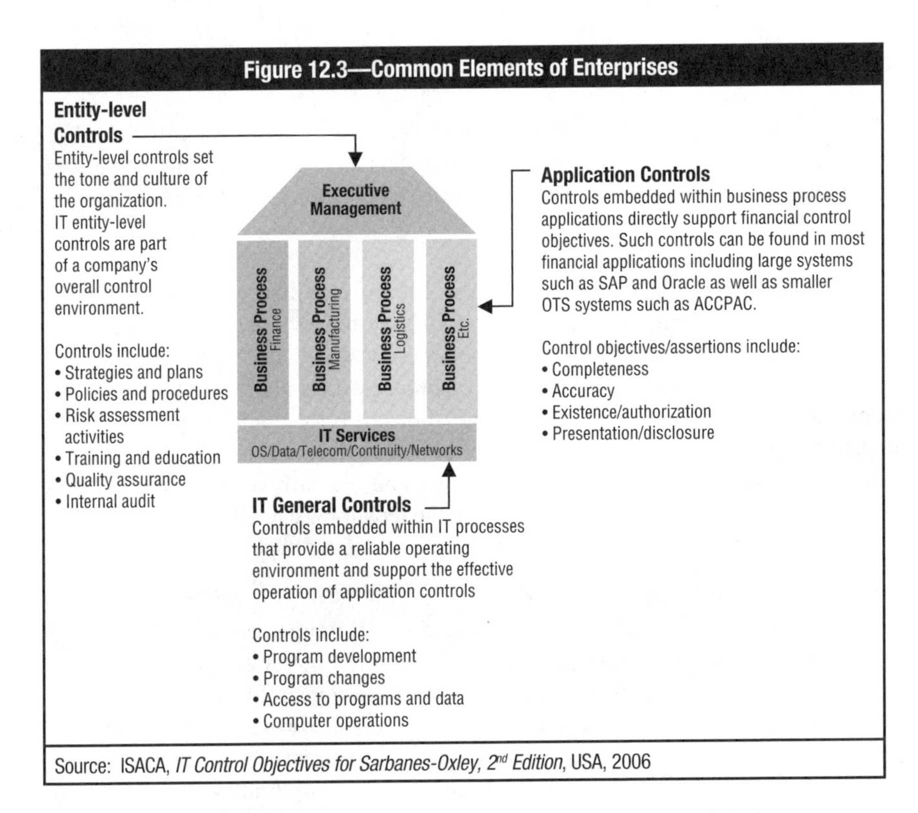

Entity-level Controls
Entity-level controls set the tone and culture of the organization. IT entity-level controls are part of a company's overall control environment.
Controls include:
• Strategies and plans
• Policies and procedures
• Risk assessment activities
• Training and education
• Quality assurance
• Internal audit
Executive Management
Business Process Finance
Business Process Manufacturing
Business Process Logistics
Business Process Etc.
IT Services
OS/Data/Telecom/Continuity/Networks
Application Controls
Controls embedded within business process applications directly support financial control objectives. Such controls can be found in most financial applications including large systems such as SAP and Oracle as well as smaller OTS systems such as ACCPAC.
Control objectives/assertions include:
• Completeness
• Accuracy
• Existence/authorization
• Presentation/disclosure
IT General Controls
Controls embedded within IT processes that provide a reliable operating environment and support the effective operation of application controls
Controls include:
• Program development
• Program changes
• Access to programs and data
• Computer operations

Figure 12.3—Common Elements of Enterprises

Source: ISACA, *IT Control Objectives for Sarbanes-Oxley, 2nd Edition*, USA, 2006

assess, monitor and manage risk and inform investors of material changes to the enterprise's risk profile should be implemented.

- **Reporting functionality**—The ability of the ERP system to extract meaningful information for internal and external reporting purposes
- **Accountability**—The ability of the ERP system to provide adequate and sufficient evidence of automated controls, such as digital sign-offs and automated reconciliations, and logs of changes to key data. This assists with providing greater visibility over SoD in the underlying business processes that support the significant accounts in the financial statements.

The potential for control automation is the largest benefit of using ERP systems to support corporate governance and financial reporting. When enterprises design any form of internal controls, they must first consider the potential for automating the required controls through the ERP system. In addition, the overall governance framework across the ERP system has the potential to be an integrated, all-in-one approach through the advent of GRC tools. As detailed in chapter 11, the comprehensive Oracle GRC tool suite can automate and provide continuous monitoring over the majority of internal controls across ERP systems such as Oracle EBS.

Using Oracle EBS Tools to Support Corporate Governance

The continuing development of the Oracle GRC tool set has greatly increased the tools available, as well as the impact they can have on supporting enterprises with their governance needs. The capabilities of the Oracle GRC tool suite are covered in chapter 11.

Regardless of what technology solution an enterprise utilizes to assist with its compliance programs (whether it is an ERP system or functionality built into systems spread across the enterprise), merely having the technology or employing consultants to configure the technology does not suggest that an enterprise is compliant with specific regulations. It is the entire process of identifying risks, defining control objectives, designing controls, implementing controls (possibly as part of an ERP system), and regularly reviewing and testing these controls that is more important than the technology itself.

Application Security Assurance

ERP systems have powerful and complex security arrangements. Testing security is not just a matter of reviewing the security matrix of users to functions. Activity groups or composite security responsibilities may be used, and unraveling these to find out who has access to what often requires automated diagnostic tools to complete the task effectively. Security exposures can also arise from the combination of security responsibilities assigned to a user through weak or poorly configured access controls. To effectively identify and assess this type of exposure, automated security diagnostic tools are required.

Automated security diagnostic tools are needed to:

- Provide a historic view, e.g., identifying when security parameters were changed.
- Unravel complex security responsibilities.
- Identify potential SoD conflicts during access creation.
- Provide dynamic documentation of user access, including who approved certain access rights granted to a user.
- Promote independence by gathering the information required for an audit/review automatically through a tool; therefore, independently gathered compared to the system administrator providing the information.
- Provide a noninvasive measurement tool (i.e., there is minimal disruption to the operation of the production system, usually through extracting and downloading the required information offline for subsequent evaluation).

Tools that are either part of or integrated with the Oracle EBS environment can be used for continuous monitoring of the adequacy of security access. Tools such as the Oracle GRC suite detailed in chapter 11 can provide real-time preventive controls so that as access is granted to new or existing users, automated checks against their segregation of duty tables are executed and (depending on configuration of tools) inappropriate access may not be obtained. This may be tailored to allow segregation of duties conflicts upon adoption of certain predefined and approved mitigating controls. In addition to being able to handle SoD conflicts, the Oracle GRC suite comes with workflow approval solutions, a full suite of reporting capabilities, configuration control monitoring, and management and business process control automation techniques. The techniques are extensive and easily tailored to a business-specific need. They also offer closed-loop remediation techniques to:

- Manage the risk and control framework.
- Identify potential control issues.
- Route the details to the appropriate individuals.
- Capture their corrective action plans.
- Monitor progress over time in addressing the issue.

Data Integrity Assurance

Despite the deployment of business process and application security controls, transactions inconsistent with management's intentions may continue to be processed. For example, while the ERP software may prevent a duplicate invoice number from being entered, the data entry operator may have entered a duplicate invoice with a different supplier name or with the invoice number in a different case (e.g., abc123 and ABC123). This may lead to a duplicate payment that may go undetected, and it is evident that duplicate payments can occur in an ERP environment.

Thus, data assurance techniques and tools such as Enterprise Transaction Controls Governor (ETCG) from the GRC suite are needed to control and test specific identified risks associated with the data in an ERP environment. If performed manually, knowledge of the ERP tables, data structures, reports, programming techniques and data extraction techniques is needed so that data are extracted in the most cost-effective manner for analysis. Tools in this area, such as TCG, are evolving, and the tools now have the ability to extract information from the modules and tables required and then perform analytics and test for data integrity and quality. For example, they can obtain information on payments and then analyze it for duplications and report back if any of these payments are suspicious.

Oracle Preventive Controls Governor (PCG) gives enterprises the ability to configure and monitor preventive controls in the system on items such as transaction entry forms.

With PCG placing and automatically enforcing controls around data entry and manipulation, data quality is increased and the chance of incorrect or inappropriate transactions being entered into the system is reduced.

Summary

The advent of ERP systems has necessitated a change in the audit approach (i.e., toward a business process focus, with greater emphasis on application security, configurable controls and continuous assurance). The audit approach is continuing to change with ERP at the heart of the controls required to meet the changing compliance landscape. The rigor of internal controls required by regulators, combined with the advent of continuous monitoring tools such as GRC, is requiring a change in the enterprise's and auditor's concepts of the audit universe. Tools such as the Oracle GRC suite provide a first step to auditors to build in the internal controls required for the future and help their enterprises maximize the very real and tangible business benefits that can be achieved with enterprise applications.

Enterprises now have an ever-increasing demand and need for quality data so that they can better understand the business. ERP systems and surrounding solutions will play a key role in providing quality data.

13. Navigator Paths

This chapter outlines, in table form, the character mode forms and menu paths, their respective GUI windows or processes, and the navigation path for Financial Accounting, Accounts Payable, the Common Country Features, and the Systems Administrator character mode form.

GL Navigator Paths

Figure 13.1 outlines the Financial Accounting character mode forms mapped to the Oracle EBS R12.1 Financials GUI window. Access to only the appropriate GL functions should be provided.

Figure 13.1—GL Navigator Paths

Character Mode Form and Menu Path	GUI Window or Process, and Navigation Path
Account Inquiry \Navigate Inquiry Account	Account Inquiry window Navigator: Inquiry→Account
Archive and Purge Options \Navigate Setup System Purge	Archive and Purge window Navigator: Setup→System→Purge
Assign Security Rules \Navigate Setup Financials Flexfields Descriptive Security Assign	Assign Security Rules window Navigator: Setup→Financials→Flexfields→Descriptive→Security→Assign Enable Descriptive Flexfield, enter search criteria, and choose the Find button.
Assign Document Sequences \Navigate Setup Financials Sequences Accounting Assign	Sequence Assignments window Navigator: Setup→Financials→Sequences→Accounting→Assign
Assign Key Flexfield Security Rules \Navigate Setup Financials Flexfields Key Security Assign	Assign Security Rules window Navigator: Setup→Financials→Flexfields→Key→Security→Assign Enable Key flexfield, enter search criteria and choose the Find button.
Assign Security Rules \Navigate Setup Financials Flexfields Validation Security Assign	Assign Security Rules window Navigator: Setup→Financials→Flexfields→Validation→Security→Assign Enable ValueSet, enter search criteria and choose the Find button.
AutoCopy \Navigate Reports AutoCopy	AutoCopy window Navigator: Reports→AutoCopy
Budget Inquiry \Navigate Inquiry Budget	Budget Inquiry window Navigator: Inquiry→Budget

Figure 13.1—GL Navigator Paths *(cont.)*	
Character Mode Form and Menu Path	**GUI Window or Process, and Navigation Path**
Budget Transfer \Navigate Budgets Enter Transfer	Budget Transfer window Navigator: Budgets➔Enter➔Transfer
Calculate Budget Amounts \Navigate Budgets Generate Formulas	Calculate Budget Amounts window Navigator: Budgets➔Generate➔Formulas
Correct Journal Import Data \Navigate Journals Import Correct	Correct Journal Import Data window Navigator: Journals➔Import➔Correct
Correct Source from Journal Import \Navigate Journals Import Correct	Correct Journal Import Data window Navigator: Journals➔Import➔Correct
Define Accounting Flexfield Combination \Navigate Setup Financials Flexfields Accounting	GL Accounts window Navigator: Setup➔Accounts➔Combinations
Define Automatic Posting Options \Navigate Setup Journal AutoPost	AutoPost Criteria Sets window Navigator: Setup➔Journal➔AutoPost
Define Budget \Navigate Budgets Define Budget	Define Budget window Navigator: Budgets➔Define➔Budget
Define Budget Formula \Navigate Budgets Define Formula	Define Budget Formula window Navigator: Budgets➔Define➔Formula
Define Budget Organization \Navigate Budgets Define Organization	Define Budget Organization window Navigator: Budgets➔Define➔Organization
Define Budgetary Control Groups \Navigate Budgets Define Controls	Budgetary Control Group window Navigator: Budgets➔Define➔Controls
Define Calendar \Navigate Setup Financials Calendar Periods	Accounting Calendar window Navigator: Setup➔Financials➔Calendar➔Accounting
Define Column Set \Navigate Reports Define ColumnSet	Column Set window Navigator: Reports➔Define➔ColumnSet
Define Consolidation \Navigate Consolidation Define	Consolidation Workbench window Navigator: Consolidation➔Workbench
Define Content Set \Navigate Reports Define ContentSet	Content Set window Navigator: Reports➔Define➔ContentSet

Figure 13.1—GL Navigator Paths *(cont.)*	
Character Mode Form and Menu Path	**GUI Window or Process, and Navigation Path**
Define Cross-Validation Rule \Navigate Setup Financials Flexfields Key Rules	Cross-Validation Rules window Navigator: Setup➔Financials➔Flexfields➔Key➔Rules
Define Currency \Navigate Setup Financials Currency Currency	Currencies window Navigator: Setup➔Currencies➔Define
Define Daily Conversion Rate Types \Navigate Setup Financials Currency RateTypes	Conversion Rate Types window Navigator: Setup➔Currencies➔Rates➔Types
Define Daily Rates \Navigate Setup Financials Currency DailyRates	Daily Rates window Navigator: Setup➔Currencies➔Rates➔Daily
Define Definition Access Sets \Navigate Setup Financials Definition Access Sets Define	Define Access Sets window Navigator: Setup➔Financials➔Definition Access Sets➔Define
Define Descriptive Flexfield Segments \Navigate Setup Financials Flexfields Descriptive Segments	Descriptive Flexfield Segments window Navigator: Setup➔Financials➔Flexfields➔ Descriptive➔Segments
Define Descriptive Security Rule \Navigate Setup Financials Flexfields Descriptive Security Define	Define Security Rules window Navigator: Setup➔Financials➔Flexfields➔Descriptive➔ Security➔Define Enable the Descriptive Flexfield, enter search criteria and choose the Find button.
Define Descriptive Segment Values \Navigate Setup Financials Flexfields Descriptive Values	Segment Values window Navigator: Setup➔Financials➔Flexfields➔Descriptive➔ Values Enter search criteria and choose the Find button.
Define Document Sequences \Navigate Setup Financials Sequences Define	Document Sequences window Navigator: Setup➔Financials➔Sequences➔Define
Define Encumbrance Types \Navigate Setup Journals Encumbrances	Encumbrance Types window Navigator: Setup➔Journal➔Encumbrances
Define Financial Report Set \Navigate Reports Define ReportSet	Financial Report Set window Navigator: Reports➔Define➔ReportSet

Figure 13.1—GL Navigator Paths *(cont.)*	
Character Mode Form and Menu Path	**GUI Window or Process, and Navigation Path**
Define Historical Rates \Navigate Setup Currencies Rates Historical	Historical Rates window Navigator: Setup➔Currencies➔Rates➔Historical
Define Journal Entry Categories \Navigate Setup Journal Categories	Journal Categories window Navigator: Setup➔Journal➔Categories
Define Journal Entry Formula \Navigate Journals Define Recurring	Define Recurring Journal Formula window Navigator: Journals➔Define➔Recurring
Define Journal Entry Sources \Navigate Setup Journal Sources	Journal Sources window Navigator: Setup➔Journal➔Sources
Define Key Flexfield Security Rule \Navigate Setup Financials Flexfields Key Security Define	Define Security Rules window Navigator: Setup➔Financials➔Flexfields➔Key➔Security➔Define Enable Key Flexfield, enter search criteria and choose the Find button.
Define Key Flexfield Segments \Navigate Setup Financials Flexfields Key Segments	Key Flexfield Segments window Navigator: Setup➔Financials➔Flexfields➔Key➔Segments
Define Key Segment Values \Navigate Setup Financials Flexfields Key Values	Segment Values window Navigator: Setup➔Financials➔Flexfields➔Key➔Values Enter search criteria and choose the Find button.
Define MassAllocations \Navigate Journals Define Allocation	Define MassAllocations window Navigator: Journals➔Define➔Allocation
Define MassBudgets \Navigate Budgets Define MassBudget	Define MassBudgets window Navigator: Budgets➔Define➔MassBudget
Define Period Rates \Navigate Setup Currencies Rates Daily	Daily Rates window Navigator: Setup➔Currencies➔Rates➔Daily
Define Period Types \Navigate Setup Financials Calendar Types	Period Types window Navigator: Setup➔Financials➔Calendar➔Types
Define Periods \Navigate Setup System Calendar Periods	Accounting Calendar window Navigator: Setup➔Financials➔Calendar➔Accounting
Define Report \Navigate Reports Define Report	Define Financial Report window Navigator: Reports➔Define➔Report

Figure 13.1—GL Navigator Paths *(cont.)*	
Character Mode Form and Menu Path	**GUI Window or Process, and Navigation Path**
Define Report Display Group \Navigate Reports Display Group	Display Group window Navigator: Reports➔Display➔Group
Define Report Display Set \Navigate Reports Display Set	Display Set window Navigator: Reports➔Display➔Set
Define Report Set \Navigate Other Report Set	Request Set window Navigator: Other➔Report➔Set
Define Rollup Groups \Navigate Setup Financials Flexfields Key Groups	Rollup Groups window Navigator: Setup➔Financials➔Flexfields➔Key➔Groups Enter criteria and choose the Find button.
Define Row Order \Navigate Reports Define Order	Row Order window Navigator: Reports➔Define➔Order
Define RowSet \Navigate Reports Define RowSet	Row Set window Navigator: Reports➔Define➔RowSet
Define Security Rule \Navigate Setup Financials Flexfields Validation Security Define	Define Security Rules window Navigator: Setup➔Financials➔Flexfields➔Validation➔ Security➔Define Enable ValueSet, enter search criteria and choose the Find button.
Define Segment Values \Navigate Setup Financials Flexfields Validation Values	Segment Values window Navigator: Setup➔Financials➔Flexfields➔Validation➔ Values Enter search criteria and choose the Find button.
Define Shorthand Aliases \Navigate Setup Financials Flexfields Key Aliases	Shorthand Aliases window Navigator: Setup➔Financials➔Flexfields➔Key➔Aliases
Define Statistical Units of Measure \Navigate Setup Accounts Units	Statistical Units of Measure window Navigator: Setup➔Accounts➔Units
Define Summary Accounts \Navigate Setup Accounts Summary	Summary Accounts window Navigator: Setup➔Accounts➔Summary
Define Suspense Accounts \Navigate Setup Accounts Suspense	Suspense Accounts window Navigator: Setup➔Accounts➔Suspense
Define ValueSet \Navigate Setup Financials Flexfields Validation Sets	Value Sets window Navigator: Setup➔Financials➔Flexfields➔Validation➔ Sets
Enter Budget Amounts \Navigate Budgets Enter Amounts	Enter Budget Amounts window Navigator: Budgets➔Enter➔Amounts
Enter Budget Journals \Navigate Budgets Enter Journals	Enter Budget Journals window Navigator: Budgets➔Enter➔Journals

Figure 13.1—GL Navigator Paths *(cont.)*	
Character Mode Form and Menu Path	**GUI Window or Process, and Navigation Path**
Enter Encumbrances \Navigate Journals Encumbrance	Enter Encumbrances window Navigator: Journals➔Encumbrance
Enter Journals \Navigate Journals Enter	Enter Journals window Navigator: Journals➔Enter
Enter Transaction Rate \Navigate Journals Enter (\Other Zoom)	Enter Journals window Navigator: Journals➔Enter
Freeze Budgets \Navigate Budgets Freeze	Freeze Budget window Navigator: Budgets➔Freeze
Generate MassAllocation Journals \Navigate Journals Generate Allocation	Generate MassAllocation Journals window Navigator: Journals➔Generate➔Allocation
Generate MassBudget Journals \Navigate Budgets Generate MassBudgets	Generate MassBudget Journals window Navigator: Budgets➔Generate➔MassBudgets
Generate Recurring Journals \Navigate Journals Generate Recurring	Generate Recurring Journals window Navigator: Journals➔Generate➔Recurring
Journal Entry Inquiry \Navigate Inquiry Journal	Journal Entry Inquiry window Navigator: Inquiry➔Journal
Open and Close Periods \Navigate Setup Open/Close	Open and Close Periods window Navigator: Setup➔Open/Close
Post Journals \Navigate Journals Post	Post Journals window Navigator: Journals➔Post
Purge Consolidation Audit Data \Navigate Consolidation Purge	Purge Consolidation Audit Data window Navigator: Consolidation➔Purge
Reports QuickPick \Navigate Other Requests	Requests window Navigator: Other➔Requests or from any window, choose View➔Requests from the menu
Request Report Set \Navigate Reports Request Financial	Run Financial Reports window Navigator: Reports➔Request➔Financial
Reverse Journals \Navigate Journals Generate Reversal	Reverse Journals window Navigator: Journals➔Generate➔Reversal
Run Consolidation \Navigate Consolidation Run	Consolidation Workbench window Navigator: Consolidation➔Workbench

Figure 13.1—GL Navigator Paths *(cont.)*	
Character Mode Form and Menu Path	**GUI Window or Process, and Navigation Path**
Run Journal Import \Navigate Journals Import Run	Import Journals window Navigator: Journals➔Import➔Run
Run Optimizer \Navigate Setup System Optimize	Submit Request window Navigator: Reports➔Request➔Standard; or Other➔Report➔Run
Run Reports \Navigate Other Report Run \Navigate Reports Request	Submit Request window Navigator: Reports➔Request➔Standard; or Other➔Report➔Run
Run Revaluation \Navigate Currency Revaluation	Revalue Balances window Navigator: Currency➔Revaluation
Run Translation \Navigate Currency Translation	Translate Balances window Navigator: Currency➔Translation
Summary Account Inquiry \Navigate Inquiry Summary	Account Inquiry window Navigator: Inquiry➔Account
Update Personal Profile Options \Navigate Other Profile	Personal Profile Values window Navigator: Other➔Profile; or Profile➔Personal Access using system administrator responsibility.
Update Storage Parameters \Navigate Setup System Storage	Storage Parameters window Navigator: Setup➔System➔Storage
Upload Budgets \Navigate Budgets Enter Upload	Upload Budgets window Navigator: Budgets➔Enter➔Upload
View Budgetary Control Transactions \Navigate Journals Enter (\Other Zoom)	Journals window Navigator: Journals➔Enter Enter search criteria, then choose Find. Select a journal batch or entry, then choose Review Batch or Review Journal to display the Journals window. Choose the More Action button and then choose the View Results button to see if a transaction failed a funds check.
View Funds Available \Navigate Inquiry Funds	Funds Available Inquiry window Navigator: Inquiry➔Funds
View Requests \Navigate Other Requests	Requests window Navigator: Other➔Requests or from any window, choose View➔Requests
Year-End Carry Forward \Navigate Journals Generate Carryforward	Year-End Carry Forward window Navigator: Journals➔Generate➔Carryforward

Expenditure Navigator Paths

Figure 13.2 outlines the Accounts Payable character mode forms mapped to the Oracle EBS R12.1 Financials GUI window. Access to only the appropriate Payables functions should be provided.

Figure 13.2—Expenditure Navigator Paths

Character Mode Form and Menu Path	GUI Window or Process, and Navigation Path
Accrual Write-offs \Navigate Tasks Accrual Write-offs	Accrual Write-offs Navigator: Invoices→Accrual Write-off
Adjust Distributions \Navigate Invoices Update Distribution	Open the Distributions window in entry mode and make adjustments. Navigator: Invoices→Entry→Invoices Choose the Distributions button.
Adjust Payment Schedule \Navigate Invoices Update PaymentSchedule	Open the Scheduled Payments window in entry mode and make adjustments. Navigator: Invoices→Entry→Invoices Choose the Scheduled Payments button.
Apply Prepayments \Navigate Payments Prepayment Apply	Apply/Unapply Prepayments window Navigator: Invoices→Entry→Invoices Query and select the invoice, and choose the Actions button. In the Invoice Actions window, check the Apply/ Unapply Prepayment checkbox and choose OK.
Assign Descriptive Security Rules \Navigate Setup Financials Flexfields Descriptive Security Assign	Assign Security Rules window Navigator: Setup→Flexfields→Validation→Security→ Assign Enable the Descriptive Flexfield, enter search criteria and choose the Find button.
Assign Key Flexfield Security Rules \Navigate Setup Financials Flexfields Key Security Assign	Assign Security Rules window Navigator: Setup→Flexfields→Validation→Security→ Assign Enable the Key Flexfield, enter search criteria and choose the Find button.
Assign Security Rules \Navigate Setup Financials Flexfields Validation Security Assign	Assign Security Rules window Navigator: Setup→Flexfields→Validation→Security→ Assign Enable ValueSet, enter search criteria and choose the Find button.
AutoClear \Navigate Controls Reconciliation AutoClear	Use Oracle Cash Management to reconcile payments in Payables.
Concurrent Requests (Pop-up window) \Help Requests	Requests window From any window, choose Requests from the View menu.

Figure 13.2—Expenditure Navigator Paths *(cont.)*	
Character Mode Form and Menu Path	**GUI Window or Process, and Navigation Path**
Control Period Statuses \Navigate Controls Payables Periods	Control Payables Periods window Navigator: Accounting➔Control Payables Periods
Create Mass Additions for Oracle Assets \Navigate Tasks Mass Additions	Submit the Mass Additions Create from the Submit Request window. Navigator: Other➔Requests➔Run
Create QuickCheck \Navigate Payments QuickCheck	Create a quick payment in the Payments window. Navigator: Payments➔Entry➔Payments
Create Recurring Payments \Navigate Invoices Recurring CreateInvoices	Use the Recurring Invoices window to create a recurring invoice template and create recurring invoices. Navigator: Invoices➔Entry➔Recurring Invoices
Define Accounting Flexfield Combination \Navigate Setup Financials Flexfields Accounting	GL Accounts window Navigator: Setup➔Flexfields➔Combinations
Define Automatic Payment Programs \Navigate Setup Payments Administrator	Automatic Payment Administrator window Navigator: Setup➔Payment➔Payment Administrator
Define Bank Codes \Navigate Setup Payments AutoClear Codes	Use Oracle Cash Management to reconcile payments in Oracle Payables.
Define Bank File Specifications \Navigate Setup Payments AutoClear File	Use Oracle Cash Management to reconcile payments in Oracle Payables.
Define Calendar \Navigate Setup Financials Calendar Periods	Accounting Calendar window Navigator: Setup➔Calendar➔Accounting➔Periods
Define Cross-Validation Rule \Navigate Setup Financials Flexfields Key Rules	Cross-Validation Rules window Navigator: Setup➔Flexfields➔Key➔Cross-Validation
Define Currency \Navigate Setup Financials Currency Define	Currencies window Navigator: Setup➔Currency➔Define
Define Daily Conversion Rate Types \Navigate Setup Financials Currency Rate Types	Conversion Rate Types window in Oracle GL Navigator: Setup➔Currency➔Rates➔Types

Figure 13.2—Expenditure Navigator Paths *(cont.)*	
Character Mode Form and Menu Path	**GUI Window or Process, and Navigation Path**
Define Daily Rates \Navigate Setup Financials Currency Daily Rates	Daily Rates window Navigator: Setup➔Currency➔Rates➔Daily
Define Descriptive Flexfield Segments \Navigate Setup Financials Flexfields Descriptive Segments	Descriptive Flexfield Segments window Navigator: Setup➔Flexfields➔Descriptive➔Segments
Define Descriptive Security Rule \Navigate Setup Financials Flexfields Descriptive Security Define	Define Security Rules window Navigator: Setup➔Flexfields➔Validation➔Security➔Define Enable Descriptive Flexfield, enter search criteria and choose the Find button.
Define Descriptive Segment Values \Navigate Setup Financials Flexfields Descriptive Values	Segment Values window Navigator: Setup➔Flexfields➔Descriptive➔Values Enable Descriptive Flexfield, enter search criteria and choose the Find button.
Define Expense Reports \Navigate Setup Invoices Xpress	Expense Report Templates window Navigator: Setup➔Invoice➔Expense Report Templates
Define Financials Options \Navigate Setup Financials Options	Financials Options window Navigator: Setup➔Options➔Financials Options
Define Flex Segments for Expense Reporting \Navigate Setup Reports ExpenseDetail	Navigator: Other➔Requests➔Run. Extract the Account Analysis report. There is no associated setup form.
Define Income Tax Regions \Navigate Setup System TaxRegions	Income Tax Regions window Navigator: Setup➔Tax➔Regions
Define Invoice Approval Workflow \Navigate Setup Invoices Approval Workflow	Invoice Approval Workflow window Navigator: Setup➔Invoice➔Approval Workflow
Define Key Flexfield Security Rule \Navigate Setup Financials Flexfields Key Security Define	Define Security Rules window Navigator: Setup➔Flexfields➔Validation➔Security➔Define Enable Key Flexfield, enter search criteria and choose the Find button.
Define Key Flexfield Segments \Navigate Setup Financials Flexfields Key Segments	Key Flexfield Segments window Navigator: Setup➔Flexfields➔Key➔Segments

Figure 13.2—Expenditure Navigator Paths *(cont.)*	
Character Mode Form and Menu Path	**GUI Window or Process, and Navigation Path**
Define Key Segment Values \Navigate Setup Financials Flexfields Key Values	Segment Values window Navigator: Setup➜Flexfields➜Key➜Values Enable Key Flexfield, enter search criteria and choose the Find button.
Define Location \Navigate Setup Organization Location	Location window Navigator: Employees➜Locations
Define Payables QuickCodes \Navigate Setup System QuickCodes Payables	Oracle Payables Lookups and Oracle Purchasing Lookups Navigator: Setup➜Lookups➜Payables; or Setup➜Lookups➜Purchasing
Define Payment Administrator \Navigate Setup Payments Administrator	Payment Administrator window Navigator: Setup➜Payment➜Payment Administrator
Define Payment Interest Rates \Navigate Setup Payments Interest	Payment Interest Rates window Navigator: Setup➜Payment➜Interest Rates
Define Payment Terms \Navigate Setup Invoices Terms	Payment Terms window Navigator: Setup➜Invoice➜Payment Terms
Define Period Types (Financials) \Navigate Setup Financials Calendar Types	Period Types window Navigator: Setup➜Calendar➜Accounting➜Types
Define Period Types (Payables) \Navigate Setup System Calendar Types	In the Special Calendar window, periods are defined when the special calendar is defined. Navigator: Setup➜Calendar➜Special Calendar
Define Periods \Navigate Setup System Calendar Periods	Accounting Calendar window Navigator: Setup➜Calendar➜Accounting➜Periods
Define QuickCodes \Navigate Setup System QuickCodes Employee	Oracle Human Resources Lookups window Navigator: Setup➜Lookups➜Employee
Define Recurring Payments \Navigate Invoices Recurring Define	Recurring Invoices window Navigator: Invoices➜Entry➜Recurring Invoices
Define Report Set \Navigate Setup Reports ReportSets	Request Set window Navigator: Other➜Requests➜Set
Define Reporting Entities \Navigate Setup Organization Reporting Entities	Reporting Entity window Navigator: Setup➜Tax➜Reporting Entities

Figure 13.2—Expenditure Navigator Paths *(cont.)*	
Character Mode Form and Menu Path	**GUI Window or Process, and Navigation Path**
Define Rollup Groups \Navigate Setup Financials Flexfields Key Groups	Rollup Groups window Navigator: Setup➔Flexfields➔Key➔Groups Enable Key Flexfield, enter search criteria and choose the Find button.
Define Security Rule \Navigate Setup Financials Flexfields Validation Security Define	Define Security Rules window Navigator: Setup➔Flexfields➔Validation Security➔Define Enable the ValueSet, enter search criteria and choose the Find button.
Define Segment Values \Navigate Setup Flexfields Validation Values	Segment Values window Navigator: Setup➔Flexfields➔Validation➔Values Enable Key Flexfield, enter search criteria and choose the Find button.
Define Shorthand Aliases \Navigate Setup Financials Flexfields Key Aliases	Shorthand Aliases window Navigator: Setup➔Flexfields➔Key➔Aliases
Define System Options and Defaults \Navigate Setup Options Payables System Setup	Payables Options window Navigator: Setup➔Options➔Payables System Setup
Define Tax Names \Navigate Setup Taxes Names	Tax Codes window Navigator: Setup➔Tax➔Codes
Define Tolerances \Navigate Setup Invoices Tolerances	Invoice Tolerances window Navigator: Setup➔Invoice➔Tolerances
Define ValueSet \Navigate Setup Financials Flexfields Validation Sets	Value Sets window Navigator: Setup➔Flexfields➔Validation➔Sets
Define Vendor QuickCodes \Navigate Setup System QuickCodes Vendor	Oracle Payables Lookups window and Oracle Purchasing Lookups window Navigator: Setup➔Lookups➔Payables/Purchasing
Define Withholding Tax Groups \Navigate Setup Taxes Groups	Withholding Tax Groups window Navigator: Setup➔Tax➔Withholding➔Groups
Distribution Inquiry \Navigate Controls Distribution Inquiry	View Journal Entry Lines window Navigator: Accounting➔Subledger Accounting➔Journal Entry Lines In the Find Journal Entry Lines window, enter search criteria and choose the Find button.
Enter Employee \Navigate Setup Organization Employees Enter	Enter Person window Navigator: Employees➔Enter Employees

Figure 13.2—Expenditure Navigator Paths *(cont.)*	
Character Mode Form and Menu Path	**GUI Window or Process, and Navigation Path**
Enter Invoices \Navigate Invoices Entry	Invoices window Navigator: Invoices→Entry→Invoices
Enter Manual Payment \Navigate Payments Manual	Payments window Navigator: Payments→Entry→Payments
Enter Prepayments \Navigate Payments Prepayment Enter	In the Invoices window, enter an invoice and use Prepayment in the Type field. Navigator: Invoices→Entry→Invoices
Enter QuickInvoices \Navigate Invoices Quick	Invoices window Navigator: Invoices→Entry→Invoices
Enter Standard Notes \Navigate Setup Invoices Notes	Standard notices can be customized. From the Invoices window, choose the Actions button. Select Print or write notes outside of Oracle Payables, and then attach the text, spreadsheet or image files to invoices.
Enter Vendor \Navigate Vendors Entry	Suppliers window Navigator: Suppliers→Entry
Enter/Adjust Manual Payment \Navigate Controls Payment AdjustPayment	Payments window Navigator: Payments→Entry→Payments
Fix Payment Distributions \Navigate Controls Payment FixDistributions	Accounting has changed in Oracle Payables.
GL Interface \Navigate Tasks GLPost	From the Submit Request window, submit the Payables transfer to the GL program. Navigator: Other→Requests→Run
Invoice Approval \Navigate Invoices Approval	Invoice Holds window Navigator: Invoices→Enter→Invoices Choose the Hold button.
Maintain Countries and Territories \Navigate Setup System Countries	Countries and Territories window Navigator: Setup→Countries
Maintain Distribution Sets \Navigate Setup Invoices DistributionSets	Distribution Sets window Navigator: Setup→Invoice→Distribution Sets
Maintain Tax Certificates and Exceptions \Navigate Setup Taxes Certificates	Withholding Tax Certificates and Exceptions window Navigator: Setup→Tax→Withholding→Certificates

Figure 13.2—Expenditure Navigator Paths *(cont.)*	
Character Mode Form and Menu Path	**GUI Window or Process, and Navigation Path**
Modify Invoice Selection \Navigate Payments Automatic Modify	Modify Payment Batch window Navigator: Payments→Entry→Payments Choose the Actions button. Select Modify Payment Batch and choose OK.
Payment Inquiry \Navigate Payments Inquiry	Payments window Navigator: Payments→Inquiry→Payments or Payment Overview window Navigator: Payments→Inquiry→Payment Overview
Reconcile Payments \Navigate Controls Reconciliation Manual	Use Oracle Cash Management to reconcile payments in Oracle Payables.
Reset Payment Batch \Navigate Controls Payment ResetPaymentBatch	Payment Batch Actions window Navigator: Payments→Entry→Payment Batches Choose the Actions button. Select Confirm Payment Batch and choose OK.
Resolve AutoClear Exceptions \Navigate Controls Reconciliation Exceptions	Use Oracle Cash Management to reconcile payments in Payables.
Run Reports \Navigate Reports Standard	Submit Request window Navigator: Other→Requests→Run
Set Up Aging Periods \Navigate Setup Reports Invoice Aging	Aging Periods window Navigator: Setup→Calendar→Aging Periods
Set Up Bank Information \Navigate Setup Banks and Bank Accounts	Banks window Navigator: Setup→Payment→Banks and Bank Accounts
Stop Payment \Navigate Payments Stop	Initiate Stop checkbox in the Payment Actions window Navigator: Payments→Entry→Payments From the Payments window, select the payment and choose Actions. In the Payment Actions window, check the Initiate Stop checkbox and choose OK.
Submit AutoApproval Process \Navigate Tasks AutoApproval	From the Submit Request window, submit the Payables approval program. Navigator: Other→Requests→Run
Submit Expense Distribution Detail Report \Navigate Reports ExpenseDetail	The Expense Distribution Detail report has been replaced by the Payables Account Analysis report. From the Submit Request window, submit the Payables Account Analysis report. Navigator: Other→Requests→Run

Figure 13.2—Expenditure Navigator Paths *(cont.)*	
Character Mode Form and Menu Path	**GUI Window or Process, and Navigation Path**
Submit Invoice Import \Navigate Tasks InvoiceImport	From the Submit Request window, submit the Payables Invoice Import program or the Payables Open Interface program. Navigator: Other→Requests→Run
Submit Purge (purge responsibility only) \Navigate Purge	Submit Purge window Use the purge responsibility. Navigator: Purge
Update Personal Profile Options \Navigate Other Profile	Personal Profile Values window Navigator: Other→Profile
Use Prepayments \Navigate Payments Prepayment Use	Apply/Unapply Prepayments window Navigator: Invoices→Entry→Invoices Query and select the invoice and choose Actions. In the Invoice Actions window, check the Apply/Unapply Prepayment checkbox and choose OK.
VendorMerge \Navigate Controls VendorMerge	Supplier Merge window Navigator: Suppliers→Supplier Merge
View Budgetary Control Transactions (zoom only)	In the Invoices window, if a transaction fails the funds check, choose Budgetary Control from the Tools menu. Navigator: Invoices→Entry→Invoices
View Employees \Navigate Setup Organization Employees View	Enter Person window Navigator: Employees→View Employees
View Invoices \Navigate Invoices Inquiry	Invoices window Navigator: Invoices→Inquiry→Invoices or Invoice Overview window Navigator: Invoices→Inquiry→Invoice Overview
View Notes (zoom only)	Notes can be written outside of Payables and then attached to invoices. Choose the paperclip icon from the tool bar.
View Purchase Order Distribution Detail (zoom only)	Purchase Order Distributions window When matching to a PO in the Match to Purchase Orders window, choose the View Purchase Orders button. In the Purchase Order Shipments window, choose the Distributions button.
View Purchase Order Header (zoom only)	Purchase Orders window Use Oracle Purchasing.
View Purchase Order Line (zoom only)	Purchase Orders window Use Oracle Purchasing.
View Purchase Orders Shipment (zoom only)	Purchase Order Shipments window When matching to a purchase order in the Match to Purchase Orders window, choose the View Purchase Orders button.

Figure 13.2—Expenditure Navigator Paths *(cont.)*

Character Mode Form and Menu Path	GUI Window or Process, and Navigation Path
View Requests \Navigate Other Concurrent	Requests window Navigator: Other→Requests→View
View Vendors \Navigate Vendors Inquiry	Suppliers window Navigator: Suppliers→Inquiry
Void Payments \Navigate Payments Void	The Void checkbox in the Payment Actions window Navigator: Payments→Entry→Payments From the Payments window, select the payment and choose the Actions button. Check the Void checkbox and choose OK.
XpenseXpress \Navigate Invoices XpenseXpress	Expense Reports window Navigator: Invoices→Entry→Expense Reports

Oracle EBS Common Country Features

Figure 13.3 shows the Oracle EBS Common Country Features character mode menu paths and form names to GUI menu paths and windows or processes.

Figure 13.3—Oracle EBS Common Country Features

Character Mode Form and Menu Path	GUI Window or Process, and Navigation Path
<Country> Local GL Define Accounting Flexfield Combination window \Navigate Setup Account Combination	Oracle GL GL Accounts window Navigator: Setup→Accounts→Combinations
<Country> Local GL Enter Journals window \Navigate Journals Enter	Oracle GL Enter Journals window Navigator: Journals→Enter
<Country> Local GL Manual Reconciliation Lines window \Navigate Journal Reconciliations Reconcile	<Country> GL Localizations Reconciliation Lines window Navigator: Journals→Reconciliations→Reconcile
<Country> Local Receivables Maintain Customer Profiles window \Navigate Setup Customer Profile Maintain	Oracle Receivables Customers window Navigator: Customers→Customers

Figure 13.3—Oracle EBS Common Country Features *(cont.)*	
Character Mode Form and Menu Path	**GUI Window or Process, and Navigation Path**
<Country> Local Receivables Define Customer Profile Class window \Navigate Setup Customer Profile Class	Oracle Receivables Customer Profile Classes window Navigator: Customers➔Profile Classes
<Country> Local Receivables Modify Interest Invoices window \Navigate InterestInvoice Modify	<Country> AR Localizations Interest Invoice Batches window Navigator: InterestInvoice➔Maintain Batches
<Country> Local Receivables <Country> Receivables window \Navigate Memo Contra	<Country> AR Localizations Contra Charges window Navigator: Contra Charging➔Contra Charging
<Country> Local Payables Define EFT System Payment Formats window \Navigate Setup Payments EFT	<Country> AP Localizations EFT System Information window Navigator: Payables Localizations➔EFT System Formats
<Country> Local Payables Maintain EFT Payment Format Information window \Navigate Setup Payment Formats	Oracle Payables Payments Formats window Navigator: Setup➔Payment➔Formats Choose View EFT Details from the Tools menu.
<Country> Local Payables Maintain Invoice EFT Information window \Navigate Invoices Entry	Oracle Payables Invoice EFT Information window Navigator: Invoices➔Entry➔Invoices Choose View EFT Details from the Tools menu.
<Country> Local Payables Maintain Payment Schedule EFT Information window \Navigate Invoices Entry	Oracle Payables Scheduled Payment EFT Information window Navigator: Invoices➔Entry➔Invoices Query the invoice. Click the Scheduled Payments button. In the Scheduled Payments window, choose View EFT Details from the Tools menu.
<Country> Local Payables Maintain Vendor Site EFT Information window \Navigate Vendor Entry	Oracle Payables Supplier Site EFT Information window Navigator: Suppliers➔Entry Query the invoice. Click the Sites button. In the Supplier Sites window, choose View EFT Details from the Tools menu.

System Administrator Character Mode

Figure 13.4 shows the system administration character mode forms and the windows or processes that have the same functionality in the GUI product. The navigation paths displayed are for the system administrator responsibility.

Figure 13.4—System Administration Character Mode Forms and the Windows/processes With the Same Functionality in the GUI Product

Character Mode Form and Menu Path	GUI Window or Process, and Navigation Path
Administer Concurrent Managers \Navigate Concurrent Manager Administer	Administer Concurrent Managers window Navigator: Concurrent→Manager→Administer
Administer Request Sets \Navigate Concurrent Sets \Navigate Report Sets	Request Set window Navigator: Concurrent→Set; or Requests→Set
Assign Descriptive Flexfield Security Rules \Navigate Security Responsibility Flexfield Descriptive Assign	Assign Security Rules window Navigator: Security→Responsibility→ValueSet→Assign In the Find window, choose Descriptive Flexfield.
Assign Key Flexfield Security Rules \Navigate Security Responsibility Flexfield Key Assign	Assign Security Rules window Navigator: Security→Responsibility→ValueSet→Assign In the Find window, choose Key Flexfield.
Assign Parameter Security Rules \Navigate Security Responsibility Report Rules Assign	Assign Security Rules window Navigator: Security→Responsibility→ValueSet→Assign In the Find window, choose Concurrent Program.
Assign Printer Drivers \Navigate Install Printer Driver Assign	Printer Drivers window Navigator: Install→Printer→Driver
Assign Security Rules \Navigate Security Responsibility ValueSet Assign	Assign Security Rules window Navigator: Security→Responsibility→ValueSet→Assign
Define Application User \Navigate Security User Define	Users window Navigator: Security→User→Define
Define Combined Specialization Rules \Navigate Concurrent Manager Rule	Combined Specialization Rules window Navigator: Concurrent→Manager→Rule
Define Concurrent Manager \Navigate Concurrent Manager Define	Concurrent Managers window Navigator: Concurrent→Manager→Define

Figure 13.4—System Administration Character Mode Forms and the Windows/processes With the Same Functionality in the GUI Product *(cont.)*	
Character Mode Form and Menu Path	**GUI Window or Process, and Navigation Path**
Define Concurrent Program \Navigate Concurrent Program Define	Concurrent Programs window Navigator: Concurrent➔Program➔Define
Define Concurrent Program Executable \Navigate Concurrent Program Executable	Concurrent Program Executable window Navigator: Concurrent➔Program➔Executable
Define Concurrent Request Types \Navigate Concurrent Program Types	Concurrent Request Types window Navigator: Concurrent➔Program➔Types
Define Cross-Validation Rule \Navigate Application Flexfield Key Cross-Validation	Cross-Validation window Navigator: Application➔Flexfield➔ Key➔Cross-Validation
Define Currency \Navigate Application Currency	Currencies window Navigator: Application➔Currency
Define Data Group \Navigate Security Oracle Data Group	Data Groups window Navigator: Security➔Oracle➔DataGroup
Define Descriptive Flexfield Security Rule \Navigate Security Responsibility Flexfield Descriptive Define	Define Security Rules window Navigator: Security➔Responsibility➔ValueSet➔Define In the Find window, choose Descriptive Flexfield.
Define Descriptive Flexfield Segments \Navigate Application Flexfield Descriptive Segments	Descriptive Flexfield Segments window Navigator: Application➔Flexfield➔Descriptive➔Segments
Define Document Categories \Navigate Application Document Categories	Document Categories window Navigator: Application➔Document➔Categories
Define Document Sequences \Navigate Application Document Define	Document Sequences window Navigator: Application➔Document➔Define
Define Key Flexfield Security Rule \Navigate Security Responsibility Flexfield	Key Define, Define Security Rules window Navigator: Security➔Responsibility➔ValueSet➔Define In the Find window, choose Key Flexfield.
Define Key Flexfield Segments \Navigate Application Flexfield Key Segments	Key Flexfield Segments window Navigator: Application➔Flexfield➔Key➔Segments

Figure 13.4—System Administration Character Mode Forms and the Windows/processes With the Same Functionality in the GUI Product *(cont.)*	
Character Mode Form and Menu Path	**GUI Window or Process, and Navigation Path**
Define Key Segment Values \Navigate Application Flexfield Key Values	Segment Values window Navigator: Application→Flexfield→Key→Values
Define Logical Databases \Navigate Concurrent Databases	 Navigator: Concurrent→Conflicts Domains
Define Menu \Navigate Application Menu	Menus window Navigator: Application→Menu
Define Parameter Security Rule \Navigate Security Responsibility Report Rules Define	Define Security Rules window Navigator: Security→Responsibility→ValueSet→Define In the Find window, choose Concurrent Program.
Define Parameter Values \Navigate Application Validation Report	Define ValueSet Values window Navigator: Application→Validation→Values In the Find window, choose ValueSet.
Define Print Style \Navigate Install Printer Style	Print Styles window Navigator: Install→Printer→Style
Define Printer Driver \Navigate Install Printer Driver	Printer Drivers window Navigator: Install→Printer→Driver
Define Printer Types \Navigate Install Printer Type	Printer Types window Navigator: Install→Printer→Types
Define Report Group \Navigate Security Responsibility Report Group	Request Groups window Navigator: Security→Responsibility→Request
Define Responsibility \Navigate Security Responsibility Define	Responsibilities window Navigator: Security→Responsibility→Define
Define Rollup Groups \Navigate Application Flexfield Key Groups	Rollup Groups window Navigator: Application→Flexfield→Key→Groups
Define Segment Values \Navigate Application Validation Values	Segment Values window Navigator: Application→Validation→Values; or Application→Flexfield→Key→Values
Define Shorthand Aliases \Navigate Application Flexfield Key Aliases	Shorthand Aliases window Navigator: Application→Flexfield→Key→Aliases
Define ValueSet \Navigate Application Validation Set	Value Sets window Navigator: Application→Validation→Set

Figure 13.4—System Administration Character Mode Forms and the Windows/processes With the Same Functionality in the GUI Product *(cont.)*

Character Mode Form and Menu Path	GUI Window or Process, and Navigation Path
Define Work Shifts \Navigate Concurrent Manager Work Shifts	Work Shifts window Navigator: Concurrent➔Manager➔WorkShifts
Monitor Application Users \Navigate Security User Monitor	Monitor Users window Navigator: Security➔User➔Monitor
Register Applications \Navigate Application Register	Applications window Navigator: Application➔Register
Register Nodes \Navigate Install Nodes	Nodes window Navigator: Install➔Nodes
Register Oracle IDs \Navigate Security Oracle Register	Oracle Users window Navigator: Security➔Oracle➔Register
Register Printers \Navigate Install Printer Register	Printers Navigator: Install➔Printer➔Register
Run Reports \Navigate Report Run	Submit Requests window Navigator: Requests➔Run
Update Personal Profile Options \Navigate Profile Personal	Personal Profile Values window Navigator: Profile➔Personal
Update System Profile Options \Navigate Profile System	System Profile Values window Navigator: Profile➔System
View Concurrent Requests \Navigate Concurrent Requests	Requests window Navigator: Concurrent➔Requests Choose the View Output button to view the request output or the View Log button to view the request log. Use the menu to choose Tools➔Manager Log to view the manager log.
View Reports \Navigate Report View	Requests window Navigator: Concurrent➔Requests Choose the View Output button to view the request output or the View Log button to view the request log.

Page intentionally left blank

Appendix 1. Frequently Asked Questions

1. **Should the Oracle EBS administrator have access to functional modules under the Oracle EBS software?**

   No. The Oracle EBS system administrator's role should be restricted to administration of the system and maintaining user access as requested by the data owners. It may be necessary for support purposes to create inquiry-only responsibilities for business analysts or other IT support staff to help diagnose module-specific issues.

2. **Should the authorization to develop queries be given freely to all users?**

   No. Incorrectly structured queries can adversely affect system performance. Further, in the Oracle EBS environment, the query requirements of users may be able to be predetermined. Therefore, queries should be created on the development client and fully tested on the integration client. Users should have access only to run, not develop, queries on the production system.

3. **Is evaluating the Oracle EBS software performance the responsibility of IS auditors?**

   In an ERP environment, the IS auditor should look beyond security and controls and suggest options for system performance improvement. The Oracle EBS software does have several system performance monitoring tools, and an auditor should point out the need for appropriate procedures for review of these measures by the system administrator.

   A few tips to ensure performance efficiency include scheduling high-volume processes (e.g., generation of a trial balance report, stock valuation, purchase orders clearing) during nonpeak periods, and restricting the number of sessions that a user can initiate.

4. **What is the most important module in Oracle EBS Financials?**

   The GL module is considered the most important module in Oracle EBS Financials because it is the basis for all other Oracle Financial modules that provide information to the GL.

5. **What is the Multi-Organization Access Control (MOAC) and what is it used for?**

   Multiple operating units and their relationships are allowed to be defined within a single installation of Oracle EBS using MOAC. As a result, this keeps each operating unit's transaction data separate and secure. MOAC allows users to access multiple operating units' data from a single responsibility. Users can access reports, concurrent programs, and all setup screens of multiple operating units from a single responsibility without switching responsibilities.

   To determine whether a system has MOAC installed, the following query can be used: select **multi_org_flag** from **fnd_product_groups**.

6. **What is the difference between fields and Flexfields?**

   A field is a position on a form that one uses to enter, view, update or delete information. A field prompt describes each field by telling what kind of information appears in the field or, alternatively, what kind of information should be entered in the field.

   A Flexfield is a flexible data field that an enterprise can customize to its business needs without programming. Security rules can be implemented to restrict data entry or reporting on specific accounting segments (e.g., cost centers, account ranges). Each security rule is composed of one or more rule elements, restricting or allowing user access to specific ranges of Flexfield values. A Key Flexfield definition is required by applications as part of setup. Key Flexfield and Descriptive Flexfield are required by applications as part of setup.

7. **What is the difference between Oracle EBS R11*i*, R12 and R12.1?**

   The key difference between Oracle EBS R11*i* and R12 is in the technology stack. Also introduced in R12 were changes in the security authorization concept to a Role-based Access Control (RBAC) model, as well as changes to the Financial Application architecture including movement from a Set of Books based architecture to Ledgers and Ledger Sets. See chapter 2 for more information and examples of the changes.

   No changes to the technology stack were made between R12 and R12.1. The key difference between these two releases is the additional functionality changes introduced to the non-Financial enterprise applications in Oracle EBS (e.g., Supply Chain Management, Procurement, Customer Relationship Management, Human Capital Management). Another change introduced in Oracle EBS R12.1 is the ability for enterprises to now implement stand-alone solutions, which are compatible with existing Oracle EBS R11*i* or R12 environments without having to perform a major upgrade to R12.1.

8. What makes the audit of the Oracle EBS environment different from the audit of legacy systems?

The essential difference between Oracle EBS and older legacy application systems is that most processing is performed online in real time, where Oracle EBS integrated modules allow for a single business event initiated in a functional area to be completed through to its final financial application, the update of the general ledger.

The impact of this difference on auditing the integrity of data produced by the Oracle EBS system is the need to place a higher degree of reliance on the controls over authorizations and access controls, as well as general computer controls supporting the Oracle EBS environment. Oracle EBS presents significant opportunities for using computer-aided audit techniques (CAATs), such as Oracle Reports, and the Advanced Analytics component of the Oracle Business Intelligence Suite.

9. How can the Oracle Governance, Risk and Compliance (GRC) tool set be used to improve audits of the Oracle EBS environment?

Oracle GRC tools can help automate the auditing and monitoring of system security and configuration controls in a preventive and detective manner. Oracle GRC tools can assist enterprises with understanding the current state of their internal controls and provide them with the ability to manage, monitor and audit the controls in a harmonized way whereby controls can be effectively tested once to satisfy many stakeholders. This removes the need for multiple audits and lessens the burden placed on frontline staff.

The latest Oracle GRC tool set (known broadly as E-GRC suite) includes the following key components which are covered in more detail in chapter 11:

- GRC Insight/Fusion GRC Intelligence
- GRC Process Management/Enterprise GRC Manager
- GRC Application Controls
  - Application Access Controls Governor (AACG)
  - Configuration Controls Governor (CCG)
  - Transaction Controls Governor (TCG)
  - Preventive Controls Governor (PCG)

10. Is there a framework to identify incompatible tasks to have proper segregation of duties (SoD)?

Oracle GRC includes the Application Access Control Governor (AACG), which provides the ability for real-time enforcement, monitoring and detailed analysis of crucial user access policies and controls including those surrounding SoD. There is a comprehensive library of standard SoD rules which can be also implemented with the tool; however, this is only a baseline and should be adapted to the specific enterprise context.

11. How are workflows and advanced analytics used in Oracle Business Intelligence Suite?

Oracle Workflow is functionality supplied within the Oracle Business Intelligence Suite, which automates and streamlines transaction flows, and manages exceptions to these flows. Oracle Workflow automatically processes and routes information of any type, according to business rules, that the user is able to configure or change to any person or system inside or outside the organization.

The Oracle Business Intelligence Suite also offers Advanced Analytics features, which allow users to obtain answers to complex data questions. Users are able to use the Query Builder function within Oracle Business Intelligence to simplify the formulation of customizable and complex business questions and present answers in an insightful way. Also, the Calculation Builder function within Oracle Business Intelligence can be used to define new business indicators for analysis, and allow users to define new models using business terms and definitions.

Appendix 2. Recommended Reading

Hagerty, J.; D. Gaughan; *The Governance, Risk Management, and Compliance (GRC) Landscape, Part 1: A Segmented Marketplace With Distinct Buyers*, AMR Research, USA, 2008

Hagerty, J.; K. Verma; D. Gaughan; *The Governance, Risk Management and Compliance (GRC) Landscape, Part 2: Software's Integral Role in GRC Automation*, AMR Research, USA, 2008

Hamerman, P. D.; *The State of ERP 2009: Market Forces Drive Specialization, Consolidation and Innovation*, Forrester, USA, 2009

Herbert, L.; *et al.*; *Successfully Managing ERP Implementation Providers*, Forrester, USA, 2008

Hurwitz, J.; *et al.*; *Service Oriented Architecture for Dummies, 2nd Edition*, John Wiley & Sons, USA, 2007

ISACA, COBIT® 4.1, USA, 2007, *www.isaca.org/cobit*

ISACA, *CobiT® Control Practices: Guidance of Achieve Control Objectives for Successful IT Governance, 2nd Edition*, USA, 2007

ISACA, *ITAF™: A Professional Practices Framework for IT Assurance*, USA, 2008

ISACA, *IT Assurance Guide: Using CobiT®*, USA, 2007

ISACA, *IT Control Objectives for Sarbanes-Oxley: The Role of IT in the Design and Implementation of Internal Control Over Financial Reporting, 2nd Edition*, USA, 2006

ISACA, *Security, Audit and Control Features: Oracle Database, 3rd Edition*, USA, 2009

Matthews, B.; *et al.*; *Installing, Upgrading and Maintaining Oracle Applications,* Barbara Matthews, USA, 2003

Oracle Corp., *Oracle® Applications Access Controls Governor User Guide, Release 8.2*, USA, 2008

Oracle Corp., *Oracle® E-Business Suite General Ledger Reference Guide, Release 12.1*, USA, 2009

Oracle Corp., *Oracle® E-Business Suite General Ledger User's Guide, Release 12.1*, USA, 2009

Oracle Corp., *Oracle® E-Business Suite Payables User's Guide, Release 12.1*, USA, 2009

Oracle Corp., *Oracle® E-Business Suite Purchasing User's Guide, Release 12.1*, USA, 2009

Oracle Corp., *Oracle® E-Business Suite System Administrator's Guide, Security Release 12.1*, USA, 2009

Oracle Corp., *Oracle® E-Business Suite User's Guide Release 12.1*, USA, 2009

Oracle Corp., *Oracle® Governance, Risk and Compliance Controls Suite User Guide, Release 7.2.3*, USA, 2008

Oracle Corp., *Oracle® Governance, Risk and Compliance Manager User's Guide, Release 7.8*, USA, 2008

Oracle Corp., *Oracle® GRC Suite, www.oracle.com/solutions/corporate_governance/grc-process-management.html*

Oracle Corp., Oracle MetaLinks: "Best Practices for Securing Oracle E-Business Suite," MetaLink Note ID 189367.1; "Oracle® List for E-Business Suite," MetaLink Note ID 278724.1, USA

Oracle Corp., *Oracle® Preventive Controls Governor User Guide, Release 7.3*, USA, 2008

Oracle Corp., *Oracle® Transaction Controls Governor User Guide, Release 7.3*, USA, 2008

Oracle Corp., *Post-Fusion Support Policy, www.oracle.com/support/library/brochure/lifetime-support-applications.pdf*

Rasmussen, M., *A Tale of Two GRC Strategies: Analysis and Comparison of Oracle's and SAP's GRC Strategies*, Forrester, USA, 2007

Shepherd, J.; *Clarifying Oracle's Fusion Strategy*, AMR Research, USA, 2006

Swanton, B., Geishecker, L., *Oracle®'s Application Strategy—What's Your Five-Year Plan*, AMR Research, USA, 2007

Thomas, Z., *Oracle® EBS Is A Strong Contender for Integrated Performance and Compensation Solutions*, Forrester, USA, 2007

Appendix 3. Oracle Financial Accounting Business Cycle Audit Plan

I. Introduction

Overview

ISACA developed *ITAF™: A Professional Practices Framework for IT Assurance* as a comprehensive and good-practice-setting model. ITAF provides standards that are designed to be mandatory, and are the guiding principles under which the IT audit and assurance profession operates. The guidelines provide information and direction for the practice of IT audit and assurance. The tools and techniques provide methodologies, tools and templates to provide direction in the application of IT audit and assurance processes.

Purpose

The audit/assurance program is a tool and template to be used as a road map for the completion of a specific assurance process. ISACA has commissioned audit/assurance programs to be developed for use by IT audit and assurance practitioners. This audit/assurance program is intended to be utilized by IT audit and assurance professionals with the requisite knowledge of the subject matter under review, as described in ITAF, section 2200—General Standards. The audit/assurance programs are part of ITAF, section 4000—IT Assurance Tools and Techniques.

Control Framework

The audit/assurance programs have been developed in alignment with the ISACA COBIT® framework—specifically, COBIT 4.1—using generally applicable and accepted good practices. They reflect ITAF, sections 3400—IT Management Processes, 3600—IT Audit and Assurance Processes, and 3800—IT Audit and Assurance Management.

Many enterprises have embraced several frameworks at an enterprise level, including the Committee of Sponsoring Organizations of the Treadway Commission (COSO) Internal Control Framework. The importance of the control framework has been enhanced due to regulatory requirements by the US Securities and Exchange Commission (SEC) as directed by the US Sarbanes-Oxley Act of 2002 and similar legislation in other countries. They seek to integrate control framework elements used by the general audit/assurance team into the IT audit and assurance framework. Since COSO is widely used, it has been selected for inclusion in this audit/assurance program. The reviewer may delete or rename columns in the audit program to align with the enterprise's control framework.

IT Governance, Risk and Control

IT governance, risk and control are critical in the performance of any assurance management process. Governance of the process under review will be evaluated as part of the policies and management oversight controls. Risk plays an important role in evaluating what to audit and how management approaches and manages risk. Both issues will be evaluated as steps in the audit/assurance program. Controls are the primary evaluation point in the process. The audit/assurance program will identify the control objectives with steps to determine control design and effectiveness.

Responsibilities of IT Audit and Assurance Professionals

IT audit and assurance professionals are expected to customize this document to the environment in which they are performing an assurance process. This document is to be used as a review tool and starting point. It may be modified by the IT audit and assurance professional; it *is not* intended to be a checklist or questionnaire. It is assumed that the IT audit and assurance professional has the necessary subject matter expertise required to conduct the work and is supervised by a professional with the Certified Information Systems Auditor (CISA) designation and necessary subject matter expertise to adequately review the work performed.

II. Using This Document

This audit/assurance program was developed to assist the audit and assurance professional in designing and executing a review. Details regarding the format and use of the document follow.

Work Program Steps

The first column of the program describes the steps to be performed. The numbering scheme used provides built-in work paper numbering for ease of cross-reference to the specific work paper for that section. IT audit and assurance professionals are encouraged to make modifications to this document to reflect the specific environment under review.

COBIT Cross-reference

The COBIT cross-reference provides the audit and assurance professional with the ability to refer to the specific COBIT control objective that supports the audit/assurance step. The COBIT control objective should be identified for each audit/assurance step in the section. Multiple cross-references are not uncommon. Processes at lower levels in the work program are too granular to be cross-referenced to COBIT. The audit/assurance program is organized in a manner to facilitate an evaluation through a structure parallel to the development process.

COBIT provides in-depth control objectives and suggested control practices at each level. As professionals review each control, they should refer to COBIT 4.1 or the *IT Assurance Guide: Using CobiT®* for good-practice control guidance.

COSO Components

As noted in the introduction, COSO and similar frameworks have become increasingly popular among audit and assurance professionals. This ties the assurance work to the enterprise's control framework. While the IT audit/assurance function has COBIT as a framework, operational audit and assurance professionals use the framework established by the enterprise. Since COSO is the most prevalent internal control framework, it has been included in this document and is a bridge to align IT audit/assurance with the rest of the audit/assurance function. Many audit/assurance organizations include the COSO control components within their report and summarize assurance activities to the audit committee of the board of directors.

For each control, the audit and assurance professional should indicate the COSO component(s) addressed. It is possible, but generally not necessary, to extend this analysis to the specific audit step level.

The original COSO internal control framework contained five components. In 2004, COSO was revised as the *Enterprise Risk Management (ERM) Integrated Framework* and extended to eight components. The primary difference between the two frameworks is the additional focus on ERM and integration into the business decision model. ERM is in the process of being adopted by large enterprises. The two frameworks are compared in **figure A3.1**.

Figure A3.1—Comparison of COSO Internal Control and ERM Integrated Frameworks

Internal Control Framework	ERM Integrated Framework
Control Environment: The control environment sets the tone of an organization, influencing the control consciousness of its people. It is the foundation for all other components of internal control, providing discipline and structure. Control environment factors include the integrity, ethical values, management's operating style, delegation of authority systems, as well as the processes for managing and developing people in the organization.	**Internal Environment:** The internal environment encompasses the tone of an organization, and sets the basis for how risk is viewed and addressed by an enterprise's people, including risk management philosophy and risk appetite, integrity and ethical values, and the environment in which they operate.

Figure A3.1—Comparison of COSO Internal Control and ERM Integrated Frameworks *(cont.)*	
Internal Control Framework	**ERM Integrated Framework**
	Objective Setting: Objectives must exist before management can identify potential events affecting their achievement. Enterprise risk management ensures that management has in place a process to set objectives and that the chosen objectives support and align with the enterprise's mission and are consistent with its risk appetite.
	Event Identification: Internal and external events affecting achievement of an enterprise's objectives must be identified, distinguishing between risks and opportunities. Opportunities are channeled back to management's strategy or objective-setting processes.
Risk Assessment: Every enterprise faces a variety of risks from external and internal sources that must be assessed. A precondition to risk assessment is establishment of objectives, and thus risk assessment is the identification and analysis of relevant risks to achievement of assigned objectives. Risk assessment is a prerequisite for determining how the risks should be managed.	**Risk Assessment:** Risks are analyzed, considering the likelihood and impact, as a basis for determining how they could be managed. Risk areas are assessed on an inherent and residual basis.
	Risk Response: Management selects risk responses—avoiding, accepting, reducing or sharing risk—developing a set of actions to align risks with the enterprise's risk tolerances and risk appetite.
Control Activities: Control activities are the policies and procedures that help ensure management directives are carried out. They help ensure that necessary actions are taken to address risks to achievement of the enterprise's objectives. Control activities occur throughout the organization, at all levels and in all functions. They include a range of activities as diverse as approvals, authorizations, verifications, reconciliations, reviews of operating performance, security of assets and segregation of duties.	**Control Activities:** Policies and procedures are established and implemented to help ensure the risk responses are effectively carried out.

Figure A3.1—Comparison of COSO Internal Control and ERM Integrated Frameworks *(cont.)*	
Internal Control Framework	**ERM Integrated Framework**
Information and Communication: Information systems play a key role in internal control systems as they produce reports, including operational, financial and compliance-related information that make it possible to run and control the business. In a broader sense, effective communication must ensure information flows down, across and up the organization. Effective communication should also be ensured with external parties, such as customers, suppliers, regulators and shareholders.	**Information and Communication:** Relevant information is identified, captured and communicated in a form and time frame that enable people to carry out their responsibilities. Effective communication also occurs in a broader sense, flowing down, across and up the enterprise.
Monitoring: Internal control systems need to be monitored—a process that assesses the quality of the system's performance over time. This is accomplished through ongoing monitoring activities or separate evaluations. Internal control deficiencies detected through these monitoring activities should be reported upstream, and corrective actions should be taken to ensure continuous improvement of the system.	**Monitoring:** The entirety of enterprise risk management is monitored and modifications made as necessary. Monitoring is accomplished through ongoing management activities, separate evaluations or both.
Information for **figure A3.1** was obtained from the COSO web site, *www.coso.org/aboutus.htm.*	

The original COSO internal control framework addresses the needs of the IT audit and assurance professional: control environment, risk assessment, control activities, information and communication, and monitoring. As such, ISACA has elected to utilize the five-component model for these audit/assurance programs. As more enterprises implement the ERM model, the additional three columns can be added, if relevant. When completing the COSO component columns, consider the definitions of the components as described in **figure A3.1**.

Reference/Hyperlink

Good practices require the audit and assurance professional to create a work paper for each line item, which describes the work performed, issues identified and conclusions. The reference/hyperlink is to be used to cross-reference the audit/assurance step to the work paper that supports it. The numbering system of this document provides a ready numbering scheme for the work papers. If desired, a link to the work paper can be pasted into this column.

Issue Cross-reference

This column can be used to flag a finding/issue that the IT audit and assurance professional wants to further investigate or establish as a potential finding. The potential findings should be documented in a work paper that indicates the disposition of the findings (formally reported, reported as a memo or verbal finding, or waived).

Comments

The comments column can be used to indicate the waiving of a step or other notations. It is not to be used in place of a work paper describing the work performed.

III. Controls Maturity Analysis

One of the consistent requests of stakeholders who have undergone IT audit/assurance reviews is a desire to understand how their performance compares to good practices. Audit and assurance professionals must provide an objective basis for the review conclusions. Maturity modeling for management and control over IT processes is based on a method of evaluating the organization, so it can be rated from a maturity level of nonexistent (0) to optimized (5). This approach is derived from the maturity model that the Software Engineering Institute (SEI) of Carnegie Mellon University defined for the maturity of software development.

The *IT Assurance: Guide Using CobiT®*, Appendix VII—Maturity Model for Internal Control, in **figure A3.2**, provides a generic maturity model showing the status of the internal control environment and the establishment of internal controls in an enterprise. It shows how the management of internal control, and an awareness of the need to establish better internal controls, typically develops from an *ad hoc* to an optimized level. The model provides a high-level guide to help COBIT users appreciate what is required for effective internal controls in IT and to help position their enterprise on the maturity scale.

Figure A3.2—Maturity Model for Internal Control

Maturity Level	Status of the Internal Control Environment	Establishment of Internal Controls
0 Non-existent	There is no recognition of the need for internal control. Control is not part of the organization's culture or mission. There is a high risk of control deficiencies and incidents.	There is no intent to assess the need for internal control. Incidents are dealt with as they arise.

Figure A3.2—Maturity Model for Internal Control *(cont.)*		
Maturity Level	**Status of the Internal Control Environment**	**Establishment of Internal Controls**
1 Initial/*ad hoc*	There is some recognition of the need for internal control. The approach to risk and control requirements is *ad hoc* and disorganized, without communication or monitoring. Deficiencies are not identified. Employees are not aware of their responsibilities.	There is no awareness of the need for assessment of what is needed in terms of IT controls. When performed, it is only on an *ad hoc* basis, at a high level and in reaction to significant incidents. Assessment addresses only the actual incident.
2 Repeatable but Intuitive	Controls are in place but are not documented. Their operation is dependent on the knowledge and motivation of individuals. Effectiveness is not adequately evaluated. Many control weaknesses exist and are not adequately addressed; the impact can be severe. Management actions to resolve control issues are not prioritized or consistent. Employees may not be aware of their responsibilities.	Assessment of control needs occurs only when needed for selected IT processes to determine the current level of control maturity, the target level that should be reached and the gaps that exist. An informal workshop approach, involving IT managers and the team involved in the process, is used to define an adequate approach to controls for the process and to motivate an agreed-upon action plan.
3 Defined	Controls are in place and adequately documented. Operating effectiveness is evaluated on a periodic basis and there is an average number of issues. However, the evaluation process is not documented. While management is able to deal predictably with most control issues, some control weaknesses persist and impacts could still be severe. Employees are aware of their responsibilities for control.	Critical IT processes are identified based on value and risk drivers. A detailed analysis is performed to identify control requirements and the root cause of gaps and to develop improvement opportunities. In addition to facilitated workshops, tools are used and interviews are performed to support the analysis and ensure that an IT process owner owns and drives the assessment and improvement process.
4 Managed and Measurable	There is an effective internal control and risk management environment. A formal, documented evaluation of controls occurs frequently. Many controls are automated and regularly reviewed. Management is likely to detect most control issues, but not all issues are routinely identified. There is consistent follow-up to address identified control weaknesses. A limited, tactical use of technology is applied to automate controls.	IT process criticality is regularly defined with full support and agreement from the relevant business process owners. Assessment of control requirements is based on policy and the actual maturity of these processes, following a thorough and measured analysis involving key stakeholders. Accountability for these assessments is clear and enforced. Improvement strategies are supported by business cases. Performance in achieving the desired outcomes is consistently monitored. External control reviews are organized occasionally.

Figure A3.2—Maturity Model for Internal Control *(cont.)*		
Maturity Level	Status of the Internal Control Environment	Establishment of Internal Controls
5 Optimized	An enterprisewide risk and control program provides continuous and effective control and risk issues resolution. Internal control and risk management are integrated with enterprise practices, supported with automated real-time monitoring with full accountability for control monitoring, risk management and compliance enforcement. Control evaluation is continuous, based on self-assessments and gap and root cause analyses. Employees are proactively involved in control improvements.	Business changes consider the criticality of IT processes and cover any need to reassess process control capability. IT process owners regularly perform self-assessments to confirm that controls are at the right level of maturity to meet business needs and they consider maturity attributes to find ways to make controls more efficient and effective. The organization benchmarks to external best practices and seeks external advice on internal control effectiveness. For critical processes, independent reviews take place to provide assurance that the controls are at the desired level of maturity and working as planned.

The maturity model evaluation is one of the final steps in the evaluation process. The IT audit and assurance professional can address the key controls within the scope of the work program and formulate an objective assessment of the maturity levels of the control practices. The maturity assessment can be a part of the audit/assurance report, and used as a metric from year to year to document progression in the enhancement of controls. However, it must be noted that the perception of the maturity level may vary between the process/IT asset owner and the auditor. Therefore, an auditor should obtain the concerned stakeholder's concurrence before submitting the final report to management.

At the conclusion of the review, once all findings and recommendations are completed, the professional assesses the current state of the COBIT control framework and assigns it a maturity level using the six-level scale. Some practitioners utilize decimals (x.25, x.5, x.75) to indicate gradations in the maturity model. As a further reference, COBIT provides a definition of the maturity designations by control objective. While this approach is not mandatory, the process is provided as a separate section at the end of the audit/assurance program for those enterprises that wish to implement it. It is suggested that a maturity assessment be made at the COBIT control level. To provide further value to the client/customer, the professional can also obtain maturity targets from the client/customer. Using the assessed and target maturity levels, the professional can create an effective graphic presentation that describes the achievement or gaps between the actual and targeted maturity goals.

IV. Assurance and Control Framework

ISACA IT Assurance Framework and Standards

ISACA has long recognized the specialized nature of IT assurance and strives to advance globally applicable standards. Guidelines and procedures provide detailed guidance on how to follow those standards. IT Audit/Assurance Standard S15 IT Controls, and IT Audit/Assurance Guideline G38 Access Controls are relevant to this audit/assurance program.

ISACA Controls Framework

COBIT is an IT governance framework and supporting tool set that allows managers to bridge the gap among control requirements, technical issues and business risks. COBIT enables clear policy development and good practice for IT control throughout enterprises.

Utilizing COBIT as the control framework on which IT audit/assurance activities are based aligns IT audit/assurance with good practices as developed by the enterprise.

Refer to ISACA's *COBIT® Control Practices: Guidance to Achieve Control Objectives for Successful IT Governance, 2nd Edition*, published in 2007, for the related control practice value and risk drivers.

V. Executive Summary of Audit/Assurance Focus

Oracle EBS Security

Since launching its first product offering approximately 30 years ago, Oracle Corp. has grown globally. In recent years, Oracle Corp. has been involved in a number of large acquisitions, including PeopleSoft, Siebel and Sun, as part of its enterprise application integration strategy that forms the core of its Fusion Middleware product range.

Oracle Corp. released Oracle EBS R12 in 2007, which introduced changes from the previous version, R11*i*.10, to the application technology platform, provided changes to the security authorization concept, and provided a new technology stack and architecture for Oracle EBS Financials. The latest release, Oracle EBS R12.1, was released in 2009 and introduced changes to the other enterprise application areas (e.g., Supply Chain Management, Procurement, Customer Relationship Management and Human Capital Management). With the ever-changing compliance landscape, a review of the enterprise Oracle EBS environment is vital to ensuring that it is secure.

Business Impact and Risk

Oracle EBS is widely used in many enterprises. Improper configuration of Oracle EBS could result in the inability of the enterprise to execute its critical processes.

Oracle EBS risks resulting from ineffective or incorrect configurations or use could result in some of the following:

- Disclosure of privileged information
- Single points of failure
- Low data quality
- Loss of physical assets
- Loss of intellectual property
- Loss of competitive advantage
- Loss of customer confidence
- Violation of regulatory requirements
- Financial loss

Objective and Scope

Objective—The objective of the Oracle EBS audit/assurance review is to provide management with an independent assessment relating to the effectiveness of configuration and security of the enterprise's Oracle EBS architecture.

Scope—The review will focus on configuration of the relevant Oracle EBS components and modules within the enterprise. The selection of the specific components and modules will be based upon the risks introduced to the enterprise by these components and modules.

Minimum Audit Skills

This review is considered highly technical. The IT audit and assurance professional must have an understanding of Oracle EBS best-practice processes and requirements, and be highly conversant in Oracle EBS tools, exposures, and functionality. It should not be assumed that an audit and assurance professional holding the CISA designation has the requisite skills to perform this review.

VI. Financial Accounting Business Cycle—Audit/Assurance Program

Audit/Assurance Program Step	COBIT Cross-reference	COSO					Reference Hyperlink	Issue Cross-reference	Comments
		Control Environment	Risk Assessment	Control Activities	Information and Communication	Monitoring			
A. Prior Audit/Examination Report Follow-up									
1. Review the prior report, if one exists, to verify completion of any agreed-upon corrections and note any remaining deficiencies.	ME1								
1.1 Determine whether: • Senior management has assigned responsibilities for information, its processing and its use • User management is responsible for providing information that supports the enterprise's objectives and policies • Information systems management is responsible for providing the information systems capabilities necessary for achievement of the defined information systems objectives and policies of the enterprise • Senior management approves plans for development and acquisition of information systems • There are procedures to ensure that the information system being developed or acquired meets user requirements • There are procedures to ensure that information systems, programs and configuration changes are tested adequately prior to implementation in a separate testing environment	AI6 DS4 ME1								

VI. Financial Accounting Business Cycle—Audit/Assurance Program *(cont.)*

Audit/Assurance Program Step	COBIT Cross-reference	COSO					Reference Hyperlink	Issue Cross-reference	Comments
		Control Environment	Risk Assessment	Control Activities	Information and Communication	Monitoring			
A. Prior Audit/Examination Report Follow-up ***(cont.)***									
1.1 *(cont.)* • All personnel involved in the system acquisition and configuration activities receive adequate training and supervision • There are procedures to ensure that information systems are implemented/configured/upgraded in accordance with the established standards • User management participates in the conversion of data from the existing system to the new system • Final approval is obtained from user management prior to going live with a new information/upgraded system • There are procedures to document and schedule all changes to information systems • There are procedures to ensure that only authorized changes are initiated • There are procedures to ensure that only authorized, tested and documented changes to information systems are accepted into the production client • There are procedures to allow for and control emergency changes • There are procedures for the approval, monitoring and control of the acquisition and upgrade of hardware and systems software									

VI. Financial Accounting Business Cycle—Audit/Assurance Program (cont.)

Audit/Assurance Program Step	COBIT Cross-reference	COSO					Reference Hyperlink	Issue Cross-reference	Comments
		Control Environment	Risk Assessment	Control Activities	Information and Communication	Monitoring			
A. Prior Audit/Examination Report Follow-up ***(cont.)***									
1.1 *(cont.)* • There is a process for monitoring the volume of named and concurrent Oracle EBS users to ensure that the license agreement is not being violated • The organization structure, established by senior management, provides for an appropriate segregation of incompatible functions • The database and application servers are located in a physically separate and protected environment (i.e., a data center) • Emergency, backup and recovery plans are documented and tested on a regular basis to ensure that they remain current and operational • Backup and recovery plans allow users of information systems to resume operations in the event of an interruption • Application controls are designed with regard to any weaknesses in segregation, security, development and processing controls that may affect the information system									

VI. Financial Accounting Business Cycle—Audit/Assurance Program (cont.)

Audit/Assurance Program Step	COBIT Cross-reference	COSO					Reference Hyperlink	Issue Cross-reference	Comments
		Control Environment	Risk Assessment	Control Activities	Information and Communication	Monitoring			
B. Preliminary Audit Steps									
1. Gain an understanding of the Oracle EBS environment.									
1.1 The same background information obtained for the Oracle EBS Security audit plan is required for and relevant to the business cycles. In particular the following information is important: • Version and release of Oracle EBS software that has been implemented • Total number of named users (for comparison with logical access security testing results) • Number of Oracle EBS Database instances • Accounting period, company codes and Chart of Accounts (COA) • Identification of the modules being used • Locally developed application programs, reports or tables created by the enterprise • Details of the risk assessment approach taken in the enterprise to identify and prioritize risks • Copies of the enterprise's key security policies and standards • Outstanding audit findings, if any, from previous years	PO2 PO3 PO4 PO6 PO9 AI2 AI6 DS2 DS5 ME1 ME2								

VI. Financial Accounting Business Cycle—Audit/Assurance Program (cont.)

Audit/Assurance Program Step	COBIT Cross-reference	COSO					Reference Hyperlink	Issue Cross-reference	Comments
		Control Environment	Risk Assessment	Control Activities	Information and Communication	Monitoring			
B. Preliminary Audit Steps ***(cont.)***									
1.2 Obtain details of the following: • The organizational management model as it relates to financial accounting activity (required when evaluating the results of access security control testing) • An interview of the systems implementation team, if possible, and the process design documentation for materials management	AI1 DS5 DS6								
2. Identify the significant risks and determine the key controls.									
2.1 Develop a high-level process flow diagram and overall understanding of the Financial Accounting cycle, including the following subprocesses: • Master data maintenance • Journal processing • Reconciliation and financial reporting	PO9 AI1 DS13								
2.2 Assess the key risks, determine key controls or control weaknesses, and test controls (refer to following sample test program and chapter 4 for techniques for testing configurable controls and logical access security) regarding the following factors: • The control culture of the enterprise (e.g., a "just-enough" control philosophy)	PO9 DS5 DS9 ME2								

VI. Financial Accounting Business Cycle—Audit/Assurance Program (cont.)

Audit/Assurance Program Step	COBIT Cross-reference	COSO					Reference Hyperlink	Issue Cross-reference	Comments
		Control Environment	Risk Assessment	Control Activities	Information and Communication	Monitoring			
B. Preliminary Audit Steps ***(cont.)***									
2.2 *(cont.)* • The need to exercise judgment to determine the key controls in the process and whether the control structure is adequate (Any weaknesses in the control structure should be reported to executive management and resolved.)									
C. Detailed Audit Steps									
1. Master Data Maintenance									
1.1 Changes made to master data are valid, complete, accurate and timely.									
1.1.1 Extract a list of the target high-risk functions/forms, and review the list to ascertain who has access privileges to the following functions/forms: • Account Generation processes—Select workflow to generate accounts • Account Hierarchy editor—Edit Account Hierarchies • Assign Reporting Ledger Sets • AutoAllocation Workbench: GL—Define, run, schedule and monitor the GL AutoAllocation process	AI2 AI6 DS6 DS11								

VI. Financial Accounting Business Cycle—Audit/Assurance Program (cont.)

Audit/Assurance Program Step	COBIT Cross-reference	COSO					Reference Hyperlink	Issue Cross-reference	Comments
		Control Environment	Risk Assessment	Control Activities	Information and Communication	Monitoring			
C. Detailed Audit Steps *(cont.)*									
1.1.1 *(cont.)* • Concurrent Program Controls—Define Concurrent Program Controls • Consolidation Workbench • Conversion Rates Types—Define Conversion Rates Types • Currencies—Currencies form • Daily Rates—Define Daily Rates • Define Content Set • Define Mass Allocations • Document Sequence Assignments—Sequential numbers: Document Sequence Assignments form • Document Sequences—Sequential numbers: Document Sequences form • Elimination Sets—Define Elimination Sets • Encumbrance Types—Define Encumbrance Types • Financial Item—Define Financial Item • Generate AutoAllocation—Enter parameters to generate AutoAllocation requests • Generate AutoAllocation—Run AutoAllocation request • Generate AutoAllocation—Run Mass Allocation request									

VI. Financial Accounting Business Cycle—Audit/Assurance Program (cont.)

Audit/Assurance Program Step	COBIT Cross-reference	COSO					Reference Hyperlink	Issue Cross-reference	Comments
		Control Environment	Risk Assessment	Control Activities	Information and Communication	Monitoring			
C. Detailed Audit Steps *(cont.)*									
1.1.1 *(cont.)* • Generate AutoAllocation—Schedule Mass Allocation request • Generate Eliminations • Generate Mass Allocations • GL Accounts—Define GL accounts • Historical Rates—Define Historical Rates • Intercompany Accounts—Define Intercompany Accounts • Intercompany Clearing Accounts—Define Intercompany Clearing Accounts • Intercompany Transaction Types—Define Intercompany Transaction Types • Mass Maintenance Workbench—Define and run Mass Maintenance requests • Mass Maintenance Workbench: Prevalidate—Submit prevalidation from Mass Maintenance Workbench • Mass Maintenance Workbench: Reversal—Submit reversal from Mass Maintenance Workbench • Mass Maintenance Workbench: Submit—Submit move/merge or mass creation from Mass Maintenance Workbench									

VI. Financial Accounting Business Cycle—Audit/Assurance Program (cont.)

Audit/Assurance Program Step	COBIT Cross-reference	COSO					Reference Hyperlink	Issue Cross-reference	Comments
		Control Environment	Risk Assessment	Control Activities	Information and Communication	Monitoring			
C. Detailed Audit Steps *(cont.)*									
1.1.1 *(cont.)* • Period Rates—Define period-end and period average rates • Period Types—Define Period Types • Profile System Values—Profile System Values form • Profile User Values—Profile User Values form • Revaluations—Process Navigator definition for revaluation purposes • Revalue Balances • Rollup Groups—Flexfield Rollup Groups: Key mode form • Ledger Sets—Define Ledger Sets • Statistical Units of Measure—Define Statistical Units of Measure • Storage Parameters—Update Storage Parameters • Subledger Import Process—Process Navigator definition for Subledger Import Process • Subsidiaries—Define Subsidiaries • Summary Accounts—Define Summary Accounts									

VI. Financial Accounting Business Cycle—Audit/Assurance Program (cont.)

Audit/Assurance Program Step	COBIT Cross-reference	COSO					Reference Hyperlink	Issue Cross-reference	Comments
		Control Environment	Risk Assessment	Control Activities	Information and Communication	Monitoring			
C. Detailed Audit Steps *(cont.)*									
1.1.1 *(cont.)* • Suspense Accounts—Define Suspense Accounts • Tax Codes • Tax Codes and Rates • Tax Options—Define Tax Options • Transaction Calendar—Define Transaction Calendar • Translate Balances—Translate Balances • Translations—Process Navigator definition for translation process									
1.1.2 Extract a list of the target high-risk functions/forms, and review the list to ascertain who has access privileges to the following functions/forms: • Archive and Purge Account Balances and Journal Entries • Assign Descriptive Flexfield Rules • Assign Flexfield Security Rules • Assign Key Flexfield Security Rules • Calendars • Column Set • Cross-Validation Rules									

VI. Financial Accounting Business Cycle—Audit/Assurance Program (cont.)

Audit/Assurance Program Step	COBIT Cross-reference	COSO					Reference Hyperlink	Issue Cross-reference	Comments
		Control Environment	Risk Assessment	Control Activities	Information and Communication	Monitoring			
C. Detailed Audit Steps ***(cont.)***									
1.1.2 *(cont.)* • Descriptive Flexfield Security Rules • Descriptive Flexfield Segments • Descriptive Flexfield Values • Flexfield Security Rules • Flexfield Value • Flexfield Value Sets • Key Flexfield Security Rules • Key Flexfield Segments • Key Flexfield Values • Mass Maintenance Workbench: Purge • Purge Consolidation Audit Data									
1.1.3 Ask management whether the organization is utilizing the Dynamic Insertion functionality during COA maintenance. Further, verify this by navigating to the Key Flexfield Segments form (Setup→Financial→Flexfields→Key→Segments) and selecting the name of the application (Oracle GL) and Flexfield (e.g., Accounting Flexfield). Check that the Allow Dynamic Inserts checkbox is enabled, if the organization is utilizing Dynamic Insertion. If this has been enabled, check that the Cross-Validate Segments checkbox is also enabled.	AI6 DS5 DS11 ME1								

VI. Financial Accounting Business Cycle—Audit/Assurance Program (cont.)

Audit/Assurance Program Step	COBIT Cross-reference	COSO					Reference Hyperlink	Issue Cross-reference	Comments
		Control Environment	Risk Assessment	Control Activities	Information and Communication	Monitoring			
C. Detailed Audit Steps *(cont.)*									
1.1.3 *(cont.)* Ask management if either of the following standard Oracle EBS reports (Requests➔Run, using system administrator responsibility) are reviewed to monitor changes to the Cross-Validation Rules: • Cross-Validation Rule Violation report • Cross-Validation Rules Listing report Determine whether management reviews these reports on a regular basis and whether exceptions are identified for follow-up and resolution. Inspect physical evidence that the reports are run, approved and retained, and that exceptions are resolved, as necessary.									
1.1.4. To review the setting of required fields, navigate to the properties of a required field through the path Help+ Diagnostics➔ Properties➔Item and confirm that the null value is set to False, which defines the field as required. Note that a password is required to view the attributes of a field. The preloaded Oracle EBS password is “apps.” If it has not been changed, security may be compromised. Request the password to test whether the client has changed the password.	DS5 DS13								

VI. Financial Accounting Business Cycle—Audit/Assurance Program (cont.)

Audit/Assurance Program Step	COBIT Cross-reference	COSO					Reference Hyperlink	Issue Cross-reference	Comments
		Control Environment	Risk Assessment	Control Activities	Information and Communication	Monitoring			
C. Detailed Audit Steps ***(cont.)***									
1.1.5. Ask management if any of the following standard Oracle EBS reports (Requests➔Run➔ Standard, using responsibility with relevant GL report set) are reviewed to monitor changes to the COA: • COA—Account Hierarchy report • COA—Detail Listing • COA—Segment Values Listing • COA—Suspense Accounts Listing Determine whether management reviews these reports on a regular basis and whether exceptions are identified for follow-up and resolution. Inspect physical evidence that the reports are run, approved and retained, and that exceptions are resolved, as necessary.	PO8 DS11 ME1								
1.2 COA master data remain current and accurate.									
1.2.1 Ask management whether there is a procedural periodic review of the management of master data to check their currency and ongoing accuracy. Inspect physical evidence that the reports are run, approved and retained, and that exceptions are resolved, as necessary.	PO4 DS3								

VI. Financial Accounting Business Cycle—Audit/Assurance Program (cont.)

Audit/Assurance Program Step	COBIT Cross-reference	COSO					Reference Hyperlink	Issue Cross-reference	Comments
		Control Environment	Risk Assessment	Control Activities	Information and Communication	Monitoring			
C. Detailed Audit Steps *(cont.)*									
2. Journal Processing									
2.1 Valid journal entries are booked to the GL.									
2.1.1 Extract a list of the target high-risk functions/ forms, and review the list to ascertain who has access privileges to the following functions/ forms: • Correct Journal Import Data • Defining Recurring Journals—Defining Recurring Journal Formula • Delete Journal Import Data • Enter Encumbrances • Enter Intercompany Transactions • Enter Journals • Entering Journals—Process Navigator Definition for Entering Journal Process • General AutoAllocation: Run Recurring Journal Request • General AutoAllocation: Schedule Recurring Journal • Generate Recurring Intercompany Transactions • Generate Recurring Journals • Import Journals • Journal Categories—Define Journal Categories	DS5 DS10								

VI. Financial Accounting Business Cycle—Audit/Assurance Program (cont.)

Audit/Assurance Program Step	COBIT Cross-reference	COSO					Reference Hyperlink	Issue Cross-reference	Comments
		Control Environment	Risk Assessment	Control Activities	Information and Communication	Monitoring			
C. Detailed Audit Steps ***(cont.)***									
2.1.1 *(cont.)* • Journal Entry Enquiry—Query Journals and Journal Batches • Journal Sources—Define Journal Sources • Recurring Intercompany Transactions—Define Recurring Intercompany Transactions • Review the configuration for AutoAllocation Rules									
2.1.2. Extract a list of the target high-risk functions/forms, and review the list to ascertain who has access privileges to the following functions/forms: • Autopost Criteria—Define Autopost Criteria • Enter Journals: Post—Post in the Enter Journals or Enter Encumbrances forms • GL Enter Employee—Enter, change or delete employees who make up the Journal Approval hierarchy • Post Journals • Run Autopost Requests Confirm with management that there is appropriate segregation of duties.	PO4 DS11 DS13								

VI. Financial Accounting Business Cycle—Audit/Assurance Program *(cont.)*

Audit/Assurance Program Step	COBIT Cross-reference	COSO					Reference Hyperlink	Issue Cross-reference	Comments
		Control Environment	Risk Assessment	Control Activities	Information and Communication	Monitoring			
C. Detailed Audit Steps ***(cont.)***									
2.1.2 *(cont.)* Navigate to the Systems Profiles window (System Administrator responsibility), Profile→System, search with jo% in the Find field. Check that the following has been set as follows: • Journals: Allow Posting During Journal Entry is set to No • Journals: Allow Preparer Approval is set to No • Journals: Find Approver Method is set to Go Up Management Chain or Go Direct or One Stop and Then Direct (depends on the organization requirement)									
2.1.3 Test the control according to point 1.1.3.	DS5 DS13								
2.1.4 Review the journal entry procedures documentation to ensure that the Oracle EBS reversing and recurring journal entry features prevent omitted or inaccurate journal entries.	P03 P07 AI3 AI6								

VI. Financial Accounting Business Cycle—Audit/Assurance Program *(cont.)*

Audit/Assurance Program Step	COBIT Cross-reference	COSO					Reference Hyperlink	Issue Cross-reference	Comments
		Control Environment	Risk Assessment	Control Activities	Information and Communication	Monitoring			
C. Detailed Audit Steps ***(cont.)***									
2.1.4 *(cont.)* Navigate to the Generate Recurring Journals screen (Journals➔Generate➔Recurring) to bring up a list of all recurring journals that have been set up, and review for evidence of performance. Navigate to the Generate Reversal Journals screen (Journals➔Generate➔Reversals) to bring up a list of all reversing journals that have been set up, and review for evidence of performance.									
2.2 Journal entries are posted only once to the GL.									
2.2.1 Ensure that Oracle EBS requires a unique combination of journal number, batch name and date for a journal entry to be created when processing journal entries in Oracle GL modules. These three fields form the key for the data table, which is verified to be unique prior to the database accepting the input. If a record already exists with the same combination of journal number, batch name and date, Oracle EBS rejects the record as a duplicate and requires the user to change one of the fields before continuing with processing.	AI6 DS11								

VI. Financial Accounting Business Cycle—Audit/Assurance Program (cont.)

Audit/Assurance Program Step	COBIT Cross-reference	COSO					Reference Hyperlink	Issue Cross-reference	Comments
		Control Environment	Risk Assessment	Control Activities	Information and Communication	Monitoring			
C. Detailed Audit Steps *(cont.)*									
2.3 All journal entries are posted to the GL.									
2.3.1 Ask management if any of the following standard Oracle EBS reports (Reports→Request→ Standard in a responsibility with the relevant GL report set) are reviewed to monitor the progress of posted journal entries against the monthly closing schedule: • Journals report • Journals—Batch Summary report • Journal—Entry report • Journal Day Ledger reports (header and line descriptions) • Journal—Line report • Journals—Tax report • Journal—Document Number report • Journal—Entered Currency report Determine whether there is a procedure for reviewing these reports. Ascertain if they are reviewed on a regular basis during the monthly closing cycle, depending upon the level of journal activity, to confirm that all scheduled journal entries and adjustments have been made at the appropriate time.	DS11 ME1								

VI. Financial Accounting Business Cycle—Audit/Assurance Program (cont.)

Audit/Assurance Program Step	COBIT Cross-reference	COSO					Reference Hyperlink	Issue Cross-reference	Comments
		Control Environment	Risk Assessment	Control Activities	Information and Communication	Monitoring			
C. Detailed Audit Steps *(cont.)*									
2.4 All journal entries are posted in the correct period.									
2.4.1 Go to the Setup➔Open/Close option in a responsibility with the ability to view Setup (e.g., GL Superuser). Review the status of the periods listed to determine which are: • Open • Future—Entry • Closed • Permanently Closed • Any	AI6								
2.4.2 Extract a list of the target high-risk functions/forms. Review the function/form listing in 2.1.2 to ascertain who has access privileges to the following functions/forms: • Open and Close Periods • Parent Close—Parent Company—Parent Navigator Definition for Period Close—Parent Company Process • Parent Close—Subsidiary Company—Process Navigator Definition for Period Close—Subsidiary Company Process	DS5								

VI. Financial Accounting Business Cycle—Audit/Assurance Program (cont.)

Audit/Assurance Program Step	COBIT Cross-reference	COSO					Reference Hyperlink	Issue Cross-reference	Comments
		Control Environment	Risk Assessment	Control Activities	Information and Communication	Monitoring			
C. Detailed Audit Steps *(cont.)*									
2.5 All journal entries are accurate and balanced.									
2.5.1 Online edits and validation tools should be tested, according to point 1.1.4, confirming that the null value is set to False (i.e., field value is required). Review QuickPick lists for fields that have been identified as having lists, and confirm that the choices are relevant. Review defaults.	DS3 DS11								
2.5.2 Determine whether management runs the COA—Suspense Accounts Listing—and reviews a list of all suspense accounts. Ask whether management reviews this on a regular basis and includes it in a formal procedure. Ask whether a staff member in the accounting department has been tasked with clearing the suspense accounts and reclassifying the expenses to the proper accounts on a regular basis.	PO10 DS11								

VI. Financial Accounting Business Cycle—Audit/Assurance Program (cont.)

Audit/Assurance Program Step	COBIT Cross-reference	COSO					Reference Hyperlink	Issue Cross-reference	Comments
		Control Environment	Risk Assessment	Control Activities	Information and Communication	Monitoring			
C. Detailed Audit Steps ***(cont.)***									
2.5.3 Observe that an error message is displayed when attempting to modify a posted journal entry. Extract a list of target high-risk functions/forms. Review the function/form listing in 2.4.2 to ascertain who has access privileges to the following functions/forms: • Enter Journals: Reverse—Reverse in the Enter Journals or Enter Encumbrances forms • Reversal Criteria—Define journal reversal criteria • Reverse Journals	DS5 DS11								
2.5.4 Observe that the Control Totals feature is being utilized. A violation message will be displayed if the journals do not balance (i.e., total debits do not equal total credits) when attempting to post.	DS5								
3. Processing Disbursements									
3.1 All valid GL account balances are included in the financial statements. Financial statements are accurate and reconcile to the GL.									
3.1.1 Extract a list of the target high-risk functions/forms, and review the list to ascertain who has access privileges to the following functions/forms: • Autocopy Report Component	PO2 DS5 DS10								

VI. Financial Accounting Business Cycle—Audit/Assurance Program (cont.)

Audit/Assurance Program Step	COBIT Cross-reference	COSO					Reference Hyperlink	Issue Cross-reference	Comments
		Control Environment	Risk Assessment	Control Activities	Information and Communication	Monitoring			
C. Detailed Audit Steps ***(cont.)***									
3.1.1 *(cont.)* • Define Financial Report • Define Financial Report Set • Process Navigator definition for Define Financial Reports process • Process Navigator definition for Extract GL Balances to FA process • Run Financial reports • Define Row Set • Define Column Set Check that the system Profile Option FSG: Enforce Segment Value Security is set to Yes if the organization is utilizing Flexfield Value Security.									
3.1.2 Extract a list of the target high-risk functions/forms, and review the list to ascertain who has access privileges to the following functions/forms: • Assign Descriptive Flexfield Rules—Flexfield Security Rules: Descriptive Mode form • Assign Flexfield Security Rules—Flexfield Security Rules form	P02 DS5 DS10								

VI. Financial Accounting Business Cycle—Audit/Assurance Program *(cont.)*

Audit/Assurance Program Step	COBIT Cross-reference	COSO					Reference Hyperlink	Issue Cross-reference	Comments
		Control Environment	Risk Assessment	Control Activities	Information and Communication	Monitoring			
C. Detailed Audit Steps ***(cont.)***									
3.1.2 *(cont.)* • Assign Key Flexfield Security Rules—Flexfield Security Rules: Key Mode form • Cross-Validation Rules—Flexfield Cross-Validation Rules form • Descriptive Flexfield Security Rules—Flexfield Security Rules: Descriptive Mode form • Descriptive Flexfield Values—Flexfield Segment Values: Descriptive Mode form • Flexfield Security Rules—Flexfield Security Rules form • Flexfield Value—Flexfield Values form • Flexfield Value Sets—Flexfield Value Sets form • Key Flexfield Security Rules—Flexfield Security Rules: Key Mode form • Key Flexfield Segments—Key Flexfield Segments form • Key Flexfield Values—Flexfield Segments Values: Key Mode form									

VI. Financial Accounting Business Cycle—Audit/Assurance Program (cont.)

Audit/Assurance Program Step	COBIT Cross-reference	COSO					Reference Hyperlink	Issue Cross-reference	Comments
		Control Environment	Risk Assessment	Control Activities	Information and Communication	Monitoring			
C. Detailed Audit Steps *(cont.)*									
3.1.3 Ask management for documentation to review for: • Policies, procedures, standards and guidance regarding segment value and financial statement updates • System-generated reports used in performing such updates and the purpose of each report • Policies, procedures, standards and guidance regarding the design of the COA • Policies, procedures, standards and guidance regarding the design and configuration of financial reports	AI7 DS13 ME1								
3.1.4 Submit the COA—Segment Values Listing report (Select Segment Account). Reports can be run using the Requests window in Reports→Request→Standard (in a responsibility with the relevant GL report set). Review the report to determine the account ranges. Compare the FSG or FA report settings to ensure that they include the complete account range extracted above and selected account ranges, if applicable.	DS1 ME1								

VI. Financial Accounting Business Cycle—Audit/Assurance Program (cont.)

Audit/Assurance Program Step	COBIT Cross-reference	COSO					Reference Hyperlink	Issue Cross-reference	Comments
		Control Environment	Risk Assessment	Control Activities	Information and Communication	Monitoring			
C. Detailed Audit Steps *(cont.)*									
3.1.5 Ask management for procedural documentation that indicates that reports identified in point 1.1.2 are cross-referenced on a regular basis to the report criteria used to produce the financial statements, to ensure that changes to the COA are complete and accurately reflected.	AI6								
3.2 All account reconciliations are performed monthly.									
3.2.1 Ask management for procedural documentation (e.g., reconciliation schedules) to review evidence that key accounts are reconciled prior to the final close. Accounts that reflect the activity of a subledger or other system (e.g., AP, AR) should be reconciled to the monthly balance reflected in the subledger before closing the period.	DS5 DS11								
3.3 Financial reports are generated and distributed in a timely manner.									
3.3.1 Ask management for procedural documentation (e.g., reconciliation schedules) to review evidence that management includes guidance as to when the different versions of the financial statements should be generated for review.	AI7 DS5								

VI. Financial Accounting Business Cycle—Audit/Assurance Program (cont.)

Audit/Assurance Program Step	COBIT Cross-reference	COSO					Reference Hyperlink	Issue Cross-reference	Comments
		Control Environment	Risk Assessment	Control Activities	Information and Communication	Monitoring			
C. Detailed Audit Steps ***(cont.)***									
3.3.2 Ask management for procedural documentation to review evidence of a monthly closing schedule and the monitoring of the schedule by an appropriate staff member.	DS9								
3.4 Account reconciliations are current.									
3.4.1 Refer to 3.1.3 and 3.3.1.	AI7 DS5 DS13 ME1								
3.5 Only authorized personnel may generate financial statements.									
3.5.1 Refer to 1.1.3 and 3.1.1.	PO2 DS5 DS10 DS13								
3.5.2 Ask management for procedural documentation to review evidence of the retention and safekeeping of interim financial reports until they are finalized.	DS3 DS13								

VI. Financial Accounting Business Cycle—Audit/Assurance Program (cont.)

Audit/Assurance Program Step	COBIT Cross-reference	COSO					Reference Hyperlink	Issue Cross-reference	Comments
		Control Environment	Risk Assessment	Control Activities	Information and Communication	Monitoring			
C. Detailed Audit Steps ***(cont.)***									
3.6 Adequate closing procedures prevent further postings to periods and accurately reflect the transactions that took place in a given accounting period.									
3.6.1 Refer to 2.4.1 and 2.4.2 in Journal Processing. Review the function/form listing in 3.1.1 in chapter 6 to ascertain who has access privilege to create, modify and delete Year-end Carry Forward—Perform Year-end Carry Forward. Review the status of the periods (i.e., that they are closed for the ledger) by navigating to the Open and Close Periods window (Setup→Open/Close). This displays all the accounting periods defined within the ledger. Note that: • Closed—A closed period can be reopened at any time. • Permanently closed—A permanently closed period cannot be reopened. Review access to the Open and Close Periods window, and ensure that it has been restricted appropriately.									

VII. Maturity Assessment

The maturity assessment is an opportunity for the reviewer to assess the maturity of the processes reviewed. Based on the results of audit/assurance review, and the reviewer's observations, a maturity level should be assigned to each of the following COBIT control practices.

COBIT Control Practice	Assessed Maturity	Target Maturity	Reference Hyperlink	Comments
AI6.1 Change Standards and Procedures 1. Develop, document and promulgate a change management framework that specifies the policies and processes, including: • Roles and responsibilities • Classification and prioritisation of all changes based on business risk • Assessment of impact • Authorisation and approval of all changes by the business process owners and IT • Tracking and status of changes • Impact on data integrity (e.g., all changes to data files being made under system and application control rather than by direct user intervention) 2. Establish and maintain version control over all changes. 3. Implement roles and responsibilities that involve business process owners and appropriate technical IT functions. Ensure appropriate segregation of duties. 4. Establish appropriate record management practices and audit trails to record key steps in the change management process. Ensure timely closure of changes. Elevate and report to management changes that are not closed in a timely fashion. 5. Consider the impact of contracted services providers (e.g., of infrastructure, application development and shared services) on the change management process. Consider integration of organisational change management processes with change management processes of service providers. Consider the impact of the organisational change management process on contractual terms and SLAs.				

VII. Maturity Assessment (cont.)

COBIT Control Practice	Assessed Maturity	Target Maturity	Reference Hyperlink	Comments
AI6.2 Impact Assessment, Prioritisation and Authorisation 1. Develop a process to allow business process owners and IT to request changes to infrastructure, systems or applications. Develop controls to ensure that all such changes arise only through the change request management process. 2. Categorise all requested changes (e.g., infrastructure, operating systems, networks, application systems, purchased/packaged application software). 3. Prioritise all requested changes. Ensure that the change management process identifies both the business and technical needs for the change. Consider legal, regulatory and contractual reasons for the requested change. 4. Assess all requests in a structured fashion. Ensure that the assessment process addresses impact analysis on infrastructure, systems and applications. Consider security, legal, contractual and compliance implications of the requested change. Consider also interdependencies amongst changes. Involve business process owners in the assessment process, as appropriate. 5. Ensure that each change is formally approved by business process owners and IT technical stakeholders, as appropriate.				
AI6.4 Change Status Tracking and Reporting 1. Ensure that a documented process exists within the overall change management process to declare, assess, authorise and record an emergency change. 2. Ensure that emergency changes are processed in accordance with the emergency change element of the formal change management process. 3. Ensure that all emergency access arrangements for changes are appropriately authorised, documented and revoked after the change has been applied. 4. Conduct a post-implementation review of all emergency changes, involving all concerned parties. The review should consider implications for aspects such as further application system maintenance, impact on development and test environments, application software development quality, documentation and manuals, and data integrity.				

VII. Maturity Assessment (cont.)

COBIT Control Practice	Assessed Maturity	Target Maturity	Reference Hyperlink	Comments
DS5.3 Identity Management 1. Establish and communicate policies and procedures to uniquely identify, authenticate and authorise access mechanisms and access rights for all users on a need-to-know/need-to-have basis, based on predetermined and preapproved roles. Clearly state accountability of any user for any action on any of the systems and/or applications involved. 2. Ensure that roles and access authorisation criteria for assigning user access rights take into account: • Sensitivity of information and applications involved (data classification) • Policies for information protection and dissemination (legal, regulatory, internal policies and contractual requirements) • Roles and responsibilities as defined within the enterprise • The need-to-have access rights associated with the function • Standard but individual user access profiles for common job roles in the organisation • Requirements to guarantee appropriate segregation of duties 3. Establish a method for authenticating and authorising users to establish responsibility and enforce access rights in line with sensitivity of information and functional application requirements and infrastructure components, and in compliance with applicable laws, regulations, internal policies and contractual agreements. 4. Define and implement a procedure for identifying new users and recording, approving and maintaining access rights. This needs to be requested by user management, approved by the system owner and implemented by the responsible security person. 5. Ensure that a timely information flow is in place that reports changes in jobs (i.e., people in, people out, people change). Grant, revoke and adapt user access rights in co-ordination with human resources and user departments for users who are new, who have left the organisation, or who have changed roles or jobs.				

VII. Maturity Assessment (cont.)

COBIT Control Practice	Assessed Maturity	Target Maturity	Reference Hyperlink	Comments
DS5.4 User Account Management 1. Ensure that access control procedures include but are not limited to: • Using unique user IDs to enable users to be linked to and held accountable for their actions • Awareness that the use of group IDs results in the loss of individual accountability and is permitted only when justified for business or operational reasons and compensated by mitigating controls. Group IDs must be approved and documented. • Checking that the user has authorisation from the system owner for the use of the information system or service, and the level of access granted is appropriate to the business purpose and consistent with the organisational security policy • A procedure to require users to understand and acknowledge their access rights and the conditions of such access • Ensuring that internal and external service providers do not provide access until authorisation procedures have been completed • Maintaining a formal record, including access levels, of all persons registered to use the service • A timely and regular review of user IDs and access rights 2. Ensure that management reviews or reallocates user access rights at regular intervals using a formal process. User access rights should be reviewed or reallocated after any job changes, such as transfer, promotion, demotion or termination of employment. Authorisations for special privileged access rights should be reviewed independently at more frequent intervals.				

VII. Maturity Assessment (cont.)

COBIT Control Practice	Assessed Maturity	Target Maturity	Reference Hyperlink	Comments
DS9.1 Configuration Repository and Baseline 1. Implement a configuration repository to capture and maintain configuration management items. The repository should include hardware; application software; middleware; parameters; documentation; procedures; and tools for operating, accessing and using the systems, services, version numbers and licencing details. 2. Implement a tool to enable the effective logging of configuration management information within a repository. 3. Provide a unique identifier to a configuration item so the item can be easily tracked and related to physical asset tags and financial records. 4. Define and document configuration baselines for components across development, test and production environments, to enable identification of system configuration at specific points in time (past, present and planned). 5. Establish a process to revert to the baseline configuration in the event of problems, if determined appropriate after initial investigation. 6. Install mechanisms to monitor changes against the defined repository and baseline. Provide management reports for exceptions, reconciliation and decision making.				

VII. Maturity Assessment (cont.)

COBIT Control Practice	Assessed Maturity	Target Maturity	Reference Hyperlink	Comments
DS9.2 Identification and Maintenance of Configuration Items 1. Define and implement a policy requiring all configuration items and their attributes and versions to be identified and maintained. 2. Tag physical assets according to a defined policy. Consider using an automated mechanism, such as barcodes. 3. Define a policy that integrates incident, change and problem management procedures with the maintenance of the configuration repository. 4. Define a process to record new, modified and deleted configuration items and their relative attributes and versions. Identify and maintain the relationships between configuration items in the configuration repository. 5. Establish a process to maintain an audit trail for all changes to configuration items. 6. Define a process to identify critical configuration items in relationship to business functions (component failure impact analysis). 7. Record all assets—including new hardware and software, procured or internally developed—within the configuration management data repository. 8. Define and implement a process to ensure that valid licences are in place to prevent the inclusion of unauthorised software.				
DS9.3 Configuration Integrity Review 1. To validate the integrity of configuration data, implement a process to ensure that configuration items are monitored. Compare recorded data against actual physical existence, and ensure that errors and deviations are reported and corrected. 2. Using automated discovery tools where appropriate, reconcile actual installed software and hardware periodically against the configuration database, licence records and physical tags. 3. Periodically review against the policy for software usage the existence of any software in violation or in excess of current policies and licence agreements. Report deviations for correction.				

Page intentionally left blank

Appendix 4. Oracle Expenditure Business Cycle Audit Plan

I. Introduction

Overview

ISACA developed *ITAF™: A Professional Practices Framework for IT Assurance* as a comprehensive and good-practice-setting model. ITAF provides standards that are designed to be mandatory, and are the guiding principles under which the IT audit and assurance profession operates. The guidelines provide information and direction for the practice of IT audit and assurance. The tools and techniques provide methodologies, tools and templates to provide direction in the application of IT audit and assurance processes.

Purpose

The audit/assurance program is a tool and template to be used as a road map for the completion of a specific assurance process. ISACA has commissioned audit/assurance programs to be developed for use by IT audit and assurance practitioners. This audit/assurance program is intended to be utilized by IT audit and assurance professionals with the requisite knowledge of the subject matter under review, as described in ITAF, section 2200—General Standards. The audit/assurance programs are part of ITAF, section 4000—IT Assurance Tools and Techniques.

Control Framework

The audit/assurance programs have been developed in alignment with the ISACA COBIT® framework—specifically, COBIT 4.1—using generally applicable and accepted good practices. They reflect ITAF, sections 3400—IT Management Processes, 3600—IT Audit and Assurance Processes, and 3800—IT Audit and Assurance Management.

Many enterprises have embraced several frameworks at an enterprise level, including the Committee of Sponsoring Organizations of the Treadway Commission (COSO) Internal Control Framework. The importance of the control framework has been enhanced due to regulatory requirements by the US Securities and Exchange Commission (SEC) as directed by the US Sarbanes-Oxley Act of 2002 and similar legislation in other countries. They seek to integrate control framework elements used by the general audit/assurance team into the IT audit and assurance framework. Since COSO is widely used, it has been selected for inclusion in this audit/assurance program. The reviewer may delete or rename columns in the audit program to align with the enterprise's control framework.

IT Governance, Risk and Control

IT governance, risk and control are critical in the performance of any assurance management process. Governance of the process under review will be evaluated as part of the policies and management oversight controls. Risk plays an important role in evaluating what to audit and how management approaches and manages risk. Both issues will be evaluated as steps in the audit/assurance program. Controls are the primary evaluation point in the process. The audit/assurance program will identify the control objectives with steps to determine control design and effectiveness.

Responsibilities of IT Audit and Assurance Professionals

IT audit and assurance professionals are expected to customize this document to the environment in which they are performing an assurance process. This document is to be used as a review tool and starting point. It may be modified by the IT audit and assurance professional; it *is not* intended to be a checklist or questionnaire. It is assumed that the IT audit and assurance professional has the necessary subject matter expertise required to conduct the work and is supervised by a professional with the Certified Information Systems Auditor (CISA) designation and necessary subject matter expertise to adequately review the work performed.

II. Using This Document

This audit/assurance program was developed to assist the audit and assurance professional in designing and executing a review. Details regarding the format and use of the document follow.

Work Program Steps

The first column of the program describes the steps to be performed. The numbering scheme used provides built-in work paper numbering for ease of cross-reference to the specific work paper for that section. IT audit and assurance professionals are encouraged to make modifications to this document to reflect the specific environment under review.

COBIT Cross-reference

The COBIT cross-reference provides the audit and assurance professional with the ability to refer to the specific COBIT control objective that supports the audit/assurance step. The COBIT control objective should be identified for each audit/assurance step in the section. Multiple cross-references are not uncommon. Processes at lower levels in the work program are too granular to be cross-referenced to COBIT. The audit/assurance program is organized in a manner to facilitate an evaluation through a structure parallel to the development process.

COBIT provides in-depth control objectives and suggested control practices at each level. As professionals review each control, they should refer to COBIT 4.1 or the *IT Assurance Guide: Using CobiT®* for good-practice control guidance.

COSO Components

As noted in the introduction, COSO and similar frameworks have become increasingly popular among audit and assurance professionals. This ties the assurance work to the enterprise's control framework. While the IT audit/assurance function has COBIT as a framework, operational audit and assurance professionals use the framework established by the enterprise. Since COSO is the most prevalent internal control framework, it has been included in this document and is a bridge to align IT audit/assurance with the rest of the audit/assurance function. Many audit/assurance organizations include the COSO control components within their report and summarize assurance activities to the audit committee of the board of directors.

For each control, the audit and assurance professional should indicate the COSO component(s) addressed. It is possible, but generally not necessary, to extend this analysis to the specific audit step level.

The original COSO internal control framework contained five components. In 2004, COSO was revised as the *Enterprise Risk Management (ERM) Integrated Framework* and extended to eight components. The primary difference between the two frameworks is the additional focus on ERM and integration into the business decision model. ERM is in the process of being adopted by large enterprises. The two frameworks are compared in **figure A4.1**.

Figure A4.1—Comparison of COSO Internal Control and ERM Integrated Frameworks

Internal Control Framework	ERM Integrated Framework
Control Environment: The control environment sets the tone of an organization, influencing the control consciousness of its people. It is the foundation for all other components of internal control, providing discipline and structure. Control environment factors include the integrity, ethical values, management's operating style, delegation of authority systems, as well as the processes for managing and developing people in the organization.	**Internal Environment:** The internal environment encompasses the tone of an organization, and sets the basis for how risk is viewed and addressed by an enterprise's people, including risk management philosophy and risk appetite, integrity and ethical values, and the environment in which they operate.

Figure A4.1—Comparison of COSO Internal Control and ERM Integrated Frameworks *(cont.)*	
Internal Control Framework	**ERM Integrated Framework**
	Objective Setting: Objectives must exist before management can identify potential events affecting their achievement. Enterprise risk management ensures that management has in place a process to set objectives and that the chosen objectives support and align with the enterprise's mission and are consistent with its risk appetite.
	Event Identification: Internal and external events affecting achievement of an enterprise's objectives must be identified, distinguishing between risks and opportunities. Opportunities are channeled back to management's strategy or objective-setting processes.
Risk Assessment: Every enterprise faces a variety of risks from external and internal sources that must be assessed. A precondition to risk assessment is establishment of objectives, and thus risk assessment is the identification and analysis of relevant risks to achievement of assigned objectives. Risk assessment is a prerequisite for determining how the risks should be managed.	**Risk Assessment:** Risks are analyzed, considering the likelihood and impact, as a basis for determining how they could be managed. Risk areas are assessed on an inherent and residual basis.
	Risk Response: Management selects risk responses—avoiding, accepting, reducing or sharing risk—developing a set of actions to align risks with the enterprise's risk tolerances and risk appetite.
Control Activities: Control activities are the policies and procedures that help ensure management directives are carried out. They help ensure that necessary actions are taken to address risks to achievement of the enterprise's objectives. Control activities occur throughout the organization, at all levels and in all functions. They include a range of activities as diverse as approvals, authorizations, verifications, reconciliations, reviews of operating performance, security of assets and segregation of duties.	**Control Activities:** Policies and procedures are established and implemented to help ensure the risk responses are effectively carried out.

Figure A4.1—Comparison of COSO Internal Control and ERM Integrated Frameworks *(cont.)*	
Internal Control Framework	**ERM Integrated Framework**
Information and Communication: Information systems play a key role in internal control systems as they produce reports, including operational, financial and compliance-related information that make it possible to run and control the business. In a broader sense, effective communication must ensure information flows down, across and up the organization. Effective communication should also be ensured with external parties, such as customers, suppliers, regulators and shareholders.	**Information and Communication:** Relevant information is identified, captured and communicated in a form and time frame that enable people to carry out their responsibilities. Effective communication also occurs in a broader sense, flowing down, across and up the enterprise.
Monitoring: Internal control systems need to be monitored—a process that assesses the quality of the system's performance over time. This is accomplished through ongoing monitoring activities or separate evaluations. Internal control deficiencies detected through these monitoring activities should be reported upstream, and corrective actions should be taken to ensure continuous improvement of the system.	**Monitoring:** The entirety of enterprise risk management is monitored and modifications made as necessary. Monitoring is accomplished through ongoing management activities, separate evaluations or both.
Information for **figure A4.1** was obtained from the COSO web site, *www.coso.org/aboutus.htm.*	

The original COSO internal control framework addresses the needs of the IT audit and assurance professional: control environment, risk assessment, control activities, information and communication, and monitoring. As such, ISACA has elected to utilize the five-component model for these audit/assurance programs. As more enterprises implement the ERM model, the additional three columns can be added, if relevant. When completing the COSO component columns, consider the definitions of the components as described in **figure A4.1**.

Reference/Hyperlink

Good practices require the audit and assurance professional to create a work paper for each line item, which describes the work performed, issues identified and conclusions. The reference/hyperlink is to be used to cross-reference the audit/assurance step to the work paper that supports it. The numbering system of this document provides a ready numbering scheme for the work papers. If desired, a link to the work paper can be pasted into this column.

Issue Cross-reference

This column can be used to flag a finding/issue that the IT audit and assurance professional wants to further investigate or establish as a potential finding. The potential findings should be documented in a work paper that indicates the disposition of the findings (formally reported, reported as a memo or verbal finding, or waived).

Comments

The comments column can be used to indicate the waiving of a step or other notations. It is not to be used in place of a work paper describing the work performed.

III. Controls Maturity Analysis

One of the consistent requests of stakeholders who have undergone IT audit/assurance reviews is a desire to understand how their performance compares to good practices. Audit and assurance professionals must provide an objective basis for the review conclusions. Maturity modeling for management and control over IT processes is based on a method of evaluating the organization, so it can be rated from a maturity level of nonexistent (0) to optimized (5). This approach is derived from the maturity model that the Software Engineering Institute (SEI) of Carnegie Mellon University defined for the maturity of software development.

The *IT Assurance: Guide Using CobiT®*, Appendix VII—Maturity Model for Internal Control, in **figure A4.2**, provides a generic maturity model showing the status of the internal control environment and the establishment of internal controls in an enterprise. It shows how the management of internal control, and an awareness of the need to establish better internal controls, typically develops from an *ad hoc* to an optimized level. The model provides a high-level guide to help COBIT users appreciate what is required for effective internal controls in IT and to help position their enterprise on the maturity scale.

Figure A4.2—Maturity Model for Internal Control

Maturity Level	Status of the Internal Control Environment	Establishment of Internal Controls
0 Non-existent	There is no recognition of the need for internal control. Control is not part of the organization's culture or mission. There is a high risk of control deficiencies and incidents.	There is no intent to assess the need for internal control. Incidents are dealt with as they arise.

Figure A4.2—Maturity Model for Internal Control *(cont.)*		
Maturity Level	**Status of the Internal Control Environment**	**Establishment of Internal Controls**
1 Initial/*ad hoc*	There is some recognition of the need for internal control. The approach to risk and control requirements is *ad hoc* and disorganized, without communication or monitoring. Deficiencies are not identified. Employees are not aware of their responsibilities.	There is no awareness of the need for assessment of what is needed in terms of IT controls. When performed, it is only on an *ad hoc* basis, at a high level and in reaction to significant incidents. Assessment addresses only the actual incident.
2 Repeatable but Intuitive	Controls are in place but are not documented. Their operation is dependent on the knowledge and motivation of individuals. Effectiveness is not adequately evaluated. Many control weaknesses exist and are not adequately addressed; the impact can be severe. Management actions to resolve control issues are not prioritized or consistent. Employees may not be aware of their responsibilities.	Assessment of control needs occurs only when needed for selected IT processes to determine the current level of control maturity, the target level that should be reached and the gaps that exist. An informal workshop approach, involving IT managers and the team involved in the process, is used to define an adequate approach to controls for the process and to motivate an agreed-upon action plan.
3 Defined	Controls are in place and adequately documented. Operating effectiveness is evaluated on a periodic basis and there is an average number of issues. However, the evaluation process is not documented. While management is able to deal predictably with most control issues, some control weaknesses persist and impacts could still be severe. Employees are aware of their responsibilities for control.	Critical IT processes are identified based on value and risk drivers. A detailed analysis is performed to identify control requirements and the root cause of gaps and to develop improvement opportunities. In addition to facilitated workshops, tools are used and interviews are performed to support the analysis and ensure that an IT process owner owns and drives the assessment and improvement process.
4 Managed and Measurable	There is an effective internal control and risk management environment. A formal, documented evaluation of controls occurs frequently. Many controls are automated and regularly reviewed. Management is likely to detect most control issues, but not all issues are routinely identified. There is consistent follow-up to address identified control weaknesses. A limited, tactical use of technology is applied to automate controls.	IT process criticality is regularly defined with full support and agreement from the relevant business process owners. Assessment of control requirements is based on policy and the actual maturity of these processes, following a thorough and measured analysis involving key stakeholders. Accountability for these assessments is clear and enforced. Improvement strategies are supported by business cases. Performance in achieving the desired outcomes is consistently monitored. External control reviews are organized occasionally.

Figure A4.2—Maturity Model for Internal Control *(cont.)*		
Maturity Level	**Status of the Internal Control Environment**	**Establishment of Internal Controls**
5 Optimized	An enterprisewide risk and control program provides continuous and effective control and risk issues resolution. Internal control and risk management are integrated with enterprise practices, supported with automated real-time monitoring with full accountability for control monitoring, risk management and compliance enforcement. Control evaluation is continuous, based on self-assessments and gap and root cause analyses. Employees are proactively involved in control improvements.	Business changes consider the criticality of IT processes and cover any need to reassess process control capability. IT process owners regularly perform self-assessments to confirm that controls are at the right level of maturity to meet business needs and they consider maturity attributes to find ways to make controls more efficient and effective. The organization benchmarks to external best practices and seeks external advice on internal control effectiveness. For critical processes, independent reviews take place to provide assurance that the controls are at the desired level of maturity and working as planned.

The maturity model evaluation is one of the final steps in the evaluation process. The IT audit and assurance professional can address the key controls within the scope of the work program and formulate an objective assessment of the maturity levels of the control practices. The maturity assessment can be a part of the audit/assurance report, and used as a metric from year to year to document progression in the enhancement of controls. However, it must be noted that the perception of the maturity level may vary between the process/IT asset owner and the auditor. Therefore, an auditor should obtain the concerned stakeholder's concurrence before submitting the final report to management.

At the conclusion of the review, once all findings and recommendations are completed, the professional assesses the current state of the COBIT control framework and assigns it a maturity level using the six-level scale. Some practitioners utilize decimals (x.25, x.5, x.75) to indicate gradations in the maturity model. As a further reference, COBIT provides a definition of the maturity designations by control objective. While this approach is not mandatory, the process is provided as a separate section at the end of the audit/assurance program for those enterprises that wish to implement it. It is suggested that a maturity assessment be made at the COBIT control level. To provide further value to the client/customer, the professional can also obtain maturity targets from the client/customer. Using the assessed and target maturity levels, the professional can create an effective graphic presentation that describes the achievement or gaps between the actual and targeted maturity goals.

IV. Assurance and Control Framework

ISACA IT Assurance Framework and Standards

ISACA has long recognized the specialized nature of IT assurance and strives to advance globally applicable standards. Guidelines and procedures provide detailed guidance on how to follow those standards. IT Audit/Assurance Standard S15 IT Controls, and IT Audit/Assurance Guideline G38 Access Controls are relevant to this audit/assurance program.

ISACA Controls Framework

COBIT is an IT governance framework and supporting tool set that allows managers to bridge the gap among control requirements, technical issues and business risks. COBIT enables clear policy development and good practice for IT control throughout enterprises.

Utilizing COBIT as the control framework on which IT audit/assurance activities are based aligns IT audit/assurance with good practices as developed by the enterprise.

Refer to ISACA's *CobiT® Control Practices: Guidance to Achieve Control Objectives for Successful IT Governance, 2nd Edition*, published in 2007, for the related control practice value and risk drivers.

V. Executive Summary of Audit/Assurance Focus

Oracle EBS Security

Since launching its first product offering approximately 30 years ago, Oracle Corp. has grown globally. In recent years, Oracle Corp. has been involved in a number of large acquisitions, including PeopleSoft, Siebel and Sun, as part of its enterprise application integration strategy that forms the core of its Fusion Middleware product range.

Oracle Corp. released Oracle EBS R12 in 2007, which introduced changes from the previous version, R11*i*.10, to the application technology platform, provided changes to the security authorization concept, and provided a new technology stack and architecture for Oracle EBS Financials. The latest release, Oracle EBS R12.1, was released in 2009 and introduced changes to the other enterprise application areas (e.g., Supply Chain Management, Procurement, Customer Relationship Management and Human Capital Management). With the ever-changing compliance landscape, a review of organizations' Oracle EBS environment is vital to ensuring that it is secure.

Business Impact and Risk

Oracle EBS is widely used in many enterprises. Improper configuration of Oracle EBS could result in the inability of the enterprise to execute its critical processes.

Oracle EBS risks resulting from ineffective or incorrect configurations or use could result in some of the following:
- Disclosure of privileged information
- Single points of failure
- Low data quality
- Loss of physical assets
- Loss of intellectual property
- Loss of competitive advantage
- Loss of customer confidence
- Violation of regulatory requirements
- Financial loss

Objective and Scope

Objective—The objective of the Oracle EBS audit/assurance review is to provide management with an independent assessment relating to the effectiveness of configuration and security of the enterprise's Oracle EBS architecture.

Scope—The review will focus on configuration of the relevant Oracle EBS components and modules within the enterprise. The selection of the specific components and modules will be based upon the risks introduced to the enterprise by these components and modules.

Minimum Audit Skills

This review is considered highly technical. The IT audit and assurance professional must have an understanding of Oracle EBS best-practice processes and requirements, and be highly conversant in Oracle EBS tools, exposures, and functionality. It should not be assumed that an audit and assurance professional holding the CISA designation has the requisite skills to perform this review.

VI. Expenditure Business Cycle—Audit/Assurance Program

Audit/Assurance Program Step	COBIT Cross-reference	COSO					Reference Hyperlink	Issue Cross-reference	Comments
		Control Environment	Risk Assessment	Control Activities	Information and Communication	Monitoring			
A. Prior Audit/Examination Report Follow-up									
1. Review prior report, if one exists, to verify completion of any agreed-upon corrections and note remaining deficiencies.	ME1								
1.1 Determine whether: • Senior management has assigned responsibilities for information, its processing and its use • User management is responsible for providing information that supports the enterprise's objectives and policies • Information systems management is responsible for providing the information systems capabilities necessary for achievement of the defined information systems objectives and policies of the enterprise • Senior management approves plans for development and acquisition of information systems • There are procedures to ensure that the information system being developed or acquired meets user requirements • There are procedures to ensure that information systems, programs and configuration changes are tested adequately prior to implementation in a separate testing environment	AI6 DS4 ME1								

VI. Expenditure Business Cycle—Audit/Assurance Program *(cont.)*

Audit/Assurance Program Step	COBIT Cross-reference	COSO					Reference Hyperlink	Issue Cross-reference	Comments
		Control Environment	Risk Assessment	Control Activities	Information and Communication	Monitoring			
A. Prior Audit/Examination Report Follow-up ***(cont.)***									
1.1 *(cont.)* • All personnel involved in the system acquisition and configuration activities receive adequate training and supervision • There are procedures to ensure that information systems are implemented/configured/upgraded in accordance with the established standards • User management participate in the conversion of data from the existing system to the new system • Final approval is obtained from user management prior to going live with a new information/upgraded system • There are procedures to document and schedule all changes to information systems • There are procedures to ensure that only authorized changes are initiated • There are procedures to ensure that only authorized, tested and documented changes to information systems are accepted into the production client • There are procedures to allow for and control emergency changes • There are procedures for the approval, monitoring and control of the acquisition and upgrade of hardware and systems software									

VI. Expenditure Business Cycle—Audit/Assurance Program (cont.)

Audit/Assurance Program Step	COBIT Cross-reference	COSO					Reference Hyperlink	Issue Cross-reference	Comments
		Control Environment	Risk Assessment	Control Activities	Information and Communication	Monitoring			
A. Prior Audit/Examination Report Follow-up *(cont.)*									
1.1 *(cont.)* • There is a process for monitoring the volume of named and concurrent Oracle EBS users to ensure that the license agreement is not being violated • The organization structure, established by senior management, provides for an appropriate segregation of incompatible functions • The database and application servers are located in a physically separate and protected environment (i.e., a data center) • Emergency, backup and recovery plans are documented and tested on a regular basis to ensure that they remain current and operational • Backup and recovery plans allow users of information systems to resume operations in the event of an interruption • Application controls are designed with regard to any weaknesses in segregation, security, development and processing controls that may affect the information system									

VI. Expenditure Business Cycle—Audit/Assurance Program (cont.)

Audit/Assurance Program Step	COBIT Cross-reference	COSO					Reference Hyperlink	Issue Cross-reference	Comments
		Control Environment	Risk Assessment	Control Activities	Information and Communication	Monitoring			
B. Preliminary Audit Steps									
1. Gain an understanding of the Oracle EBS environment.									
1.1 The same background information obtained for the Oracle EBS Security audit plan is required for and relevant to the business cycles. In particular, the following information is important: • Version and release of Oracle EBS software that has been implemented • Total number of named users (for comparison with logical access security testing results) • Number of Oracle EBS database instances • Accounting period, company codes and Chart of Accounts (COA) • Identification of the modules being used • Locally developed application programs, reports or tables created by the enterprise • Details of the risk assessment approach taken in the enterprise to identify and prioritize risks • Copies of the enterprise's key security policies and standards • Outstanding audit findings, if any, from previous years	PO2 PO3 PO4 PO6 PO9 AI2 AI6 DS2 DS5 ME1 ME2								

VI. Expenditure Business Cycle—Audit/Assurance Program (cont.)

Audit/Assurance Program Step	COBIT Cross-reference	COSO					Reference Hyperlink	Issue Cross-reference	Comments
		Control Environment	Risk Assessment	Control Activities	Information and Communication	Monitoring			
B. Preliminary Audit Steps ***(cont.)***									
1.2 Obtain details of the following: • The organizational management model as it relates to expenditure activity, e.g., purchasing organization unit structure in Oracle EBS software and purchasing/accounts payable organization chart (required when evaluating the results of access security control testing) • An interview of the systems implementation team, if possible, and the process design documentation for materials management	AI1 DS5 DS6								
2. Identify the significant risks and determine the key controls.									
2.1 Develop a high-level process flow diagram and overall understanding of the Expenditure processing cycle, including the following subprocesses: • Master data maintenance • Purchasing • Invoice processing • Processing disbursements	PO9 AI1 DS13								

VI. Expenditure Business Cycle—Audit/Assurance Program (cont.)

Audit/Assurance Program Step	COBIT Cross-reference	COSO					Reference Hyperlink	Issue Cross-reference	Comments
		Control Environment	Risk Assessment	Control Activities	Information and Communication	Monitoring			
B. Preliminary Audit Steps *(cont.)*									
2.2 Assess the key risks, determine key controls or control weaknesses, and test controls (refer to sample testing program below and chapter 4 for techniques for testing configurable controls and logical access security) regarding the following factors: • The control culture of the enterprise (e.g., a "just-enough" control philosophy) • The need to exercise judgment to determine the key controls in the process and whether the controls structure is adequate (Any weaknesses in the control structure should be reported to executive management and resolved.)	P09 DS5 DS9 ME2								
C. Detailed Audit Steps									
1. Master Data Maintenance									
1.1 Changes made to master data are valid, complete, accurate and timely.									
1.1.1 Review any relevant documentation relating to policies, procedures, standards and guidance for requesting changes to master data, and assess the existence of and adherence to these policies.	AI6								

VI. Expenditure Business Cycle—Audit/Assurance Program (cont.)

Audit/Assurance Program Step	COBIT Cross-reference	COSO					Reference Hyperlink	Issue Cross-reference	Comments
		Control Environment	Risk Assessment	Control Activities	Information and Communication	Monitoring			
C. Detailed Audit Steps *(cont.)*									
1.1.1 *(cont.)* Review evidence of the generation and comparison of online reports of master data changes back to source documentation (sample basis). Determine whether the recorded changes to master data are compared to authorized source documents and/or a manual log of requested changes, to ensure that they were input accurately and promptly.									
1.1.2 Review organizational policy and process design specifications regarding access to maintain master data. Test user access via a list extracted from Oracle EBS of the target high-risk functions/forms (refer to chapter 4 on how to test user access), to ascertain who has access privileges to the following functions/forms: • Suppliers • Merge Suppliers • Summary Approved Suppliers List • Suppliers Statuses	AI4 DS5								

VI. Expenditure Business Cycle—Audit/Assurance Program (cont.)

Audit/Assurance Program Step	COBIT Cross-reference	COSO					Reference Hyperlink	Issue Cross-reference	Comments
		Control Environment	Risk Assessment	Control Activities	Information and Communication	Monitoring			
C. Detailed Audit Steps *(cont.)*									
1.1.2 *(cont.)* • Define Approved Suppliers List • Setup Approved Suppliers List • Supplier Item Catalog									
1.1.3 Extract a list of supplier account names by running either the New Supplier/New Supplier Site Listing or the Suppliers report. Review a sample for compliance with the enterprise's naming convention. View or search the list for potential duplicates.	DS13								
1.2 Inventory master data remain current and pertinent.									
1.2.1 Review policies and procedures relating to periodic reviews of supplier master file data. Determine whether the appropriate members of management can produce a list of reports or the equivalent, and regularly perform a review of same. Confirm evidence of their review of the data on a rotating basis for currency and ongoing pertinence. Oracle EBS standard reports could include:	AI4 DS11								

VI. Expenditure Business Cycle—Audit/Assurance Program *(cont.)*

Audit/Assurance Program Step	COBIT Cross-reference	COSO					Reference Hyperlink	Issue Cross-reference	Comments
		Control Environment	Risk Assessment	Control Activities	Information and Communication	Monitoring			
C. Detailed Audit Steps *(cont.)*									
1.2.1 *(cont.)* • New Supplier/New Supplier Site Listing • Suppliers report • Supplier Audit report • Supplier Merge report • Supplier Paid Invoice History report • Supplier Payment History report									
2. Purchasing									
2.1 Purchase order (PO) entry and changes are valid, complete, accurate and timely.									
2.1.1 Determine whether the ability to create, change or cancel purchase requisitions, POs and outline agreements (standing POs) is restricted to authorized personnel by testing access to the following transactions: • Create, modify and cancel the purchase requisition • Autocreate purchase orders • Create, modify and delete purchase orders • Modify certain Oracle EBS setup parameters • Perform the purging operations	PO9 PO10 DS5 DS10								

VI. Expenditure Business Cycle—Audit/Assurance Program (cont.)

Audit/Assurance Program Step	COBIT Cross-reference	COSO					Reference Hyperlink	Issue Cross-reference	Comments
		Control Environment	Risk Assessment	Control Activities	Information and Communication	Monitoring			
C. Detailed Audit Steps *(cont.)*									
2.1.1 *(cont.)* Review a list of the target high-risk functions/ forms to determine who has access privileges to the following functions/forms: • Approval Assignments • Approval Groups • Autocreate Documents • Buyers • Control Purchase Orders • Control Requisitions • Define Position Hierarchy • Define Purchasing Lookup Codes • Document Types • Lookups: Common • OM: Order Purge • PO Copy Document • PO Creation—Process Navigator Definition for PO Creation Process • PO Preferences • PO Summary: Create New PO • PO Summary: Create New Release • PO Summary: Open Document									

VI. Expenditure Business Cycle—Audit/Assurance Program (cont.)

Audit/Assurance Program Step	COBIT Cross-reference	COSO					Reference Hyperlink	Issue Cross-reference	Comments
		Control Environment	Risk Assessment	Control Activities	Information and Communication	Monitoring			
C. Detailed Audit Steps ***(cont.)***									
2.1.1 *(cont.)* • Procure to Pay—Process Navigator Definition for Procurement to Pay process • Profile System Values—Profile System Values form • Profile User Values • PO Summary • Purchase Orders • POs: Suppliers Item Catalog • Purchasing Options • Purchasing Lookups • Purge • Requisition Preferences • Requisition Summary • Requisition Templates • Requisitions • Requisitions: Supplier Item Catalog • Templates									

VI. Expenditure Business Cycle—Audit/Assurance Program (cont.)

Audit/Assurance Program Step	COBIT Cross-reference	COSO					Reference Hyperlink	Issue Cross-reference	Comments
		Control Environment	Risk Assessment	Control Activities	Information and Communication	Monitoring			
C. Detailed Audit Steps *(cont.)*									
2.1.2 Determine whether document security is configured to set security levels and access levels by document type to authorized personnel. To view, navigate to the Document Types window (Setup➔Purchasing➔Document Types) and select and review the security level and access level applied against each of the existing documents. Review a list of the target high-risk functions/forms to determine who has access privileges to the Document Types function.	PO9 DS5								
2.1.3 Obtain a sufficient understanding of the system configuration to assess the adequacy of the release strategy as defined and implemented by the enterprise, as well as the functioning and effectiveness of established policies, procedures, standards and guidance. Review the following system settings to obtain an understanding of the way the system has been configured: • Approval Groups (Setup➔Approvals➔Approval Groups) • Approval Assignments (Setup➔Approvals➔Approval Assignments)	DS5 DS9 DS13 ME1								

VI. Expenditure Business Cycle—Audit/Assurance Program (cont.)

Audit/Assurance Program Step	COBIT Cross-reference	COSO					Reference Hyperlink	Issue Cross-reference	Comments
		Control Environment	Risk Assessment	Control Activities	Information and Communication	Monitoring			
C. Detailed Audit Steps *(cont.)*									
2.1.4 Determine that Cross-Validation is enabled by navigating to the Key Flexfield Segments form (Setup➔Financials➔Flexfields➔ Key➔Segments) and selecting the name of the application (Oracle GL) and Flexfield (Accounting Flexfield). Ensure that the Allow Dynamic Inserts and the Cross-Validate Segments checkboxes are enabled if the enterprise is utilizing Dynamic Insertion.	DS11								
2.1.5 Discuss with the client and determine whether Flexfield Value Security has been enabled and error messages are defined by client personnel that are meaningful to assist the user in understanding the reasons for being restricted from entering a value.	DS3								
2.1.6 Determine whether management monitors receiving adjustments to POs through reviewing standard Oracle Reports on a periodic basis. Standard Oracle Reports that can be reviewed include: • Receipt Adjustments reports									

VI. Expenditure Business Cycle—Audit/Assurance Program *(cont.)*

Audit/Assurance Program Step	COBIT Cross-reference	COSO					Reference Hyperlink	Issue Cross-reference	Comments
		Control Environment	Risk Assessment	Control Activities	Information and Communication	Monitoring			
C. Detailed Audit Steps *(cont.)*									
2.1.6 *(cont.)* • Receiving Exceptions reports • Unordered Receipts report These reports can be run through the Report window (Reports➔Run) Inspect physical evidence that these reports are reviewed on a regular basis and that exceptions are identified for follow-up and resolution.	DS11 ME1								
2.1.7 Determine whether management reviews the status of purchase orders on a timely basis by reviewing standard Oracle Reports, including: • Expected Receipts report • Blanket and Planned PO Status report • Open Purchase Orders report • Overdue Vendor Shipments report • Purchase Requisition Status report Inspect physical evidence that these reports are reviewed on a regular basis and that exceptions are identified for follow-up and resolution.	AI6								

VI. Expenditure Business Cycle—Audit/Assurance Program (cont.)

Audit/Assurance Program Step	COBIT Cross-reference	COSO					Reference Hyperlink	Issue Cross-reference	Comments
		Control Environment	Risk Assessment	Control Activities	Information and Communication	Monitoring			
C. Detailed Audit Steps *(cont.)*									
2.1.8 Determine which suppliers and items should be reflected on the system. Compare the ASL on the system to authorized documentation. Use the following path to navigate to the ASL: Supplier Base→Approved Suppliers List.	DS3								
2.1.9 Review the sequential numbering setting to determine that it has been checked for each document type. To view this setting, navigate to the Purchasing Options window via: Setup→Organizations→Purchasing Options→Numbering Tab. Determine whether management monitors the sequential numbering of transactions through the following standard Oracle Reports: • Purchasing and Requisitions Activity registers • Canceled Purchase Order and Requisitions reports • Purchase Order Detail report Inspect physical evidence that these reports are reviewed on a regular basis and that exceptions are identified for follow-up and resolution.	ME1								

VI. Expenditure Business Cycle—Audit/Assurance Program *(cont.)*

Audit/Assurance Program Step	COBIT Cross-reference	COSO					Reference Hyperlink	Issue Cross-reference	Comments
		Control Environment	Risk Assessment	Control Activities	Information and Communication	Monitoring			
C. Detailed Audit Steps ***(cont.)***									
2.1.10 Determine whether management has a procedure in place to monitor overrides of established PO prices, terms and conditions. Ascertain whether the Purchase Price Variance report is used to assist in monitoring. Inspect physical evidence that these reports are reviewed on a regular basis and that exceptions are identified for follow-up and resolution.	PO10 DS11								
2.2 Goods are received only for valid purchase orders and goods receipts are recorded completely, accurately and promptly.									
2.2.1 Examine the following reports and verify that they are used to identify long-outstanding goods receipt notes, POs and/or invoices: • Invoice Aging report • Invoice on Hold report • Open Purchase Order by Cost Center report • Receiving Exceptions report Ascertain from management if there are any reasons for any long-outstanding items on the report.	DS9								

VI. Expenditure Business Cycle—Audit/Assurance Program *(cont.)*

Audit/Assurance Program Step	COBIT Cross-reference	COSO					Reference Hyperlink	Issue Cross-reference	Comments
		Control Environment	Risk Assessment	Control Activities	Information and Communication	Monitoring			
C. Detailed Audit Steps *(cont.)*									
2.2.2 Extract a list of the target high-risk functions/ forms. Review this list to ascertain who has access privileges to the following: • Corrections (receiving) • Receiving Options • Returns (receiving)	P09								
2.3 Defective goods are returned to suppliers in a timely manner.									
2.3.1 Ascertain from management the mechanisms used to block processing and for returning rejected goods to suppliers. Determine if there are any long-outstanding materials pending return to suppliers/receipt of appropriate credits. Test user access to transactions for invoice processing.	DS3								

VI. Expenditure Business Cycle—Audit/Assurance Program (cont.)

Audit/Assurance Program Step	COBIT Cross-reference	COSO					Reference Hyperlink	Issue Cross-reference	Comments
		Control Environment	Risk Assessment	Control Activities	Information and Communication	Monitoring			
C. Detailed Audit Steps *(cont.)*									
3. Invoice Processing									
3.1 Amounts posted to accounts payable represent goods or services received.									
3.1.1 Test user access to transactions for invoice processing: • Create, modify and delete invoices • Modify open and close periods in the Accounting Calendar • Create, modify and delete credit memos/debit memos • Modify certain Oracle EBS setup parameters relating to accounts payable • Create, modify and delete Distribution Sets Extract and review a list of the target high-risk functions/forms to ascertain who has access privileges to the following: • AP Accounting Flexfield Combinations GUI • AP Accounting Periods • Calendars—Define Periods • Close Accounts Payable Period—Process Navigator Definition for Close Accounts Payable Period Process	P09 AI7 DS2 DS5 DS13								

VI. Expenditure Business Cycle—Audit/Assurance Program *(cont.)*

Audit/Assurance Program Step	COBIT Cross-reference	COSO					Reference Hyperlink	Issue Cross-reference	Comments
		Control Environment	Risk Assessment	Control Activities	Information and Communication	Monitoring			
C. Detailed Audit Steps ***(cont.)***									
3.1.1 *(cont.)* • Distribution Sets • Invoice Accounting • Invoice Actions • Invoice Apply Prepayments • Invoice Approvals • Invoice Batch Summary • Invoice Batches • Invoice Cancel • Invoice Distributions • Invoice Distributions Prorate • Invoice Distributions Reverse • Invoice Fundscheck • Invoice Gateway • Invoice Holds • Invoice Match • Invoice Overview • Invoice Payments • Invoice Payments Schedules • Invoice Payments Workbench • Invoice Print • Invoice Release Holds • Invoices • Invoices Summary • Open and Close Periods • Payables Options • Payment Terms									

VI. Expenditure Business Cycle—Audit/Assurance Program (cont.)

Audit/Assurance Program Step	COBIT Cross-reference	COSO					Reference Hyperlink	Issue Cross-reference	Comments
		Control Environment	Risk Assessment	Control Activities	Information and Communication	Monitoring			
C. Detailed Audit Steps *(cont.)*									
3.1.1 *(cont.)* • Period Close—Payment Company—Process Navigator Definition for Period Close—Parent Company Process • Period Close—Subsidiary Company Process Navigator Definition for Period Close—Subsidiary Company Process • Period Types—Define Period Types • Returns									
3.2 Accounts payable amounts are calculated completely and accurately and recorded in a timely manner.									
3.2.1 To review the setting of required fields, navigate to the properties of a required field through the Help➔Diagnostics➔Properties➔Item path, and confirm that the null value is set to False, which defines the field as required. A password is required to view the attributes of a field. The preloaded Oracle EBS password is “apps” and is case-sensitive. If it has not been changed, security may be compromised. Request the password to test to see if the client has changed the password.	DS5 DS13								

VI. Expenditure Business Cycle—Audit/Assurance Program (cont.)

Audit/Assurance Program Step	COBIT Cross-reference	COSO					Reference Hyperlink	Issue Cross-reference	Comments
		Control Environment	Risk Assessment	Control Activities	Information and Communication	Monitoring			
C. Detailed Audit Steps *(cont.)*									
3.2.2 Review the Distribution Set listing (Reports→Run) for a complete list of all the Distribution Sets that have been defined by the client.	DS11								
3.2.3 Determine whether the following reports are executed and reviewed periodically to facilitate the accurate processing of supplier invoices and credit memos: • Expense Export report • Invoice Audit Listing • Invoice Audit report • Invoice on Hold report • Credit Memo Matching report • Uninvoiced Receipts report • Unordered Receipts report Check that there are appropriate procedures in place to investigate unmatched POs. In particular, long-outstanding items should be followed up and cleared.	PO6 PO8 DS13								

VI. Expenditure Business Cycle—Audit/Assurance Program (cont.)

Audit/Assurance Program Step	COBIT Cross-reference	COSO					Reference Hyperlink	Issue Cross-reference	Comments
		Control Environment	Risk Assessment	Control Activities	Information and Communication	Monitoring			
C. Detailed Audit Steps ***(cont.)***									
3.2.4 Review the procedures for the regular reconciliation of key/large supplier statements to the AP subledger. Inspect the procedures documentation and a sample of reconciliation documentation for the period under review for evidence of performance.	DS9								
3.2.5 Tolerance limits for price variances and message settings for invoice verification (online matching) should be checked as follows: • Navigate to the Purchasing Options window (Setup Organizations→Purchasing Options Document Control Section) • Check that the Price Tolerance % field is populated and the Enforce Price Tolerance checkbox is marked. All other checkboxes should be marked in accordance with the organizational policy.	AI2								

VI. Expenditure Business Cycle—Audit/Assurance Program (cont.)

Audit/Assurance Program Step	COBIT Cross-reference	COSO					Reference Hyperlink	Issue Cross-reference	Comments
		Control Environment	Risk Assessment	Control Activities	Information and Communication	Monitoring			
C. Detailed Audit Steps ***(cont.)***									
3.2.6 Review the Unposted Invoice Sweep report to determine which invoices do not belong to the period being closed. Select Invoice Sweep reports and Posting Hold reports throughout the period of intended reliance, and determine that management has reviewed such reports prior to closing the periods. For items that have been marked as requiring action, determine that such action has been completed in accordance with management's intention.	P06								
3.2.7 Extract a list of the target high-risk functions/forms. Review the list to ascertain who has access privileges to the following: • Conversion Rate Types • Currencies—Currencies form • Daily Rates—Define Daily Rates • Exchange Rate Adjustment	P09								

VI. Expenditure Business Cycle—Audit/Assurance Program (cont.)

Audit/Assurance Program Step	COBIT Cross-reference	COSO					Reference Hyperlink	Issue Cross-reference	Comments
		Control Environment	Risk Assessment	Control Activities	Information and Communication	Monitoring			
C. Detailed Audit Steps *(cont.)*									
3.3 Credit notes and other adjustments are calculated completely and accurately and recorded in a timely manner.									
3.3.1 Review the access to the functions/forms for invoicing and credit and debit memos listed in testing technique point 3.1.1 to confirm that access is appropriately restricted. Credit and debit notes are invoice types and should be restricted in the same manner.	AI7 DS5								
3.3.2 Determine that matching settings have been checked. Do this by navigating to the Purchasing Options window. Check the Match Approval Level setting.	DS9								
3.3.3 Determine whether management reviews the Payables Account Analysis report to identify nonsystem debits to accounts payable. Determine whether there is a procedure to review this report. Ascertain whether these reports are reviewed on a regular basis and that exceptions are identified for follow-up and resolution. Inspect physical evidence that the reports are run, approved and retained, and that exceptions are resolved, as necessary.	PO7 AI4 DS3								

VI. Expenditure Business Cycle—Audit/Assurance Program (cont.)

Audit/Assurance Program Step	COBIT Cross-reference	COSO					Reference Hyperlink	Issue Cross-reference	Comments
		Control Environment	Risk Assessment	Control Activities	Information and Communication	Monitoring			
C. Detailed Audit Steps ***(cont.)***									
4. Processing Disbursements									
4.1 Disbursements are made only for goods and services received, and are calculated accurately, recorded and distributed to the appropriate suppliers in a timely manner.									
4.1.1 Review users with access to: • Create, modify and void/release payments and payment batches • Modify open and closed periods in the Accounting Calendar • Release holds on payments Extract a list of the target high-risk functions/forms. Review this list to ascertain who has access privileges to the following: • AP Accounting Periods • Calendars—Define Periods • Close Accounts Payable Period—Process Navigator Definition for Close Accounts Payable Period process • Open and Close Periods • Payment Accounting • Payment Actions • Payment Batch Sets	PO8 PO9 AI6 DS5								

VI. Expenditure Business Cycle—Audit/Assurance Program (cont.)

Audit/Assurance Program Step	COBIT Cross-reference	COSO					Reference Hyperlink	Issue Cross-reference	Comments
		Control Environment	Risk Assessment	Control Activities	Information and Communication	Monitoring			
C. Detailed Audit Steps *(cont.)*									
4.1.1 *(cont.)* • Payment Batches • Payment Batches Actions • Payment Batches Cancel • Payment Batches Confirm • Payment Batches Format • Payment Batches Modify • Payment Batches Positive Pay • Payment Batches Print • Payment Batches Summary • Payment Batches to Payment • Payment Formats • Payment Invoice Rates • Payment Invoices • Payment Overview • Payment Print Check • Payment Programs • Payment Void • Payments • Payments Summary • Period Close—Parent Company—Process Navigator Definition for Period Close—Parent Company process									

VI. Expenditure Business Cycle—Audit/Assurance Program (cont.)

Audit/Assurance Program Step	COBIT Cross-reference	COSO					Reference Hyperlink	Issue Cross-reference	Comments
		Control Environment	Risk Assessment	Control Activities	Information and Communication	Monitoring			
C. Detailed Audit Steps *(cont.)*									
4.1.1 *(cont.)* • Period Close—Subsidiary Company process • Period Types—Define Period Types									
4.1.2 Inspect the procedure documentation for reviewing supporting documentation before approving payments. Review a sample for evidence of performance and a sample for canceled supporting documentation for the period.	PO8								
4.1.3 The following Oracle Reports (Reports→Run) facilitate the accurate processing of payments: • Cash Requirements report • Discounts Available report • Preliminary Payment register • Payment Batch Control report Determine whether management reviews these reports on a regular basis, identifies exceptions for follow-up and resolves them, as appropriate. Inspect physical evidence that the reports are run, approved and retained, and that exceptions are resolved, as necessary.	DS7 DS9								

VI. Expenditure Business Cycle—Audit/Assurance Program *(cont.)*

Audit/Assurance Program Step	COBIT Cross-reference	COSO					Reference Hyperlink	Issue Cross-reference	Comments
		Control Environment	Risk Assessment	Control Activities	Information and Communication	Monitoring			
C. Detailed Audit Steps ***(cont.)***									
4.1.4 Review the procedures in place for the periodic review of returned paid checks for unauthorized signatures, alterations and/or endorsements. Inspect the procedures documentation for evidence of performance and a sample of returned paid checks for the period under review.	PO8 DS12								
4.1.5 The following Oracle Reports help manage the sequential integrity of payment documents: • Payables Posted Payment Register • Audit Report By Document Number Determine whether management reviews these reports on a regular basis, identifies exceptions for follow-up and resolves them, as appropriate. Inspect physical evidence that the reports are run, approved and retained, and that exceptions are resolved, as necessary.	PO6 DS9 DS12								

VI. Expenditure Business Cycle—Audit/Assurance Program *(cont.)*

Audit/Assurance Program Step	COBIT Cross-reference	COSO					Reference Hyperlink	Issue Cross-reference	Comments
		Control Environment	Risk Assessment	Control Activities	Information and Communication	Monitoring			
C. Detailed Audit Steps *(cont.)*									
4.1.6 The Invoice Aging report may be used to review an aged accounts payable analysis. Determine whether management reviews this report on a regular basis, identifies exceptions for follow-up and resolves them, as appropriate. Inspect physical evidence that the report is run, approved and retained, and that exceptions are resolved, as necessary.	DS13								
4.1.7 Obtain and review a list of users who have access to the drives and directories where the pay files are located. Review the process for archiving and purging these files. Users should not have write access to these files. Read access should also be restricted.	DS5								

VII. Maturity Assessment

The maturity assessment is an opportunity for the reviewer to assess the maturity of the processes reviewed. Based on the results of audit/assurance review, and the reviewer's observations, a maturity level should be assigned to each of the following COBIT control practices.

COBIT Control Practice	Assessed Maturity	Target Maturity	Reference Hyperlink	Comments
AI6.1 Change Standards and Procedures 1. Develop, document and promulgate a change management framework that specifies the policies and processes, including: • Roles and responsibilities • Classification and prioritisation of all changes based on business risk • Assessment of impact • Authorisation and approval of all changes by the business process owners and IT • Tracking and status of changes • Impact on data integrity (e.g., all changes to data files being made under system and application control rather than by direct user intervention) 2. Establish and maintain version control over all changes. 3. Implement roles and responsibilities that involve business process owners and appropriate technical IT functions. Ensure appropriate segregation of duties. 4. Establish appropriate record management practices and audit trails to record key steps in the change management process. Ensure timely closure of changes. Elevate and report to management changes that are not closed in a timely fashion. 5. Consider the impact of contracted services providers (e.g., of infrastructure, application development and shared services) on the change management process. Consider integration of organisational change management processes with change management processes of service providers. Consider the impact of the organisational change management process on contractual terms and SLAs.				

VII. Maturity Assessment (cont.)

COBIT Control Practice	Assessed Maturity	Target Maturity	Reference Hyperlink	Comments
AI6.2 Impact Assessment, Prioritisation and Authorisation 1. Develop a process to allow business process owners and IT to request changes to infrastructure, systems or applications. Develop controls to ensure that all such changes arise only through the change request management process. 2. Categorise all requested changes (e.g., infrastructure, operating systems, networks, application systems, purchased/packaged application software). 3. Prioritise all requested changes. Ensure that the change management process identifies both the business and technical needs for the change. Consider legal, regulatory and contractual reasons for the requested change. 4. Assess all requests in a structured fashion. Ensure that the assessment process addresses impact analysis on infrastructure, systems and applications. Consider security, legal, contractual and compliance implications of the requested change. Consider also interdependencies amongst changes. Involve business process owners in the assessment process, as appropriate. 5. Ensure that each change is formally approved by business process owners and IT technical stakeholders, as appropriate.				
AI6.4 Change Status Tracking and Reporting 1. Ensure that a documented process exists within the overall change management process to declare, assess, authorise and record an emergency change. 2. Ensure that emergency changes are processed in accordance with the emergency change element of the formal change management process. 3. Ensure that all emergency access arrangements for changes are appropriately authorised, documented and revoked after the change has been applied. 4. Conduct a post-implementation review of all emergency changes, involving all concerned parties. The review should consider implications for aspects such as further application system maintenance, impact on development and test environments, application software development quality, documentation and manuals, and data integrity.				

VII. Maturity Assessment (cont.)

COBIT Control Practice	Assessed Maturity	Target Maturity	Reference Hyperlink	Comments
DS5.3 Identity Management 1. Establish and communicate policies and procedures to uniquely identify, authenticate and authorise access mechanisms and access rights for all users on a need-to-know/need-to-have basis, based on predetermined and preapproved roles. Clearly state accountability of any user for any action on any of the systems and/or applications involved. 2. Ensure that roles and access authorisation criteria for assigning user access rights take into account: • Sensitivity of information and applications involved (data classification) • Policies for information protection and dissemination (legal, regulatory, internal policies and contractual requirements) • Roles and responsibilities as defined within the enterprise • The need-to-have access rights associated with the function • Standard but individual user access profiles for common job roles in the organisation • Requirements to guarantee appropriate segregation of duties 3. Establish a method for authenticating and authorising users to establish responsibility and enforce access rights in line with sensitivity of information and functional application requirements and infrastructure components, and in compliance with applicable laws, regulations, internal policies and contractual agreements. 4. Define and implement a procedure for identifying new users and recording, approving and maintaining access rights. This needs to be requested by user management, approved by the system owner and implemented by the responsible security person. 5. Ensure that a timely information flow is in place that reports changes in jobs (i.e., people in, people out, people change). Grant, revoke and adapt user access rights in co-ordination with human resources and user departments for users who are new, who have left the organisation, or who have changed roles or jobs.				

VII. Maturity Assessment (cont.)

COBIT Control Practice	Assessed Maturity	Target Maturity	Reference Hyperlink	Comments
DS5.4 User Account Management 1. Ensure that access control procedures include but are not limited to: • Using unique user IDs to enable users to be linked to and held accountable for their actions • Awareness that the use of group IDs results in the loss of individual accountability and is permitted only when justified for business or operational reasons and compensated by mitigating controls. Group IDs must be approved and documented. • Checking that the user has authorisation from the system owner for the use of the information system or service, and the level of access granted is appropriate to the business purpose and consistent with the organisational security policy • A procedure to require users to understand and acknowledge their access rights and the conditions of such access • Ensuring that internal and external service providers do not provide access until authorisation procedures have been completed • Maintaining a formal record, including access levels, of all persons registered to use the service • A timely and regular review of user IDs and access rights 2. Ensure that management reviews or reallocates user access rights at regular intervals using a formal process. User access rights should be reviewed or reallocated after any job changes, such as transfer, promotion, demotion or termination of employment. Authorisations for special privileged access rights should be reviewed independently at more frequent intervals.				

VII. Maturity Assessment (cont.)

COBIT Control Practice	Assessed Maturity	Target Maturity	Reference Hyperlink	Comments
DS9.1 Configuration Repository and Baseline 1. Implement a configuration repository to capture and maintain configuration management items. The repository should include hardware; application software; middleware; parameters; documentation; procedures; and tools for operating, accessing and using the systems, services, version numbers and licencing details. 2. Implement a tool to enable the effective logging of configuration management information within a repository. 3. Provide a unique identifier to a configuration item so the item can be easily tracked and related to physical asset tags and financial records. 4. Define and document configuration baselines for components across development, test and production environments, to enable identification of system configuration at specific points in time (past, present and planned). 5. Establish a process to revert to the baseline configuration in the event of problems, if determined appropriate after initial investigation. 6. Install mechanisms to monitor changes against the defined repository and baseline. Provide management reports for exceptions, reconciliation and decision making.				

VII. Maturity Assessment (cont.)

COBIT Control Practice	Assessed Maturity	Target Maturity	Reference Hyperlink	Comments
DS9.2 Identification and Maintenance of Configuration Items 1. Define and implement a policy requiring all configuration items and their attributes and versions to be identified and maintained. 2. Tag physical assets according to a defined policy. Consider using an automated mechanism, such as barcodes. 3. Define a policy that integrates incident, change and problem management procedures with the maintenance of the configuration repository. 4. Define a process to record new, modified and deleted configuration items and their relative attributes and versions. Identify and maintain the relationships between configuration items in the configuration repository. 5. Establish a process to maintain an audit trail for all changes to configuration items. 6. Define a process to identify critical configuration items in relationship to business functions (component failure impact analysis). 7. Record all assets—including new hardware and software, procured or internally developed—within the configuration management data repository. 8. Define and implement a process to ensure that valid licences are in place to prevent the inclusion of unauthorised software.				
DS9.3 Configuration Integrity Review 1. To validate the integrity of configuration data, implement a process to ensure that configuration items are monitored. Compare recorded data against actual physical existence, and ensure that errors and deviations are reported and corrected. 2. Using automated discovery tools where appropriate, reconcile actual installed software and hardware periodically against the configuration database, licence records and physical tags. 3. Periodically review against the policy for software usage the existence of any software in violation or in excess of current policies and licence agreements. Report deviations for correction.				

Page intentionally left blank

Appendix 5. Oracle Security Administration Audit Plan

I. Introduction

Overview

ISACA developed *ITAF™: A Professional Practices Framework for IT Assurance* as a comprehensive and good-practice-setting model. ITAF provides standards that are designed to be mandatory, and are the guiding principles under which the IT audit and assurance profession operates. The guidelines provide information and direction for the practice of IT audit and assurance. The tools and techniques provide methodologies, tools and templates to provide direction in the application of IT audit and assurance processes.

Purpose

The audit/assurance program is a tool and template to be used as a road map for the completion of a specific assurance process. ISACA has commissioned audit/assurance programs to be developed for use by IT audit and assurance practitioners. This audit/assurance program is intended to be utilized by IT audit and assurance professionals with the requisite knowledge of the subject matter under review, as described in ITAF, section 2200—General Standards. The audit/assurance programs are part of ITAF, section 4000—IT Assurance Tools and Techniques.

Control Framework

The audit/assurance programs have been developed in alignment with the ISACA COBIT® framework—specifically, COBIT 4.1—using generally applicable and accepted good practices. They reflect ITAF, sections 3400—IT Management Processes, 3600—IT Audit and Assurance Processes, and 3800—IT Audit and Assurance Management.

Many enterprises have embr-aced several frameworks at an enterprise level, including the Committee of Sponsoring Organizations of the Treadway Commission (COSO) Internal Control Framework. The importance of the control framework has been enhanced due to regulatory requirements by the US Securities and Exchange Commission (SEC) as directed by the US Sarbanes-Oxley Act of 2002 and similar legislation in other countries. They seek to integrate control framework elements used by the general audit/assurance team into the IT audit and assurance framework. Since COSO is widely used, it has been selected for inclusion in this audit/assurance program. The reviewer may delete or rename columns in the audit program to align with the enterprise's control framework.

IT Governance, Risk and Control

IT governance, risk and control are critical in the performance of any assurance management process. Governance of the process under review will be evaluated as part of the policies and management oversight controls. Risk plays an important role in evaluating what to audit and how management approaches and manages risk. Both issues will be evaluated as steps in the audit/assurance program. Controls are the primary evaluation point in the process. The audit/assurance program will identify the control objectives with steps to determine control design and effectiveness.

Responsibilities of IT Audit and Assurance Professionals

IT audit and assurance professionals are expected to customize this document to the environment in which they are performing an assurance process. This document is to be used as a review tool and starting point. It may be modified by the IT audit and assurance professional; it *is not* intended to be a checklist or questionnaire. It is assumed that the IT audit and assurance professional has the necessary subject matter expertise required to conduct the work and is supervised by a professional with the Certified Information Systems Auditor (CISA) designation and necessary subject matter expertise to adequately review the work performed.

II. Using This Document

This audit/assurance program was developed to assist the audit and assurance professional in designing and executing a review. Details regarding the format and use of the document follow.

Work Program Steps

The first column of the program describes the steps to be performed. The numbering scheme used provides built-in work paper numbering for ease of cross-reference to the specific work paper for that section. IT audit and assurance professionals are encouraged to make modifications to this document to reflect the specific environment under review.

COBIT Cross-reference

The COBIT cross-reference provides the audit and assurance professional with the ability to refer to the specific COBIT control objective that supports the audit/assurance step. The COBIT control objective should be identified for each audit/assurance step in the section. Multiple cross-references are not uncommon. Processes at lower levels in the work program are too granular to be cross-referenced to COBIT. The audit/assurance program is organized in a manner to facilitate an evaluation through a structure parallel to the development process.

COBIT provides in-depth control objectives and suggested control practices at each level. As professionals review each control, they should refer to COBIT 4.1 or the *IT Assurance Guide: Using CobiT®* for good-practice control guidance.

COSO Components

As noted in the introduction, COSO and similar frameworks have become increasingly popular among audit and assurance professionals. This ties the assurance work to the enterprise's control framework. While the IT audit/assurance function has COBIT as a framework, operational audit and assurance professionals use the framework established by the enterprise. Since COSO is the most prevalent internal control framework, it has been included in this document and is a bridge to align IT audit/assurance with the rest of the audit/assurance function. Many audit/assurance organizations include the COSO control components within their report and summarize assurance activities to the audit committee of the board of directors.

For each control, the audit and assurance professional should indicate the COSO component(s) addressed. It is possible, but generally not necessary, to extend this analysis to the specific audit step level.

The original COSO internal control framework contained five components. In 2004, COSO was revised as the *Enterprise Risk Management (ERM) Integrated Framework* and extended to eight components. The primary difference between the two frameworks is the additional focus on ERM and integration into the business decision model. ERM is in the process of being adopted by large enterprises. The two frameworks are compared in **figure A5.1**.

Figure A5.1—Comparison of COSO Internal Control and ERM Integrated Frameworks

Internal Control Framework	ERM Integrated Framework
Control Environment: The control environment sets the tone of an organization, influencing the control consciousness of its people. It is the foundation for all other components of internal control, providing discipline and structure. Control environment factors include the integrity, ethical values, management's operating style, delegation of authority systems, as well as the processes for managing and developing people in the organization.	**Internal Environment:** The internal environment encompasses the tone of an organization, and sets the basis for how risk is viewed and addressed by an enterprise's people, including risk management philosophy and risk appetite, integrity and ethical values, and the environment in which they operate.

Figure A5.1—Comparison of COSO Internal Control and ERM Integrated Frameworks *(cont.)*	
Internal Control Framework	**ERM Integrated Framework**
	Objective Setting: Objectives must exist before management can identify potential events affecting their achievement. Enterprise risk management ensures that management has in place a process to set objectives and that the chosen objectives support and align with the enterprise's mission and are consistent with its risk appetite.
	Event Identification: Internal and external events affecting achievement of an enterprise's objectives must be identified, distinguishing between risks and opportunities. Opportunities are channeled back to management's strategy or objective-setting processes.
Risk Assessment: Every enterprise faces a variety of risks from external and internal sources that must be assessed. A precondition to risk assessment is establishment of objectives, and thus risk assessment is the identification and analysis of relevant risks to achievement of assigned objectives. Risk assessment is a prerequisite for determining how the risks should be managed.	**Risk Assessment:** Risks are analyzed, considering the likelihood and impact, as a basis for determining how they could be managed. Risk areas are assessed on an inherent and residual basis.
	Risk Response: Management selects risk responses—avoiding, accepting, reducing or sharing risk—developing a set of actions to align risks with the enterprise's risk tolerances and risk appetite.
Control Activities: Control activities are the policies and procedures that help ensure management directives are carried out. They help ensure that necessary actions are taken to address risks to achievement of the enterprise's objectives. Control activities occur throughout the organization, at all levels and in all functions. They include a range of activities as diverse as approvals, authorizations, verifications, reconciliations, reviews of operating performance, security of assets and segregation of duties.	**Control Activities:** Policies and procedures are established and implemented to help ensure the risk responses are effectively carried out.

Figure A5.1—Comparison of COSO Internal Control and ERM Integrated Frameworks *(cont.)*	
Internal Control Framework	**ERM Integrated Framework**
Information and Communication: Information systems play a key role in internal control systems as they produce reports, including operational, financial and compliance-related information that make it possible to run and control the business. In a broader sense, effective communication must ensure information flows down, across and up the organization. Effective communication should also be ensured with external parties, such as customers, suppliers, regulators and shareholders.	**Information and Communication:** Relevant information is identified, captured and communicated in a form and time frame that enable people to carry out their responsibilities. Effective communication also occurs in a broader sense, flowing down, across and up the enterprise.
Monitoring: Internal control systems need to be monitored—a process that assesses the quality of the system's performance over time. This is accomplished through ongoing monitoring activities or separate evaluations. Internal control deficiencies detected through these monitoring activities should be reported upstream, and corrective actions should be taken to ensure continuous improvement of the system.	**Monitoring:** The entirety of enterprise risk management is monitored and modifications made as necessary. Monitoring is accomplished through ongoing management activities, separate evaluations or both.
Information for **figure A5.1** was obtained from the COSO web site, *www.coso.org/aboutus.htm.*	

The original COSO internal control framework addresses the needs of the IT audit and assurance professional: control environment, risk assessment, control activities, information and communication, and monitoring. As such, ISACA has elected to utilize the five-component model for these audit/assurance programs. As more enterprises implement the ERM model, the additional three columns can be added, if relevant. When completing the COSO component columns, consider the definitions of the components as described in **figure A5.1**.

Reference/Hyperlink

Good practices require the audit and assurance professional to create a work paper for each line item, which describes the work performed, issues identified and conclusions. The reference/hyperlink is to be used to cross-reference the audit/assurance step to the work paper that supports it. The numbering system of this document provides a ready numbering scheme for the work papers. If desired, a link to the work paper can be pasted into this column.

Issue Cross-reference

This column can be used to flag a finding/issue that the IT audit and assurance professional wants to further investigate or establish as a potential finding. The potential findings should be documented in a work paper that indicates the disposition of the findings (formally reported, reported as a memo or verbal finding, or waived).

Comments

The comments column can be used to indicate the waiving of a step or other notations. It is not to be used in place of a work paper describing the work performed.

III. Controls Maturity Analysis

One of the consistent requests of stakeholders who have undergone IT audit/assurance reviews is a desire to understand how their performance compares to good practices. Audit and assurance professionals must provide an objective basis for the review conclusions. Maturity modeling for management and control over IT processes is based on a method of evaluating the organization, so it can be rated from a maturity level of nonexistent (0) to optimized (5). This approach is derived from the maturity model that the Software Engineering Institute (SEI) of Carnegie Mellon University defined for the maturity of software development.

The *IT Assurance: Guide Using CobiT®*, Appendix VII—Maturity Model for Internal Control, in **figure A5.2**, provides a generic maturity model showing the status of the internal control environment and the establishment of internal controls in an enterprise. It shows how the management of internal control, and an awareness of the need to establish better internal controls, typically develops from an ad hoc to an optimized level. The model provides a high-level guide to help COBIT users appreciate what is required for effective internal controls in IT and to help position their enterprise on the maturity scale.

Figure A5.2—Maturity Model for Internal Control

Maturity Level	Status of the Internal Control Environment	Establishment of Internal Controls
0 Non-existent	There is no recognition of the need for internal control. Control is not part of the organization's culture or mission. There is a high risk of control deficiencies and incidents.	There is no intent to assess the need for internal control. Incidents are dealt with as they arise.

Figure A5.2—Maturity Model for Internal Control *(cont.)*		
Maturity Level	**Status of the Internal Control Environment**	**Establishment of Internal Controls**
1 Initial/ad hoc	There is some recognition of the need for internal control. The approach to risk and control requirements is *ad hoc* and disorganized, without communication or monitoring. Deficiencies are not identified. Employees are not aware of their responsibilities.	There is no awareness of the need for assessment of what is needed in terms of IT controls. When performed, it is only on an *ad hoc* basis, at a high level and in reaction to significant incidents. Assessment addresses only the actual incident.
2 Repeatable but Intuitive	Controls are in place but are not documented. Their operation is dependent on the knowledge and motivation of individuals. Effectiveness is not adequately evaluated. Many control weaknesses exist and are not adequately addressed; the impact can be severe. Management actions to resolve control issues are not prioritized or consistent. Employees may not be aware of their responsibilities.	Assessment of control needs occurs only when needed for selected IT processes to determine the current level of control maturity, the target level that should be reached and the gaps that exist. An informal workshop approach, involving IT managers and the team involved in the process, is used to define an adequate approach to controls for the process and to motivate an agreed-upon action plan.
3 Defined	Controls are in place and adequately documented. Operating effectiveness is evaluated on a periodic basis and there is an average number of issues. However, the evaluation process is not documented. While management is able to deal predictably with most control issues, some control weaknesses persist and impacts could still be severe. Employees are aware of their responsibilities for control.	Critical IT processes are identified based on value and risk drivers. A detailed analysis is performed to identify control requirements and the root cause of gaps and to develop improvement opportunities. In addition to facilitated workshops, tools are used and interviews are performed to support the analysis and ensure that an IT process owner owns and drives the assessment and improvement process.
4 Managed and Measurable	There is an effective internal control and risk management environment. A formal, documented evaluation of controls occurs frequently. Many controls are automated and regularly reviewed. Management is likely to detect most control issues, but not all issues are routinely identified. There is consistent follow-up to address identified control weaknesses. A limited, tactical use of technology is applied to automate controls.	IT process criticality is regularly defined with full support and agreement from the relevant business process owners. Assessment of control requirements is based on policy and the actual maturity of these processes, following a thorough and measured analysis involving key stakeholders. Accountability for these assessments is clear and enforced. Improvement strategies are supported by business cases. Performance in achieving the desired outcomes is consistently monitored. External control reviews are organized occasionally.

Figure A5.2—Maturity Model for Internal Control *(cont.)*		
Maturity Level	**Status of the Internal Control Environment**	**Establishment of Internal Controls**
5 Optimized	An enterprisewide risk and control program provides continuous and effective control and risk issues resolution. Internal control and risk management are integrated with enterprise practices, supported with automated real-time monitoring with full accountability for control monitoring, risk management and compliance enforcement. Control evaluation is continuous, based on self-assessments and gap and root cause analyses. Employees are proactively involved in control improvements.	Business changes consider the criticality of IT processes and cover any need to reassess process control capability. IT process owners regularly perform self-assessments to confirm that controls are at the right level of maturity to meet business needs and they consider maturity attributes to find ways to make controls more efficient and effective. The organization benchmarks to external best practices and seeks external advice on internal control effectiveness. For critical processes, independent reviews take place to provide assurance that the controls are at the desired level of maturity and working as planned.

The maturity model evaluation is one of the final steps in the evaluation process. The IT audit and assurance professional can address the key controls within the scope of the work program and formulate an objective assessment of the maturity levels of the control practices. The maturity assessment can be a part of the audit/assurance report, and used as a metric from year to year to document progression in the enhancement of controls. However, it must be noted that the perception of the maturity level may vary between the process/IT asset owner and the auditor. Therefore, an auditor should obtain the concerned stakeholder's concurrence before submitting the final report to management.

At the conclusion of the review, once all findings and recommendations are completed, the professional assesses the current state of the COBIT control framework and assigns it a maturity level using the six-level scale. Some practitioners utilize decimals (x.25, x.5, x.75) to indicate gradations in the maturity model. As a further reference, COBIT provides a definition of the maturity designations by control objective. While this approach is not mandatory, the process is provided as a separate section at the end of the audit/assurance program for those enterprises that wish to implement it. It is suggested that a maturity assessment be made at the COBIT control level. To provide further value to the client/customer, the professional can also obtain maturity targets from the client/customer. Using the assessed and target maturity levels, the professional can create an effective graphic presentation that describes the achievement or gaps between the actual and targeted maturity goals.

IV. Assurance and Control Framework

ISACA IT Assurance Framework and Standards

ISACA has long recognized the specialized nature of IT assurance and strives to advance globally applicable standards. Guidelines and procedures provide detailed guidance on how to follow those standards. IT Audit/Assurance Standard S15 IT Controls, and IT Audit/Assurance Guideline G38 Access Controls are relevant to this audit/assurance program.

ISACA Controls Framework

COBIT is an IT governance framework and supporting tool set that allows managers to bridge the gap among control requirements, technical issues and business risks. COBIT enables clear policy development and good practice for IT control throughout enterprises.

Utilizing COBIT as the control framework on which IT audit/assurance activities are based aligns IT audit/assurance with good practices as developed by the enterprise.

Refer to ISACA's *CobiT® Control Practices: Guidance to Achieve Control Objectives for Successful IT Governance, 2nd Edition*, published in 2007, for the related control practice value and risk drivers.

V. Executive Summary of Audit/Assurance Focus

Oracle EBS Security

Since launching its first product offering approximately 30 years ago, Oracle Corp. has grown globally. In recent years, Oracle Corp. has been involved in a number of large acquisitions, including PeopleSoft, Siebel and Sun, as part of its enterprise application integration strategy that forms the core of its Fusion Middleware product range.

Oracle Corp. released Oracle EBS R12 in 2007, which introduced changes from the previous version, R11*i*.10, to the application technology platform, provided changes to the security authorization concept, and provided a new technology stack and architecture for Oracle EBS Financials. The latest release, Oracle EBS R12.1, was released in 2009 and introduced changes to the other enterprise application areas (e.g., Supply Chain Management, Procurement, Customer Relationship Management and Human Capital Management). With the ever-changing compliance landscape, a review of organizations' Oracle EBS environment is vital to ensuring that it is secure.

Business Impact and Risk

Oracle EBS is widely used in many enterprises. Improper configuration of Oracle EBS could result in the inability of the enterprise to execute its critical processes.

Oracle EBS risks resulting from ineffective or incorrect configurations or use could result in some of the following:

- Disclosure of privileged information
- Single points of failure
- Low data quality
- Loss of physical assets
- Loss of intellectual property
- Loss of competitive advantage
- Loss of customer confidence
- Violation of regulatory requirements
- Financial loss

Objective and Scope

Objective—The objective of the Oracle EBS audit/assurance review is to provide management with an independent assessment relating to the effectiveness of configuration and security of the enterprise's Oracle EBS architecture.

Scope—The review will focus on configuration of the relevant Oracle EBS components and modules within the enterprise. The selection of the specific components and modules will be based upon the risks introduced to the enterprise by these components and modules.

Minimum Audit Skills

This review is considered highly technical. The IT audit and assurance professional must have an understanding of Oracle EBS best-practice processes and requirements, and be highly conversant in Oracle EBS tools, exposures, and functionality. It should not be assumed that an audit and assurance professional holding the CISA designation has the requisite skills to perform this review.

VI. Oracle Security Administration—Audit/Assurance Program

Audit/Assurance Program Step	COBIT Cross-reference	COSO					Reference Hyperlink	Issue Cross-reference	Comments
		Control Environment	Risk Assessment	Control Activities	Information and Communication	Monitoring			
A. Prior Audit/Examination Report Follow-up									
1. Review prior report, if one exists, to verify completion of any agreed-upon corrections and note remaining deficiencies.	ME1								
1.1 Determine whether: • Senior management has assigned responsibilities for information, its processing and its use • User management is responsible for providing information that supports the enterprise's objectives and policies • Information systems management is responsible for providing the information systems capabilities necessary for achievement of the defined information systems objectives and policies of the enterprise • Senior management approves plans for development and acquisition of information systems • There are procedures to ensure that the information system being developed or acquired meets user requirement • There are procedures to ensure that information systems, programs and configuration changes are tested adequately prior to implementation in a separate testing environment	AI6 DS4 ME1								

VI. Oracle Security Administration—Audit/Assurance Program *(cont.)*

Audit/Assurance Program Step	COBIT Cross-reference	COSO					Reference Hyperlink	Issue Cross-reference	Comments
		Control Environment	Risk Assessment	Control Activities	Information and Communication	Monitoring			
A. Prior Audit/Examination Report Follow-up *(cont.)*									
1.1 *(cont.)* • All personnel involved in the system acquisition and configuration activities receive adequate training and supervision • There are procedures to ensure that information systems are implemented/configured/upgraded in accordance with the established standards • User management participate in the conversion of data from the existing system to the new system • Final approval is obtained from user management prior to going live with a new information/ upgraded system • There are procedures to document and schedule all changes to information systems • There are procedures to ensure that only authorized changes are initiated • There are procedures to ensure that only authorized, tested and documented changes to information systems are accepted into the production client • There are procedures to allow for and control emergency changes • There are procedures for the approval, monitoring and control of the acquisition and upgrade of hardware and systems software									

VI. Oracle Security Administration—Audit/Assurance Program (cont.)

Audit/Assurance Program Step	COBIT Cross-reference	COSO					Reference Hyperlink	Issue Cross-reference	Comments
		Control Environment	Risk Assessment	Control Activities	Information and Communication	Monitoring			
A. Prior Audit/Examination Report Follow-up *(cont.)*									
1.1 *(cont.)* • There is a process for monitoring the volume of named and concurrent Oracle EBS users to ensure that the license agreement is not being violated • The organization structure, established by senior management, provides for an appropriate segregation of incompatible functions • The database and application servers are located in a physically separate and protected environment (i.e., a data center) • Emergency, backup and recovery plans are documented and tested on a regular basis to ensure they remain current and operational • Backup and recovery plans allow users of information systems to resume operations in the event of an interruption • Application controls are designed with regard to any weaknesses in segregation, security, development and processing controls that may affect the information system • The organization has created any locally developed application programs, reports or tables. If so, determine how these programs, exports or tables are used. Depending on the importance/extent of use, review and document the development and change management process surrounding the creation/modification of these programs, reports or tables.									

VI. Oracle Security Administration—Audit/Assurance Program (cont.)

Audit/Assurance Program Step	COBIT Cross-reference	COSO					Reference Hyperlink	Issue Cross-reference	Comments
		Control Environment	Risk Assessment	Control Activities	Information and Communication	Monitoring			
B. Preliminary Audit Steps									
1. Gain an understanding of the Oracle EBS environment.									
1.1 The same background information obtained for the Oracle EBS Security audit plan is required for and relevant to the business cycles. In particular the following information is important: • Version and release of Oracle EBS software that has been implemented • Total number of named users (for comparison with logical access security testing results) • Number of Oracle Database instances • Location of the servers and the related LAN/ WAN connections (need to verify security and controls, including environmental, surrounding the hardware and the network security controls surrounding the connectivity). If possible, obtain copies of the network topology diagrams. • Listing of business partners, related organizations and remote locations that are permitted to connect to the Oracle EBS environment • Various means used to connect to the Oracle EBS environment (e.g., dial-up, remote access server, Internet transaction server) and the network diagram, if available • Accounting period, company codes and COA • Identification of the modules being used	PO2 PO3 PO4 PO6 PO9 AI2 AI6 DS2 DS5 ME1 ME2								

VI. Oracle Security Administration—Audit/Assurance Program (cont.)

Audit/Assurance Program Step	COBIT Cross-reference	COSO					Reference Hyperlink	Issue Cross-reference	Comments
		Control Environment	Risk Assessment	Control Activities	Information and Communication	Monitoring			
B. Preliminary Audit Steps ***(cont.)***									
1.1 *(cont.)* • Locally developed application programs, reports or tables created by the organization • Details of the risk assessment approach taken in the organization to identify and prioritize risks • Copies of the organization's key security policies and standards around access administration, as well as information regarding the awareness programs that have been delivered to staff on the key security policies and standards. Consider specifically the frequency of delivery and any statistics on the extent of coverage (e.g., what percentage of staff has received the awareness training). • Sensitivity classification • Logical and physical access control requirements • Network security requirements, including requirements for encryption and firewalls • Platform security requirements (e.g., configuration requirements) • With regard to maintenance of authorizations and profiles, determine whether job roles including the related transactions have been defined and documented, and whether procedures for maintaining (creating/changing/deleting) activity groups exist and are followed • Outstanding audit findings, if any, from previous years									

VI. Oracle Security Administration—Audit/Assurance Program (cont.)

Audit/Assurance Program Step	COBIT Cross-reference	COSO					Reference Hyperlink	Issue Cross-reference	Comments
		Control Environment	Risk Assessment	Control Activities	Information and Communication	Monitoring			
B. Preliminary Audit Steps ***(cont.)***									
2. Identify the significant risks and determine the key controls.									
2.1 Obtain details of the risk assessment approach taken in the organization to identify and prioritize risks.	PO9								
2.2 Obtain copies of and review: • Completed risk assessments impacting the Oracle EBS environment • Approved requests to deviate from security policies and standards Assess the impact of the above documents on the planning of the Oracle EBS audit. If a recent implementation/upgrade has taken place, obtain a copy of the security implementation plan. Assess whether: • The plan took into account the protection of critical objects and segregation of duties • An appropriate naming convention (e.g., for suppliers) has been developed to help security maintenance and to comply with required Oracle EBS naming conventions	PO9 ME1								

VI. Oracle Security Administration—Audit/Assurance Program (cont.)

Audit/Assurance Program Step	COBIT Cross-reference	COSO					Reference Hyperlink	Issue Cross-reference	Comments
		Control Environment	Risk Assessment	Control Activities	Information and Communication	Monitoring			
C. Detailed Audit Steps									
1. Security Administration									
1.1 Oracle EBS security/control parameters are defined adequately.									
1.1.1 Check that the password parameters are in line with the enterprise's policies on password strength: • These password System Profile Options are: – Sign-On: Password Length – Sign-On: Password Failure Limit – Sign-On: Password Hard to Guess – Sign-On: Password No Reuse • Check that all user accounts have the password expiration set to the enterprise policy (e.g., 90 days). • Verify that the timeout/background disconnect interval in the System Profile Option ICX Session Timeout is set to the enterprise policy (e.g., 1800 seconds/30 minutes).	AI1 DS5								

VI. Oracle Security Administration—Audit/Assurance Program (cont.)

Audit/Assurance Program Step	COBIT Cross-reference	COSO					Reference Hyperlink	Issue Cross-reference	Comments
		Control Environment	Risk Assessment	Control Activities	Information and Communication	Monitoring			
C. Detailed Audit Steps *(cont.)*									
1.2 The security administration function and user access change management procedures exist.									
1.2.1. Ask operational management about the existence and effectiveness of a data administration group. Verify whether there is a security administration group/administrator that controls access to Oracle EBS.	PO2 DS5 DS10								
1.2.2 Review documentation on policies/ procedures, organizational structure of the IS group, security administrator job descriptions and mission statements.									
1.2.3 Using the system administrator responsibility, go to the User Maintenance page and Roles and Responsibilities. This will allow you to determine what user roles have been set up in the system.									
1.2.4 Also using the system administrator responsibility, use the Submit Requests form to obtain the Active Responsibilities report to determine whether the client is using a security administrator responsibility and who is assigned the responsibility.									

VI. Oracle Security Administration—Audit/Assurance Program (cont.)

Audit/Assurance Program Step	COBIT Cross-reference	COSO					Reference Hyperlink	Issue Cross-reference	Comments
		Control Environment	Risk Assessment	Control Activities	Information and Communication	Monitoring			
C. Detailed Audit Steps ***(cont.)***									
1.2.5 Obtain and review the Function Security Function report to determine the configuration of the security administrator responsibility.									
1.2.6 Review the User Access Change Management process. Determine what assessment is made on assignment of user access levels, and review the approval and update procedures.									
1.2.7 Review evidence that changes to access authority are tested by security administrators. Through observation, witness a sample of access requests being tested by the system security administrators.									
1.3 Access to programs, data and other information resources is restricted and maintained.									
1.3.1 Ask management about the logical tools and techniques that are implemented by the enterprise, and obtain and review evidence that corroborates their use, such as: • Information system plans • Security policies, procedures, standards and guidance	DS4 DS5 DS9 DS10 DS11 ME1								

VI. Oracle Security Administration—Audit/Assurance Program *(cont.)*

Audit/Assurance Program Step	COBIT Cross-reference	COSO					Reference Hyperlink	Issue Cross-reference	Comments
		Control Environment	Risk Assessment	Control Activities	Information and Communication	Monitoring			
C. Detailed Audit Steps ***(cont.)***									
1.3.1 *(cont.)* • Security and internal control frameworks • Minutes of planning/steering committee meetings • Review and approval analysis documentation • Oracle Reports									
1.3.2 With regard to management reviews, inspect physical evidence that reviews and approvals of the implementation and configuration of Oracle EBS security techniques occur. Furthermore, inspect evidence that any issues identified from the reviews are followed up and resolved.									
1.3.3 With regard to information security techniques, review system configuration settings and security reports to ensure that control activities have been performed in accordance with established policies and procedures, for example:									

VI. Oracle Security Administration—Audit/Assurance Program (cont.)

Audit/Assurance Program Step	COBIT Cross-reference	COSO					Reference Hyperlink	Issue Cross-reference	Comments
		Control Environment	Risk Assessment	Control Activities	Information and Communication	Monitoring			
C. Detailed Audit Steps ***(cont.)***									
1.3.3 *(cont.)* • Oracle User Profile Option Values report for settings on: - Sign-On: Notification - Sign-On: Audit Level - Audit Trail Activate • Oracle Sign-On Audit reports: - Sign-On: Audit Concurrent Requests - Sign-On: Audit Forms - Sign-On: Audit Responsibilities - Sign-On: Audit Unsuccessful Logins - Sign-On: Audit Users									
1.3.4 With regard to default passwords, test the default user IDs and passwords for the following by attempting to log in to Oracle Application ID/password: • SYSADMIN/SYSADMIN • SYS/Change_On_Install • SYSTEM/Manager • APPS/APPS • GL/GL • AP/AP • AR/AR • PO/PO									

VI. Oracle Security Administration—Audit/Assurance Program (cont.)

Audit/Assurance Program Step	COBIT Cross-reference	COSO					Reference Hyperlink	Issue Cross-reference	Comments
		Control Environment	Risk Assessment	Control Activities	Information and Communication	Monitoring			
C. Detailed Audit Steps ***(cont.)***									
1.3.5 Check that the SYSADMIN account is disabled by setting the effective date equal to the expiration date in the user profile. System administrators should be signing on with their unique user IDs and assigned to the system administrator responsibility. Furthermore, audit trails should be reviewed to check for any use of the SYSADMIN account.									
1.3.6 With regard to password controls, using the Submit Requests form, view user Profile Options by printing the User Profile Option Values report. Review the profiles for the Sign-On Password Length Profile Option. If no value is entered, the minimum length defaults to five. Alternatively, view the Profile Options using the Find System Profile Values window.									
1.3.7 With regard to password area (user window), check that the number of accesses is set to 1. This will enforce that users change passwords the first time they sign on. Check that the password settings are configured as follows: • Passwords must be at least five characters long and may extend up to 100 characters									

VI. Oracle Security Administration—Audit/Assurance Program (cont.)

Audit/Assurance Program Step	COBIT Cross-reference	COSO					Reference Hyperlink	Issue Cross-reference	Comments
		Control Environment	Risk Assessment	Control Activities	Information and Communication	Monitoring			
C. Detailed Audit Steps ***(cont.)***									
1.3.7 *(cont.)* • The username must contain alphanumeric characters (A through Z, 0 through 9) • Passwords are not displayed when entered									
1.3.8 The initial password should be set to expire after one access and thereafter every 30 days.									
1.3.9 Relevant security tables that should be analyzed as part of a security access review are: • Oracle Interface tables, AP_INVOICES_INTERFACE • Oracle Base tables, for the modules (for example, HR, AP, Purchasing) • Oracle Event tables • Tables and columns containing common sensitive data									
1.3.10 With regard to encryption of sensitive information, ask IT management or the security administrator about the: • Use of Oracle EBS-certified encryption tools and techniques									

VI. Oracle Security Administration—Audit/Assurance Program (cont.)

Audit/Assurance Program Step	COBIT Cross-reference	COSO					Reference Hyperlink	Issue Cross-reference	Comments
		Control Environment	Risk Assessment	Control Activities	Information and Communication	Monitoring			
C. Detailed Audit Steps *(cont.)*									
1.3.10 *(cont.)* • Policies/procedures regarding encryption of sensitive data over internal/external networks • Policies/procedures regarding encryption of sensitive data over interfaces, import settings and export settings • Policies/procedures regarding encryption of sensitive data over other instances of Oracle EBS • Policies/procedures regarding confidentiality controls such as logs of SQL statements • Specifications of required cryptographic techniques and strengths • Design of cryptographic infrastructure • Selection of appropriate encryption software and modules • Use of private and public key cryptographic solutions • Selection of suitable certification authority; key management, including generation, distribution, storage, entry, use and archiving; and protection of encryption hardware and software • Reports and other information used, including how they are used									

VI. Oracle Security Administration—Audit/Assurance Program (cont.)

Audit/Assurance Program Step	COBIT Cross-reference	COSO					Reference Hyperlink	Issue Cross-reference	Comments
		Control Environment	Risk Assessment	Control Activities	Information and Communication	Monitoring			
C. Detailed Audit Steps ***(cont.)***									
1.3.11 Review evidence that corroborates the responses to the previous inquiries by examining the corresponding documentation.									
1.3.12 With regard to privileged responsibilities, ask IT management or the security administrator about the: • Policies/procedures related to the limitation of usage • Organizational structure of the Security Administration function • Monitoring and review of privileged responsibilities • Steps involved, including the procedures for assignment of privileged access • Reports, such as security administration reports for privileged responsibilities, and other information used, including how they are used									
1.3.13 Review evidence that corroborates the responses to the previous inquiries by examining the corresponding documentation.									

VI. Oracle Security Administration—Audit/Assurance Program (cont.)

Audit/Assurance Program Step	COBIT Cross-reference	COSO					Reference Hyperlink	Issue Cross-reference	Comments
		Control Environment	Risk Assessment	Control Activities	Information and Communication	Monitoring			
C. Detailed Audit Steps *(cont.)*									
1.3.14 With regard to sensitive data access, ask IT management or the security administrator about the: • Policies/procedures related to the fact that authorized access to sensitive data is audited • Audit logs, which are reviewed regularly to assess whether the access and use of such data was appropriate • Steps involved, such as configurations and audit capabilities, generation of regular logs, responsibility to perform a review and resolve concerns, criteria used to review access logs, independent data to which management compares the logs, and level of detail in the access log • Reports and other information used, e.g., copies of logs of access to sensitive data, a list of sensitive data items and their corresponding classification, job descriptions and responsibilities, and documentation of action taken to resolve inappropriate access or use of sensitive data, including how they are used • Procedures performed when inappropriate access is encountered									

VI. Oracle Security Administration—Audit/Assurance Program (cont.)

Audit/Assurance Program Step	COBIT Cross-reference	COSO					Reference Hyperlink	Issue Cross-reference	Comments
		Control Environment	Risk Assessment	Control Activities	Information and Communication	Monitoring			
C. Detailed Audit Steps ***(cont.)***									
1.3.15 Review evidence that corroborates the responses to the inquiries above by examining the corresponding documentation: • Policies/procedures related to the inspection of logs and reports to identify and prevent unauthorized access to Oracle Applications • Steps involved (consider frequency of inspection of logs) • Action to be taken immediately to prevent unauthorized access and, subsequently, to identify the perpetrator of the unauthorized access attempt and assessment of the adequacy of the security configuration • Security violation reports and other information used, including how they are used • Procedures performed when exceptions, misstatements or unusual items are encountered • Extent of unauthorized access									

VI. Oracle Security Administration—Audit/Assurance Program (cont.)

Audit/Assurance Program Step	COBIT Cross-reference	COSO					Reference Hyperlink	Issue Cross-reference	Comments
		Control Environment	Risk Assessment	Control Activities	Information and Communication	Monitoring			
C. Detailed Audit Steps ***(cont.)***									
1.3.16 Review evidence that corroborates the responses to the previous inquiries by examining the corresponding documentation.									
1.4 Users perform compatible functions, and Oracle EBS responsibilities match the user's organizational duties.									
1.4.1 Ask management about how user IDs, roles and responsibilities are defined. Inquire as to whether segregation of duties is considered when defining or changing the access of users.	P04 P07 DS5								
1.4.2 Review evidence that segregation of duties has been addressed when assigning user access or creating responsibilities so that users are unable to initiate, record, approve and post a transaction (e.g., responsibilities matrix).									
1.4.3 Ensure that system administrators and application developers are not assigned business function menus or forms.									

VI. Oracle Security Administration—Audit/Assurance Program (cont.)

Audit/Assurance Program Step	COBIT Cross-reference	COSO					Reference Hyperlink	Issue Cross-reference	Comments
		Control Environment	Risk Assessment	Control Activities	Information and Communication	Monitoring			
C. Detailed Audit Steps ***(cont.)***									
1.5 System and start-up profiles are defined adequately, the ability to change profiles is restricted, and changes are monitored.									
1.5.1 Ascertain the extent to which Oracle EBS profile settings have been utilized at the user, responsibility, application and site levels. The site profile should include the following: • Set the Sign-On Notifications to Yes • Set the Sign-On Password Length to 5	AI2								
1.6 Access to system output is restricted.									
1.6.1 Inquire as to the manner in which system output is controlled so that the intended recipient receives the information produced. Review the assignment of the printer locations. Observe that report printing is controlled by the Oracle EBS user.	DS11 DS12								
1.7 User IDs, responsibilities and preferences are documented adequately.									
1.7.1 Verify with the security administration function and inspect evidence that user IDs, responsibilities and profiles documentation exists and a process for maintenance of the documentation is in place.	A13 A14								

VI. Oracle Security Administration—Audit/Assurance Program (cont.)

Audit/Assurance Program Step	COBIT Cross-reference	COSO					Reference Hyperlink	Issue Cross-reference	Comments
		Control Environment	Risk Assessment	Control Activities	Information and Communication	Monitoring			
C. Detailed Audit Steps ***(cont.)***									
1.8 Remote access by external vendors is controlled adequately.									
1.8.1 Speak with the security administration function to determine how vendor remote access is managed. Review the remote access setting to ensure that access is controlled and logged.	DS5								
1.9 Access to transport changes to the production environment is controlled; as a result, modifications are consistent with management's intentions.									
1.9.1 Ask the development team and the security administrator about the process involved for application change management. In addition, review: • Policies and procedures for controlling access to the test and production environments • Access control listings for the test and production environments • Lists of users • Violation reports	AI6								

VI. Oracle Security Administration—Audit/Assurance Program (cont.)

Audit/Assurance Program Step	COBIT Cross-reference	COSO					Reference Hyperlink	Issue Cross-reference	Comments
		Control Environment	Risk Assessment	Control Activities	Information and Communication	Monitoring			
C. Detailed Audit Steps *(cont.)*									
1.10 Only authorized and accurate changes are made to the data dictionary.									
1.10.1 Gain an understanding of management's policies and procedures regarding the review of data dictionary changes.	PO6 AI4 DS5								
1.10.2 Review any policies/procedures and assess the adequacy, taking into account the frequency with which a review is performed, the level of detail in the reports, other independent data to which management compares the reports, the likelihood that the people performing the review will be able to identify exception items, and the nature of exception items that they can be expected to identify.									
1.10.3 Review the listing of users with access to view and modify the data dictionary, and discuss with the database administrator whether these users are appropriate.									

VII. Maturity Assessment

The maturity assessment is an opportunity for the reviewer to assess the maturity of the processes reviewed. Based on the results of audit/assurance review, and the reviewer's observations, a maturity level should be assigned to each of the following COBIT control practices.

COBIT Control Practice	Assessed Maturity	Target Maturity	Reference Hyperlink	Comments
AI6.1 Change Standards and Procedures 1. Develop, document and promulgate a change management framework that specifies the policies and processes, including: • Roles and responsibilities • Classification and prioritisation of all changes based on business risk • Assessment of impact • Authorisation and approval of all changes by the business process owners and IT • Tracking and status of changes • Impact on data integrity (e.g., all changes to data files being made under system and application control rather than by direct user intervention) 2. Establish and maintain version control over all changes. 3. Implement roles and responsibilities that involve business process owners and appropriate technical IT functions. Ensure appropriate segregation of duties. 4. Establish appropriate record management practices and audit trails to record key steps in the change management process. Ensure timely closure of changes. Elevate and report to management changes that are not closed in a timely fashion. 5. Consider the impact of contracted services providers (e.g., of infrastructure, application development and shared services) on the change management process. Consider integration of organisational change management processes with change management processes of service providers. Consider the impact of the organisational change management process on contractual terms and SLAs.				

VII. Maturity Assessment (cont.)

COBIT Control Practice	Assessed Maturity	Target Maturity	Reference Hyperlink	Comments
AI6.2 Impact Assessment, Prioritisation and Authorisation 1. Develop a process to allow business process owners and IT to request changes to infrastructure, systems or applications. Develop controls to ensure that all such changes arise only through the change request management process. 2. Categorise all requested changes (e.g., infrastructure, operating systems, networks, application systems, purchased/packaged application software). 3. Prioritise all requested changes. Ensure that the change management process identifies both the business and technical needs for the change. Consider legal, regulatory and contractual reasons for the requested change. 4. Assess all requests in a structured fashion. Ensure that the assessment process addresses impact analysis on infrastructure, systems and applications. Consider security, legal, contractual and compliance implications of the requested change. Consider also interdependencies amongst changes. Involve business process owners in the assessment process, as appropriate. 5. Ensure that each change is formally approved by business process owners and IT technical stakeholders, as appropriate.				
AI6.4 Change Status Tracking and Reporting 1. Ensure that a documented process exists within the overall change management process to declare, assess, authorise and record an emergency change. 2. Ensure that emergency changes are processed in accordance with the emergency change element of the formal change management process. 3. Ensure that all emergency access arrangements for changes are appropriately authorised, documented and revoked after the change has been applied. 4. Conduct a post-implementation review of all emergency changes, involving all concerned parties. The review should consider implications for aspects such as further application system maintenance, impact on development and test environments, application software development quality, documentation and manuals, and data integrity.				

VII. Maturity Assessment (cont.)

COBIT Control Practice	Assessed Maturity	Target Maturity	Reference Hyperlink	Comments
DS5.3 Identity Management 1. Establish and communicate policies and procedures to uniquely identify, authenticate and authorise access mechanisms and access rights for all users on a need-to-know/need-to-have basis, based on predetermined and preapproved roles. Clearly state accountability of any user for any action on any of the systems and/or applications involved. 2. Ensure that roles and access authorisation criteria for assigning user access rights take into account: • Sensitivity of information and applications involved (data classification) • Policies for information protection and dissemination (legal, regulatory, internal policies and contractual requirements) • Roles and responsibilities as defined within the enterprise • The need-to-have access rights associated with the function • Standard but individual user access profiles for common job roles in the organisation • Requirements to guarantee appropriate segregation of duties 3. Establish a method for authenticating and authorising users to establish responsibility and enforce access rights in line with sensitivity of information and functional application requirements and infrastructure components, and in compliance with applicable laws, regulations, internal policies and contractual agreements. 4. Define and implement a procedure for identifying new users and recording, approving and maintaining access rights. This needs to be requested by user management, approved by the system owner and implemented by the responsible security person. 5. Ensure that a timely information flow is in place that reports changes in jobs (i.e., people in, people out, people change). Grant, revoke and adapt user access rights in co-ordination with human resources and user departments for users who are new, who have left the organisation, or who have changed roles or jobs.				

VII. Maturity Assessment (cont.)

COBIT Control Practice	Assessed Maturity	Target Maturity	Reference Hyperlink	Comments
DS5.4 User Account Management 1. Ensure that access control procedures include but are not limited to: • Using unique user IDs to enable users to be linked to and held accountable for their actions • Awareness that the use of group IDs results in the loss of individual accountability and is permitted only when justified for business or operational reasons and compensated by mitigating controls. Group IDs must be approved and documented. • Checking that the user has authorisation from the system owner for the use of the information system or service, and the level of access granted is appropriate to the business purpose and consistent with the organisational security policy • A procedure to require users to understand and acknowledge their access rights and the conditions of such access • Ensuring that internal and external service providers do not provide access until authorisation procedures have been completed • Maintaining a formal record, including access levels, of all persons registered to use the service • A timely and regular review of user IDs and access rights 2. Ensure that management reviews or reallocates user access rights at regular intervals using a formal process. User access rights should be reviewed or reallocated after any job changes, such as transfer, promotion, demotion or termination of employment. Authorisations for special privileged access rights should be reviewed independently at more frequent intervals.				

VII. Maturity Assessment (cont.)

COBIT Control Practice	Assessed Maturity	Target Maturity	Reference Hyperlink	Comments
DS9.1 Configuration Repository and Baseline 1. Implement a configuration repository to capture and maintain configuration management items. The repository should include hardware; application software; middleware; parameters; documentation; procedures; and tools for operating, accessing and using the systems, services, version numbers and licencing details. 2. Implement a tool to enable the effective logging of configuration management information within a repository. 3. Provide a unique identifier to a configuration item so the item can be easily tracked and related to physical asset tags and financial records. 4. Define and document configuration baselines for components across development, test and production environments, to enable identification of system configuration at specific points in time (past, present and planned). 5. Establish a process to revert to the baseline configuration in the event of problems, if determined appropriate after initial investigation. 6. Install mechanisms to monitor changes against the defined repository and baseline. Provide management reports for exceptions, reconciliation and decision making.				

VII. Maturity Assessment (cont.)

COBIT Control Practice	Assessed Maturity	Target Maturity	Reference Hyperlink	Comments
DS9.2 Identification and Maintenance of Configuration Items 1. Define and implement a policy requiring all configuration items and their attributes and versions to be identified and maintained. 2. Tag physical assets according to a defined policy. Consider using an automated mechanism, such as barcodes. 3. Define a policy that integrates incident, change and problem management procedures with the maintenance of the configuration repository. 4. Define a process to record new, modified and deleted configuration items and their relative attributes and versions. Identify and maintain the relationships between configuration items in the configuration repository. 5. Establish a process to maintain an audit trail for all changes to configuration items. 6. Define a process to identify critical configuration items in relationship to business functions (component failure impact analysis). 7. Record all assets—including new hardware and software, procured or internally developed—within the configuration management data repository. 8. Define and implement a process to ensure that valid licences are in place to prevent the inclusion of unauthorised software.				
DS9.3 Configuration Integrity Review 1. To validate the integrity of configuration data, implement a process to ensure that configuration items are monitored. Compare recorded data against actual physical existence, and ensure that errors and deviations are reported and corrected. 2. Using automated discovery tools where appropriate, reconcile actual installed software and hardware periodically against the configuration database, licence records and physical tags. 3. Periodically review against the policy for software usage the existence of any software in violation or in excess of current policies and licence agreements. Report deviations for correction.				

Page intentionally left blank

Appendix 6. Oracle E-Business Suite Security Audit ICQs

The following internal control questionnaires (ICQs) provide suggested control objectives/questions for conducting an audit of the two business cycles covered in this book (Financial Accounting and Expenditure) and of the Oracle EBS Security component. It also provides references to the relevant COBIT 4.1 control objectives.

Financial Accounting Business Cycle

Control Objectives/Questions	Response			Comments	COBIT References
	Yes	No	N/A		
1. Master Data Maintenance					
1.1 Changes made to master data are valid, complete, accurate and timely.					
1.1.1 Does relevant management, other than the initiators, check online reports of master data additions and changes back to source documentation on a sample basis?					DS11
1.1.2 Is access to create and change Chart of Accounts master data restricted to authorized individuals?					DS5 DS11
1.1.3 Have configurable controls been designed into the process to maintain the integrity of master data?					DS9 DS11
1.2 Chart of Accounts master data remain current and accurate.					
1.2.1 Does management periodically review master data to check their accuracy?					DS11
2. Journal Processing					
2.1 Valid journal entries are booked to the GL.					
2.1.1 Has access to the entry, import, definition, set and generation of journal entries in the Oracle GL module been appropriately restricted?					P02
2.1.2 Has access to approve and post journal entries in Oracle GL been appropriately restricted? In addition, has segregation of duties been considered to prevent the ability to both create and post Oracle EBS journal vouchers?					P04

Financial Accounting Business Cycle (cont.)

Control Objectives/Questions	Response			Comments	COBIT References
	Yes	No	N/A		
2. Journal Processing *(cont.)*					
2.1.3 Is the Cross-Validation rule being utilized within Oracle EBS to prevent inaccurate journals?					AI5 DS5
2.1.4 Are the recurring and reversing journal entry features within Oracle EBS being utilized to prevent omitted or inaccurate journal entries?					AI6
2.2 Journal entries are posted only once to the GL.					
2.2.1 Is Oracle EBS configuration in place to ensure that each journal has a unique journal number, batch name or date for a journal to post to the ledger?					DS9
2.3 All journal entries are posted to the GL.					
2.3.1 Are monthly reconciliations in place from the journal entries against the monthly closing schedule?					PO4
2.4 All journal entries are posted to the correct period.					
2.4.1 Have the periods been defined within Oracle EBS?					
2.4.2 Is access to change the status of periods within Oracle EBS restricted to authorized personnel only?					DS5
2.5 All journal entries are accurate and balanced.					
2.5.1 Have edit and validation checks been configured within the system to ensure the integrity of the data?					
2.5.2 Have configurable controls been designed into the process to maintain the integrity of the account balances (e.g., prevention of posting of out-of-balance journal entries)?					PO9
2.5.3 Is access to reverse journal functions restricted to authorized personnel only?					DS5
3. Reconciliation and Financial Reporting Risks					
3.1 All valid GL account balances are included in the financial statements. Financial statements are accurate and reconcile to the GL.					
3.1.1 Is the ability to create, modify or generate financial statements/ reports restricted appropriately?					DS11

Financial Accounting Business Cycle (cont.)

Control Objectives/Questions	Response			Comments	COBIT References
	Yes	No	N/A		
3. Reconciliation and Financial Reporting Risks ***(cont.)***					
3.1.2 Is the ability to add, change or delete Key Flexfield segments, Key Flexfield values, Value Sets and account combinations restricted appropriately?					DS11 DS13
3.1.3 Are new Accounting Flexfield segment values inserted into existing ranges and financial reports, and configured with broadly defined account ranges so new accounts are included in appropriate financial reports?					AI6
3.1.4 Are reconciliations performed between the financial statements and the GL account balances at month end? This will assist in ensuring that all of the account numbers are included.					DS11
3.1.5 Are there any manual controls in place for changes made to the Chart of Accounts?					AI6
3.2 All account reconciliations are performed monthly.					
3.2.1 Are there monthly reconciliation schedules that are retained and monitored each month to ensure that all accounts are reconciled correctly? Who performs this monitoring and review of reconciliations?					ME1
3.3 Financial reports are generated and distributed in a timely manner.					
3.3.1 Does management monitor that financial reports are being generated in a timely manner during the closing cycle? How do they ensure that all reports have been generated?					ME1
3.3.2 Is there a matrix or schedule of who should be receiving individual financial reports and by what date they should be received throughout the closing?					AI2

Financial Accounting Business Cycle (cont.)

Control Objectives/Questions	Response			Comments	COBIT References
	Yes	No	N/A		
3. Reconciliation and Financial Reporting Risks *(cont.)*					
3.4 Account reconciliations are current.					
3.4.1 Is the ability to input, change, cancel or release credit notes restricted to authorized personnel?					AI6
3.5 Only authorized personnel may generate financial statements.					
3.5.1 Refer to 3.1.3 and 3.3.1.					
3.5.2 Are printed financial results securely stored to ensure that confidential information is not leaked prior to its release date?					
3.6 Adequate closing procedures prevent further postings to that period and accurately reflect the transactions that took place in a given accounting period.					
3.6.1 How is the close of the period monitored? Are the closing periods formalized and monitored to prevent any further postings to that period?					ME1
3.6.2 Has Oracle EBS been configured to prevent posting to closed or future periods?					
3.6.3 Is the ability to open/close accounting periods appropriately restricted?					

Expenditure Business Cycle

Control Objectives/Questions	Response			Comments	COBIT References
	Yes	No	N/A		
1. Master Data Maintenance					
1.1 Changes made to master data are valid, complete, accurate and timely.					
1.1.1 Does relevant management, other than the initiators, check online reports of master data additions and changes back to source documentation on a sample basis?					PO4 DS11
1.1.2 Is access to create and change vendor and pricing master data restricted to authorized individuals?					DS5 DS11
1.1.3 Have configurable controls been designed into the process to maintain the integrity of master data?					DS9

Expenditure Business Cycle (cont.)

Control Objectives/Questions	Response			Comments	COBIT References
	Yes	No	N/A		
1. Master Data Maintenance *(cont.)*					
1.1.4 Is a naming convention used for vendor names (e.g., as per letterhead) to minimize the risk of establishing duplicated vendor master records?					DS2
1.2 Inventory master data remain current and pertinent.					
1.2.1 Does management periodically review master data to check their accuracy?					DS11
2. Purchasing					
2.1 Purchase order entry and changes are valid, complete, accurate and timely.					
2.1.1 Is the ability to create, change or cancel purchase requisitions, purchase orders and outline agreements (standing purchase orders) restricted to authorized personnel?					AI6 DS5
2.1.2 Does the Oracle EBS source list functionality allow specified materials to be purchased only from vendors included in the source list for the specified material?					AI1 DS2
2.1.3 Is the Oracle EBS release strategy used to authorize purchase requisitions, purchase orders, outline agreements (standing purchase orders) and unusual purchases (for example, capital outlays)?					AI6
2.2 Goods are received only for valid purchase orders, and goods receipts are recorded completely, accurately and in a timely manner.					
2.2.1 When goods received are matched to open purchase orders, are receipts with no purchase order or those that exceed the purchase order quantity by more than an established amount investigated? Does management review exception reports of goods not received on time for recorded purchases?					AI4 DS3

Expenditure Business Cycle (cont.)

Control Objectives/Questions	Response			Comments	COBIT References
	Yes	No	N/A		
2. Purchasing *(cont.)*					
2.2.2 Is the ability to input, change or cancel goods received transactions restricted to authorized inbound logistics/raw materials personnel?					AI6
2.3 Defective goods are returned to suppliers in a timely manner.					
2.3.1 Are rejected raw materials adequately segregated from other raw materials in a quality assurance bonding area, and are they regularly monitored to ensure timely return to suppliers?					ME1
3. Invoice Processing					
3.1 Amounts posted to accounts payable represent goods or services received.					
3.1.1 Is the ability to input, change, cancel or release vendor invoices for payment restricted to authorized personnel? Is the ability to input vendor invoices that do not have a purchase order and/or goods receipt as support further restricted to authorized personnel?					DS5
3.2 Accounts payable amounts are calculated completely and accurately and recorded in a timely manner.					
3.2.1 Is the Oracle EBS software configured to perform a three-way match?					P09
3.2.2 Is the Oracle EBS software configured with quantity and price tolerance limits?					P09
3.2.3 Is the GR/IR account regularly reconciled?					DS1
3.2.4 Are reports of outstanding purchase orders regularly reviewed?					ME1
3.2.5 Does the Oracle EBS software restrict the ability to modify the exchange rate table to authorized personnel? Does management approve values in the centrally maintained exchange rate table?					AI2 DS9 DS11

Expenditure Business Cycle (cont.)

Control Objectives/Questions	Response			Comments	COBIT References
	Yes	No	N/A		
3. Invoice Processing ***(cont.)***					
3.2.5 *(cont.)* Does the Oracle EBS software automatically calculate foreign currency translation, based on values in the centrally maintained exchange rate table?					
3.3 Credit notes and other adjustments are calculated completely and accurately and recorded in a timely manner.					
3.3.1 Is the ability to input, change, cancel or release credit notes restricted to authorized personnel?					P06 DS5
4. Processing Disbursements					
4.1 Disbursements are made only for goods and services received and are calculated, recorded and distributed to the appropriate suppliers accurately in a timely manner.					
4.1.1 Does management approve the Oracle EBS payment run parameter specification?					DS9
4.1.2 Does the Oracle EBS software restrict to authorized personnel the ability to release invoices that have been blocked for payment, either for an individual invoice or for a specified vendor?					DS5

Security Administration

Control Objectives/Questions	Response			Comments	COBIT References
	Yes	No	N/A		
1. Security Administration					
1.1 Oracle EBS security/control parameters are defined adequately.					
1.1.1 Does relevant management, other than the initiators, check online reports of master data additions and changes back to source documentation on a sample basis?					DS11
1.2 Security administration function and user access change management procedures exist.					
1.2.1 Is there an organizational structure for security administration?					DS5 DS11

Security Administration (cont.)

Control Objectives/Questions	Response			Comments	COBIT References
	Yes	No	N/A		
1. Security Administration *(cont.)*					
1.2.2 Are Oracle EBS responsibilities related to information security administration defined and assigned?					
1.2.3 Have Oracle EBS security administration policies and procedures been developed and implemented? Are administrators aware of the policy, and do they adhere to it?					AI4
1.2.4 Have user access change management procedures been formalized and documented?					AI6
1.3 Access to programs, data and other information resources is restricted and monitored.					
1.3.1 Are logical security tools and techniques implemented and configured to enable restriction of access to programs, data and other information resources? Does management review and approve such implementation and configuration of Oracle EBS security techniques?					P09 DS5
1.3.2 Does the enterprise activate information security techniques such as Oracle Alerts, user auditing or other monitoring reports to record and report security events or access to sensitive information as defined in information security policies? Are reports generated and regularly reviewed, and is necessary action taken when required?					DS12 ME1
1.3.3 Have the default Oracle EBS asswords been changed for system administrator and supervisor access?					DS5
1.3.4 To what are the Oracle EBS password policy settings configured? Have password standards been developed and enforced by making users security-aware?					P09 DS9

Security Administration (cont.)

Control Objectives/Questions	Response			Comments	COBIT References
	Yes	No	N/A		
1. Security Administration ***(cont.)***					
1.3.5 Do users have a unique user identifier within Oracle Applications?					DS5
1.3.6 Does the enterprise encrypt sensitive data?					DS5
1.3.7 Is access to privileged responsibilities limited to appropriate personnel and monitored and reviewed by management?					DS5 ME1
1.3.8 Is access to sensitive data restricted to authorized users only? Does the enterprise review access privileges on a regular basis? Has the enterprise utilized the Oracle EBS facility to record authorized access to specific data items via the Sign-On: Audit Level Profile Option setting?					DS5
1.3.9 Are unauthorized attempts to log in to Oracle Applications logged and reviewed on a regular basis?					DS5
1.3.10 Is access to the GRC tool set adequately restricted and monitored so that reliance can be placed on the information obtained out of the tools?					
1.3.11 Are the GRC tools and the controls within appropriately configured to provide reliable and accurate reports?					
1.4 Users perform compatible functions, and Oracle EBS responsibilities match the user's organizational duties.					
1.4.1 Has segregation of duties been appropriately assigned within the enterprise?					PO4
1.4.2 Do the responsibilities match the organizational functions performed by the system users?					PO4

Security Administration (cont.)

Control Objectives/Questions	Response			Comments	COBIT References
	Yes	No	N/A		
1. Security Administration *(cont.)*					
1.5 System and start-up profiles are defined adequately, the ability to change profiles is restricted, and changes are monitored.					
1.5.1 Are the system and start-up profiles appropriately defined? Has the enterprise restricted access to the system and start-up profiles and ensured that changes are monitored?					AI6 ME1
1.6 Access to system output is restricted.					
1.6.1 Does the enterprise manage system output so that all expected information is produced and received by the intended recipient?					DS11
1.7 User IDs, responsibilities and preferences are documented adequately.					
1.7.1 Has the enterprise documented user IDs, responsibilities and preferences?					PO4
1.8 Remote access by external vendors is controlled adequately.					
1.8.1 Has the enterprise restricted remote access to external vendors to the test and development environments only? Furthermore, is access granted continuously or activated only upon request? Does the enterprise log remote access?					AI7 DS5
1.9 Access to transport changes to the production environment is controlled; as a result, modifications are consistent with management's intentions.					
1.9.1 Does the enterprise appropriately segregate access to the test and production environments?					DS5
1.10 Only authorized and accurate changes are made to the data dictionary.					
1.10.1 Has access to the data dictionary been restricted to the DBA?					DS5
1.10.2 What is the process for making changes to the data dictionary? Do requests for changes need to be formally approved and reviewed by management on a regular basis?					PO9 AI6

Appendix 7. COBIT Control Objectives

This chart indicates the COBIT 4.1 control objectives used in the reference columns of the Oracle EBS audit plans and ICQs.

Domain	Process	
Plan and Organise	P01	Define a strategic IT plan.
	P02	Define the information architecture.
	P03	Determine technological direction.
	P04	Define the IT processes, organisation and relationships.
	P05	Manage the IT investment.
	P06	Communicate management aims and direction.
	P07	Manage IT human resources.
	P08	Manage quality.
	P09	Assess and manage IT risks.
	P010	Manage projects.
Acquire and Implement	AI1	Identify automated solutions.
	AI2	Acquire and maintain application software.
	AI3	Acquire and maintain technology infrastructure.
	AI4	Enable operation and use.
	AI5	Procure IT resources.
	AI6	Manage changes.
	A17	Install and accredit solutions and changes.
Deliver and Support	DS1	Define and manage service levels.
	DS2	Manage third-party services.
	DS3	Manage performance and capacity.
	DS4	Ensure continuous service.
	DS5	Ensure systems security.
	DS6	Identify and allocate costs.
	DS7	Educate and train users.
	DS8	Manage service desk and incidents.
	DS9	Manage the configuration.
	DS10	Manage problems.
	DS11	Manage data.
	DS12	Manage the physical environment.
	DS13	Manage operations.

Domain	Process	
Monitor and Evaluate	ME1	Monitor and evaluate IT performance.
	ME2	Monitor and evaluate internal control.
	ME3	Ensure compliance with external requirements.
	ME4	Provide IT governance.

Glossary

Account Hierarchy—An Oracle EBS feature used to perform summary-level funds checking. An Account Hierarchy lets Oracle Purchasing and Oracle GL quickly determine the Summary Accounts into which detail accounts roll up.

Accounting Flexfield—The code used to identify a GL account in Oracle EBS. Each Accounting Flexfield segment value corresponds to a summary or rollup account within the Chart of Accounts. The number of segments can be defined in the Accounting Flexfield, as well as the length, name and order of each of the segments. Oracle GL displays the Accounting Flexfield in a pop-up window.

Accounting Flexfield segment—One of up to 30 different sections of the Accounting Flexfield, which together make up the GL account code. Each segment is separated from the other segments by a symbol chosen (such as "-," "/," ".," or "\"). Each segment typically represents an element of the business structure, such as company, cost center or account. A segment cannot exceed 25 characters in size.

Accounting Flexfield segment value—A series of characters and a description that define a unique value for a particular Value Set. For example, 0100 in the region Value Setting may represent the Eastern region, or 510 in the product Value Setting may represent computer products.

Accounting Flexfield structure—The account code structure defined to fit the specific needs of the organization. The number of segments and the length, name and order of each segment are chosen in the Accounting Flexfield structure.

Accounting Flexfield Value Set—A group of values and attributes of the values, e.g., the value length and value type that are assigned to the Accounting Flexfield segment to identify a particular element of the business, such as company, division, region or product

Adjusting period—Can be contained in the calendar in addition to, or instead of, "real" accounting periods. The real accounting periods must not overlap, and there cannot be any gaps between real accounting periods. Adjusting accounting periods can overlap with other accounting periods. For example, a period called DEC-10 can be defined; it includes 01-DEC-2010 through 31-DEC-2010. An adjusting period called DEC31-10 can also be defined; it includes only one day: 31-DEC-2010 through 31-DEC-2010. Note: Most Oracle Corp. feeder systems, such as Oracle Inventory, Oracle Payables, Oracle Purchasing and Oracle Receivables, do not use adjusting periods. Journals can be imported only into nonadjusting periods in Oracle GL.

Allocation entry—A recurring journal entry used to allocate revenues or costs. For example, an allocation entry could be defined to allocate costs to each department, based on headcount.

Auto offset—An Oracle GL feature that automatically determines the offset (or credit) entry for the allocation entry. Auto offset automatically calculates the net of all previous journal lines in the allocation entry, reverses the sign and generates the contra amount.

Average exchange rate—An exchange rate that is the average rate for an entire accounting period. Oracle GL automatically translates revenue and expense account balances using period average rates in accordance with FASB 52 (US). For companies in highly inflationary economies, Oracle GL uses average exchange rates to translate the nonhistorical revenue and expense accounts in accordance with FASB 8 (US). It is also known as the period average exchange rate.

Balances table—An Oracle GL database table that stores account balances called GL_BALANCES

Balancing segment—An Accounting Flexfield segment that is defined so that Oracle GL automatically balances all journal entries for each value of this segment. For example, if the company segment is a balancing segment, Oracle GL ensures that, within every journal entry, the total debits to company 01 equal the total credits to company 01. This would be the Accounting Flexfield segment to which the GL will always balance, i.e., for every value within the segment, the GL will make sure that the debits equal the credits. Normally, this relates to the segment that houses COMPANY(s). A Chart of Accounts must contain one, and only one, of these segment types.

Budget—Estimated cost and revenue amounts for a given range of periods and Ledgers. There can be multiple budget versions for the same Ledger.

Budget formula—A mathematical expression used to calculate budget amounts based on actual results, other budget amounts and statistics. With budget formulas, budgets using complex equations, calculations and allocations can be automatically created.

Budget hierarchy—A group of budgets linked together at different levels, such that the budgeting authority of a lower-level budget is controlled by an upper-level budget

Budget Interface table—An Oracle GL database table that stores information needed for budget upload

Budget organization—An entity (department, cost center, division or other group) responsible for entering and maintaining budget data. Budget organizations are defined for the company, and the appropriate Accounting Flexfields are assigned to each budget organization.

Budget rules—A variety of shorthand techniques that can be used to speed manual budget entry. With budget rules, a total amount can be divided evenly among budget periods, a given amount can be repeated in each budget period, or budget amounts can be derived from the account balances entered.

Budget upload—The ability to transfer budget information from a spreadsheet to Oracle GL. For example, with the spreadsheet interface, budget information can be uploaded from the spreadsheet to Oracle GL.

Budgetary Account—An account segment value (such as 6110) that is assigned one of the two budgetary account types. Budgetary accounts are used to record the movement of funds through the budget process from appropriation to expended appropriation.

Budgetary Account type—Either of the two account types: Budgetary DR or Budgetary CR

Budgetary Accounting Flexfield—An Accounting Flexfield that contains a budgetary account

Budgetary Control—An Oracle EBS feature used to control actual and anticipated expenditures against a budget. If Budgetary Control is enabled, funds can be checked online for transactions, and funds can be reserved for transactions by creating encumbrances. Oracle EBS automatically calculates funds available (budget less encumbrances less actual expenditures) when funds are reserved for a transaction. Oracle EBS notifies the user online if there are insufficient funds available for the transaction.

Chart of Accounts (COA) structure—A classification of the Accounting Flexfield segment values so that a particular range of values has a common characteristic. For example, 1000 to 1999 might be the range of segment values for assets in the account segment of the Accounting Flexfield.

Column Set—A Financial Statement Generator report component that can be built within Oracle GL by defining all of the columns in the report. For each column, the format and content can be controlled, including column headings, spacing and size, calculations, units of measure, and precision. A typical Column Set includes a header column for headings and subheadings, currency assignments, amount types, and calculation column totals. A Column Set can also be defined with each column representing a different company to enhance consolidation reporting.

Consolidation—An Oracle GL feature that allows the results of multiple companies to be combined, even if they are in different ledger sets with different currencies, calendars and Chart of Accounts

Content Set—A report component that can be built within Oracle GL that defines the information in each report and the printing sequence of reports. For example, a departmental Content Set can be defined that prints one report for each department.

Context field prompt—A question or prompt to which a user enters a response, called a context field value. When Oracle EBS displays a Descriptive Flexfield pop-up window, it displays the context field prompt after it displays any global segments that have been defined. Each Descriptive Flexfield can have up to one context field prompt.

Context field value—A response to the context field prompt. The response is composed of a series of characters and a description. The response and description together provide a unique value for the context prompt, such as 1500, Journal Batch ID, or 2000, Budget Formula Batch ID. The context field value determines which additional Descriptive Flexfield segments appear.

Context segment value—A response to the context-sensitive segment. The response is composed of a series of characters and a description. The response and description together provide a unique value for the context-sensitive segment, such as "Redwood Shores, Oracle Corporation Headquarters" or "Minneapolis, Merrill Aviation's Hub."

Context-sensitive segment—A Descriptive Flexfield segment that appears in a second pop-up window when a response is entered to the context field prompt. For each context response, multiple context segments can be defined, and the sequence of the context segments in the second pop-up window can be controlled. Each context-sensitive segment typically prompts one item of information related to the context response.

Conversion—This option converts foreign currency transactions to the functional currency.

Corporate exchange rate—An exchange rate that can be used optionally to perform foreign currency conversion. The corporate exchange rate is generally a standard market rate determined by senior financial management for use throughout the organization.

Cost center segment—Usually relates to the various departments within an organization, i.e., sales, transportation. Oracle GL does not use this segment for any specific purpose. This type of segment is not required for a Chart of Accounts.

Cross-Validation Rules—Rules that define valid combinations of segment values that a user can enter in an Accounting Flexfield. Cross-Validation Rules restrict users from entering invalid combinations of Accounting Flexfield segment values.

Dependent segment—An Accounting Flexfield segment in which the available values depend on values entered in a previous segment, called the independent segment. For example, the dependent segment "Subaccount 0001" might mean "Bank of Alaska" when combined with the independent segment "Account 1100, Cash," but the same "Subaccount 0001" might mean "Building #3" when combined with "Account 1700, Fixed Assets."

Descriptive Flexfield—A field that the organization can extend to capture extra information that is otherwise not tracked by Oracle EBS. A Descriptive Flexfield appears on the form as a single-character, unnamed field. The organization can customize this field to capture additional information that is necessary and unique to the business.

Detail Budget—A budget for which authority is controlled by another budget

Display Group—A range of rows or columns in the Row Set or Column Set for which the display in the report is to be controlled. A Display Group is assigned to a Display Set where it is specified whether to display or hide the rows or columns.

Display Set—A Financial Statement Generator report component that is built within Oracle GL to control the display of ranges of rows and/or columns in a report, without reformatting the report or losing header information. A Display Set can be defined for reports that use specific row and Column Sets, or a generic Display Set can be defined for any report, regardless of its row and Column Set.

Dual Currency—An Oracle GL feature that allows reporting in the functional currency and in one or more foreign currencies

Dynamic Insertion—A feature specific to Accounting Flexfields that allows entering and defining new combinations of segment values directly into a flexfield pop-up window. The new combination must satisfy any cross-validation rules before the flexfield accepts the new combination. The organization can decide if an Accounting Flexfield supports Dynamic Insertion. If an Accounting Flexfield does not support Dynamic Insertion, only new combinations of segment values using the Define Accounting Flexfield Combination form can be entered.

Encumbrance Accounting—An Oracle EBS feature used to automatically create encumbrances for requisitions, purchase orders and invoices. The Budgetary Control feature uses Encumbrance Accounting to reserve funds for budgets. If only Encumbrance Accounting is enabled, encumbrances can be created automatically or manually; however, funds cannot be checked online, nor does Oracle EBS verify available funds for the transaction. See also Budgetary Control.

Exchange rate—A rate that represents the amount of one currency that can be exchanged for another at a particular point in time. Oracle EBS uses the daily, periodic and historical exchange rates maintained to perform foreign currency conversion, revaluation and translation.

Export—A utility that allows data from an Oracle EBS table to be copied to a file in the current directory. The export utility is part of the Oracle EBS Relational Database Management System.

Export file—The file that the export utility creates in the directory. The export file should be given a name that helps identify the data in the table. For example, fiscal year 2010 for a Fremont Ledger might be called the export file FY10FR.dmp. Export files must have the extension .dmp at the end of each file name.

Feeder program—A custom program written to transfer transaction information from an original system into Oracle EBS interface tables. The type of feeder program written depends on the environment from which data are imported.

Financial Statement Generator (FSG)—A powerful and flexible tool that can be used to build custom reports without programming. Reports can be defined online with complete control over the rows, columns and contents of the report.

Fiscal year—Any yearly accounting period without regard to its relationship to a calendar year

Flexfield—Oracle EBS fields made up of segments. Each segment has an assigned name and a set of valid values. Oracle EBS uses Flexfields to capture information about the organization. There are two types of Flexfields: Key Flexfields and Descriptive Flexfields.

Foreign currency conversion—A process that allows a foreign currency journal entry to be converted into the functional currency. Oracle GL automatically performs conversion whenever a journal entry in a currency other than the functional currency is entered. Oracle GL multiplies the daily exchange rate defined or the exchange rate entered to convert amounts to the functional currency. The results of foreign currency conversion can be viewed on the Enter Journals form.

Foreign currency journal entry—A journal entry in which transactions are recorded in a foreign currency. Oracle GL automatically converts foreign currency amounts into the functional currency using an exchange rate specified.

Foreign currency revaluation—A process that allows assets and liabilities denominated in a foreign currency to be revalued using a period-end (usually a balance sheet date) exchange rate. Oracle GL automatically revalues foreign assets and liabilities using the period-end exchange rate specified. Revaluation gains and losses result from fluctuations in an exchange rate, between a transaction date and a balance sheet date. Oracle GL automatically creates a journal entry in accordance with FASB 52 (US) to adjust the unrealized gain/loss account when a revaluation is run.

Foreign currency translation—A process that allows the functional currency account balances to be restated into a reporting currency. Oracle GL multiplies the average, periodic or historical rate defined by the functional currency account balances to perform foreign currency translation.

Functional currency—The principal currency used to record transactions and maintain accounting data within Oracle GL. The functional currency is generally the currency in which most of the business transactions are performed. The functional currency is specified for each Ledger in the Define Ledger form.

GL_JE_ BATCHES—The table that the Oracle GL application uses to store journal entry batch information, such as name, status, and batch debit and credit totals

GL_JE_ HEADERS—The table that the Oracle GL application uses to store journal entry header information, such as name, category, date and currency

GL_INTERFACE—The table that the Oracle GL application uses to run Journal Import. The subledgers must insert records into the Oracle GL GL_INTERFACE table. Journal import validates the data in this table to ensure that only valid journal entries are created in Oracle GL.

GL_JE_LINES—The table that the Oracle GL application uses to store journal entry line information, such as journal entry line number and Accounting Flexfield debit and credit amounts

Import—A utility that allows data from an export file to be brought into an Oracle EBS table. The import utility is part of the Oracle EBS Relational Database Management System. It is used when restoring archived data.

Import journal entry—A journal entry from a non-Oracle application, such as accounts payable, accounts receivable and fixed assets. Journal Import is used to import these journal entries from the feeder systems. See also **GL_INTERFACE**.

Journal Details tables—Database tables GL_JE_BATCHES, GL_JE_HEADERS and GL_JE_LINES that store journal details

Journal entry—A debit or credit to a GL account. See also **Manual journal entry**.

Journal entry category—A category that Oracle GL uses to describe the purpose or type of journal entry. Standard journal entry categories include accruals, payments and vouchers.

Journal entry source—The source by which Oracle GL identifies and differentiates the origin of journal entries. Standard journal entry sources include payables, payroll, personnel and receivables.

Journal Import—An Oracle GL program that creates journal entries from transaction data stored in the Oracle GL GL_INTERFACE table. Journal entries are created and stored in GL_BATCHES, GL_HEADERS and GL_LINES.

Key Flexfield—An Oracle EBS feature used to build custom fields used for entering and displaying information relating to the business. The Oracle GL Accounting Flexfield is a Key Flexfield.

Ledger—A company or group of companies within Oracle GL that shares a common Accounting Flexfield structure, calendar and functional currency

Manual journal entry—A journal entry entered at a computer terminal. Manual journal entries can include regular, statistical, intercompany and foreign currency entries.

Mass allocations—A single journal entry formula that allocates revenues and expenses across a group of cost centers, departments, divisions, etc. For example, employee benefit costs can be allocated to each department based on headcount.

Mass Budgeting—A feature that allows a complete budget to be built using simple formulas based on actual results, other budget amounts and statistics. For example, next year's budget could be drafted using last year's actual results plus 10 percent or some other growth factor. With Mass Budgeting, one rule can be applied to a range of accounts.

Master budget—A budget that controls the authority of other budgets

Natural account segment—Relates to the "natural" GL account types, i.e., asset, liability, owner's equity, revenue or expense. A Chart of Accounts must contain one, and only one, of these segment types.

Net allocation—Allocation in which the net of all allocations is posted to an allocated-out account

Organization—A business unit such as a company, division or department. Organization can refer to a complete company or to divisions within a company. Typically, an organization (or a similar term) is defined as part of the Accounting Flexfield when Oracle EBS is implemented.

Parent segment value—An Accounting Flexfield segment value that references a number of other segment values, called child segment values. Oracle GL uses parent segment values for creating Summary Accounts, for reporting on summary balances, and in mass allocations and Mass Budgeting. Parent segment values can be created for independent segments, but not for dependent segments.

Period Type—Used to define the Accounting Calendar

Profile Option—A set of changeable options that affects the way applications run. In general, Profile Options can be set at one or more of the following levels: Site, Application, Responsibility and User.

Project segment—The Accounting Flexfield is set up by defining the individual segments of the GL account code. A project segment that is used to enter the project identifier can be defined. All key attributes of the segment are defined, including field length, position of the segment within the Accounting Flexfield, prompt, type of characters (numeric or alphanumeric) and default value (optional).

Project segment value—The identifier (project name, number or code) used to designate each project. After a project segment in the Accounting Flexfield is defined, a project in Oracle GL is set up by simply defining a project segment value, e.g., a project name (ALPHA), a project number (583) or a project code (D890).

Proprietary Account—An account segment value (such as 3500) that is assigned one of the five Proprietary Account types

Proprietary Account Type—Any of the five account types: asset, liability, owner's equity, revenue and expense

Proprietary Accounting Flexfield—An Accounting Flexfield that contains a Proprietary Account

Recurring Journal Entry—A journal entry is defined once; then, upon request, Oracle GL uses it to repeat the journal entry for each accounting period. Recurring journal entries are used to define automatic consolidating and eliminating entries. It is also known as recurring formula.

Report—A combination of at least a Row Set and Column Set, and optionally a Content Set, Display Group, Row Order and runtime options, such as currency and override segment name, that can be defined and named. When submitting financial statement requests, this name is entered and Oracle GL automatically enters the report components and runtime options by specifying the accounting period.

Report component—An element of a Financial Statement Generator report that defines the format and content of a report. Report components include Row Sets, Column Sets, Content Sets, Row Orders and Display Sets. Report components are grouped together in different ways to create custom reports.

Report headings—Provide general information about the contents of the report

Report parameter—Options for sorting, formatting, selecting and summarizing the information in a report

Report Set—A grouping of reports submitted at the same time to run as one transaction. A report set allows the same set of reports to be submitted regularly without having to specify each report individually. For example, a report set can be defined that prints all of the regular month-end management reports.

Reporting currency—The currency used for financial reporting. If the reporting currency is not the same as the functional currency, foreign currency translation can be used to restate the account balances in the reporting currency.

Reporting hierarchies—Summary relationships within an Accounting Flexfield segment that let detailed values of that segment be grouped to prepare summary reports. Summary (parent) values are defined that reference the detailed (child) values of that segment.

Responsibility—A level of authority within Oracle GL. Each responsibility provides a user with access to a menu and a Ledger. One or more responsibilities can be assigned to each user. Responsibilities allow control security in Oracle GL.

Responsibility report—A financial statement containing information organized by management responsibility. For example, a Responsibility report for a cost center contains information for that specific cost center, a Responsibility report for a division manager contains information for all organizational units within that division, and so on. A manager typically receives reports for the organizational unit(s) (such as cost center, department, division and group) for which he/she is responsible.

Reversing journal entry—A journal entry that Oracle GL creates by reversing an existing journal entry. Any journal entry can be reversed and posted to any open accounting period.

Rollup Group—A collection of parent segment values for a given segment. Rollup Groups are used to define Summary Accounts based on parents in the group. Letters as well as numbers can be used to name a Rollup Group. Rollup Groups are only used to create Summary Accounts whose balances can be reported and viewed online. Rollup Groups are assigned to parent segment values.

Row Order—A report component that is used to modify the order of detail rows and Accounting Flexfield segments in a report. Rows can be ranked in ascending or descending order based on the amounts in a particular column and/or by sorting the Accounting Flexfield segments by segment value or segment value description. Display options can be specified, depending on the row-ranking method chosen. For example, if total sales are to be reviewed in descending order by product, the rows can be ranked in descending order by the total sales column, and the segments can be rearranged so that the product appears first on the report.

Row Set—A Financial Statement Generator report component that is built within Oracle GL by defining all of the lines in a report. For each row, the format and content can be controlled, including line descriptions, indentations, spacing, page breaks, calculations, units of measure, precision and so on. A typical row set includes row labels, Accounting Flexfields and calculation rows for totals. For example, a standard income statement Row Set or a standard balance sheet Row Set can be defined.

Rule numbers—A sequential step in a calculation. Rule numbers are used to specify the order in which it is desired that Oracle GL processes the factors used in the budget and actual formulas.

Shorthand Flexfield entry—A quick way to enter Key Flexfield data using shorthand aliases (names) that represent valid Flexfield combinations or patterns of valid segment values. The organization can specify which Flexfields use the shorthand Flexfield entry, and define shorthand aliases for those Flexfields to represent complete or partial sets of Key Flexfield segment values.

Skeleton entry—A recurring journal entry whose amounts change each accounting period. A recurring journal entry is defined without amounts, and then the appropriate amounts are entered into each accounting period. For example, a skeleton entry might be defined to record depreciation in the same accounts every month, but with different amounts due to additions and retirements.

Spot exchange rate—A daily exchange rate that is used to perform foreign currency conversion. The spot exchange rate is generally a quoted market rate, which applies to the immediate delivery of one currency for another.

Spreadsheet interface—A program that uploads the actual or budget data from a spreadsheet into Oracle GL

Standard entry—A recurring journal entry, the amount of which is the same for each accounting period. For example, a standard entry might be defined for fixed accruals, such as rent, interest and audit fees.

Statistical journal entry—A journal entry in which nonfinancial information is entered, such as headcount, production units and sales units

Statistics—Accounting information (other than currency amounts) used to manage the business operations, e.g., information on headcount, production units and sales units. With Oracle GL, budget and actual statistics can be maintained, and these statistics can be used with budget rules and formulas.

Summary Account—An Accounting Flexfield whose balance represents the sum of other Accounting Flexfield balances. Summary Accounts can be used for faster reporting and inquiry as well as in formulas and allocations.

Value Set—A group of values and related attributes that are assigned to an Accounting Flexfield segment or to a Descriptive Flexfield segment. Each Value Set contains a set of values with the same maximum length, validation type, alphanumeric option and so on.

Weighted-average exchange rate—An exchange rate that Oracle GL automatically calculates by multiplying journal amounts for a specific Accounting Flexfield by the translation rate that applies to each journal amount. Whether the rate that applies to each journal amount is based on the inverse of the daily conversation rate or an exception rate entered manually can be chosen. Oracle GL uses the weighted-average rate, instead of the period-end, average or historical rates, to translate balances for those Accounting Flexfields to which a weighted-average rate type is assigned.

Index

Q

R

S

T

U

V

W

ISACA Professional Guidance Publications

Many ISACA publications contain detailed assessment questionnaires and work programs. Please visit *www.isaca.org/bookstore* or e-mail *bookstore@isaca.org* for more information.

Frameworks
- CobiT® 4.1, 2007
- *Enterprise Value: Governance of IT Investments: The Val IT™ Framework 2.0*, 2008
- *ITAF™: A Professional Practices Framework for IT Assurance*, 2008
- *The Risk IT Framework*, 2009

COBIT-related Publications
- *Aligning CobiT® 4.1, ITIL V3 and ISO/IEC 27002 for Business Benefit*, 2008
- *Building the Business Case for CobiT® and Val IT™: Executive Briefing*, 2009
- *CobiT® and Application Controls*, 2009
- *CobiT® Control Practices: Guidance to Achieve Control Objectives for Successful IT Governance, 2nd Edition*, 2007
- *CobiT® Mapping: Mapping of CMMI® for Development V1.2 With CobiT® 4.0*, 2007
- *CobiT® Mapping: Mapping of FFIEC With CobiT® 4.1*, 2010
- *CobiT® Mapping: Mapping of ISO/IEC 17799:2000 With CobiT®, 2nd Edition*, 2006
- *CobiT® Mapping: Mapping of ISO/IEC 17799:2005 With CobiT® 4.0*, 2006
- *CobiT® Mapping: Mapping of ITIL With CobiT® 4.0*, 2007
- *CobiT® Mapping: Mapping of ITIL V3 With CobiT® 4.1*, 2008
- *CobiT® Mapping: Mapping of NIST SP800-53 With CobiT® 4.1*, 2007
- *CobiT® Mapping: Mapping of PMBOK® With CobiT® 4.0*, 2006
- *CobiT® Mapping: Mapping of SEI's CMM® for Software With CobiT® 4.0*, 2006
- *CobiT® Mapping: Mapping of TOGAF 8.1 With CobiT® 4.0*, 2007
- *CobiT® Mapping: Overview of International IT Guidance, 2nd Edition*, 2006
- *CobiT® Quickstart™, 2nd Edition*, 2007
- *CobiT® Security Baseline™, 2nd Edition*, 2007
- *CobiT® User Guide for Service Managers*, 2009
- *Implementing and Continually Improving IT Governance*, 2009
- *IT Assurance Guide: Using CobiT®*, 2007
- *SharePoint® Deployment and Governance: Using CobiT® 4.1*, 2010

Risk IT-related Publication
- *The Risk IT Practitioner Guide*, 2009

Val IT-related Publications

- *The Business Case: Using Val IT™ 2.0*, 2010
- *Enterprise Value: Getting Started With Value Management*, 2008
- *Value Management Guidance for Assurance Professionals: Using Val IT™ 2.0*, 2010

Executive and Management Guidance

- *An Executive View of IT Governance*, 2008
- *An Introduction to the Business Model for Information Security*, 2009
- *Board Briefing on IT Governance, 2nd Edition*, 2003
- *Defining Information Security Management Position Requirements: Guidance for Executives and Managers*, 2008
- *Identifying and Aligning Business Goals and IT Goals: Full Research Report*, 2008
- *Information Security Governance: Guidance for Boards of Directors and Executive Management, 2nd Edition*, 2006
- *Information Security Governance: Guidance for Information Security Managers*, 2008
- *Information Security Governance—Top Actions for Security Managers*, 2005
- *ITGI Enables ISO/IEC 38500:2008 Adoption*, 2009
- *IT Governance and Process Maturity*, 2008
- IT Governance Domain Practices and Competencies:
  - *Governance of Outsourcing*, 2005
  - *Information Risks: Whose Business Are They?*, 2005
  - *IT Alignment: Who Is in Charge?*, 2005
  - *Measuring and Demonstrating the Value of IT*, 2005
  - *Optimising Value Creation From IT Investments*, 2005
- IT Governance Roundtables:
  - *Defining IT Governance*, 2008
  - *IT Staffing Challenges*, 2008
  - *Unlocking Value*, 2009
  - *Value Delivery*, 2008
- *Managing Information Integrity: Security, Control and Audit Issues*, 2004
- *Understanding How Business Goals Drive IT Goals*, 2008
- *Unlocking Value: An Executive Primer on the Critical Role of IT Governance*, 2008

Practitioner Guidance

- Audit/Assurance Programs:
  - *Change Management Audit/Assurance Program*, 2009
  - *Generic Application Audit/Assurance Program*, 2009
  - *Identity Management Audit/Assurance Program*, 2009
  - *IT Continuity Planning Audit/Assurance Program*, 2009
  - *Network Perimeter Security Audit/Assurance Program*, 2009
  - *Outsourced IT Environments Audit/Assurance Program*, 2009